Human Skeletal Remains

MANUALS ON ARCHEOLOGY • 2

HUMAN SKELETAL REMAINS

THIRD EDITION

EXCAVATION, ANALYSIS, INTERPRETATION

DOUGLAS H. UBELAKER

SMITHSONIAN INSTITUTION

TARAXACUM WASHINGTON

THE MANUALS ON ARCHEOLOGY

are guides to the excavation, preservation, classification, analysis, description, and interpretation of special categories of archeological remains. The authors have drawn upon their experience and the contributions of other experts to formulate instructions for dealing with delicate, complicated or unobtrusive kinds of phenomena. Their intent is to enable those who are not specialists in particular subject matter to classify and describe finds in a useful manner and to observe details easily overlooked during excavation or analysis. The manuals will be revised periodically to incorporate new information. Suggestions for improving the clarity or completeness of the instructions, illustrations, and data are welcome, as well as recommendations of topics appropriate to the series.

Taraxacum, 1227 30th Street, Washington, D.C. 20007

First published 1978; Second printing 1980; Revised edition 1984;
Second edition 1989; Second printing 1991; Third printing 1994; Fourth printing 1996; Fifth printing 1998;
Third edition 1999.
ISBN 0-9602822-7-0

Library of Congress Catalog Card Number 88- 50355

Designed by John B. Goetz
Cover motif by George Robert Lewis
Printed in the United States of America

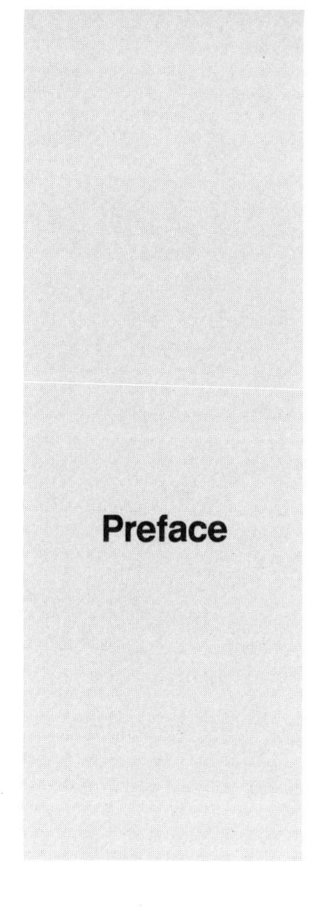

Preface

When the idea for this series originated in 1975, I agreed that a manual on human skeletal remains was needed and that I was in as good a position as anyone to write it, but I felt pressed by research obligations and growing museum responsibilities. The proposition was tempting, but I hesitated to commit the time. During the months that followed, I studied two series of skeletons from archeological sites in the New World. Both had been excavated by non-professionals over extended periods with limited funding. The field notes were largely confined to a dialogue of who-did-what-when. Descriptions of artifact positions and dimensions were copious, but data on the skeletons, which were the dominant feature at both sites, were almost non-existent. Several burials labelled secondary interments were accompanied by sketches of primary flexed skeletons. The photographs revealed little about burial position because the soil had not been properly removed. Not all the bones had been saved and those from different individuals were sometimes co-mingled, apparently reflecting the belief that fragmentary post-cranial remains cannot provide useful information.

Unfortunately, cases like those I encountered are common. Many professional archeologists lack familiarity with the human skeleton and do not record the kinds and amounts of data that specialists in human skeletal biology require for interpretation. The situation is more serious when human remains are encountered by police investigators or others lacking archeological training. In such cases, not only are field observations usually minimal, but frequently many bones are not saved, further limiting the potential for identification and interpretation.

Since these situations stem primarily from unawareness of what is needed, I decided to prepare a manual that would (1) enhance appreciation of the kinds of data human skeletons can offer and (2) provide guidelines for excavating, processing, and analyzing human remains. I have tried to be clear and to illustrate complicated and unfamiliar features. Because anyone dealing with skeletal remains should employ correct terminology, I have not attempted to avoid technical terms. Those not defined in the text or the Glossary can be found in a dictionary. Many of the analyses require equipment or expertise not possessed by most of those who will use this manual. I decided to give brief descriptions to show some of the inferences potentially obtainable from specimens properly collected. There are too few specialists to collect as well as analyze the material, and we can use all the help we can get. If the goals of this manual are met, collections such as those that provoked me to write it should become the exception rather than the rule.

Among the many individuals who have assisted me in preparing the manuscript, several deserve special recognition. All original art work was done by George Robert Lewis, Scientific Illustrator in the Smithsonian's Department of Anthropology. Victor Krantz of the Smithsonian's Division of Photographic Services prepared the original photographs, often on short notice. Smithsonian research colleagues, Stephanie Damadio and Ann Palkovich Shaw, read the manuscript and offered useful suggestions. Betty J. Meggers provided the original inspiration and considerable encouragement along the way. The ideas (and deficiencies) have remained my own; however, the organization and wording have profited considerably from her editorial efforts.

The following individuals have contributed illustrations or other materials: Donald J. Ortner, Smithsonian Institution; William M. Bass, University of Tennessee; T. D. Stewart, Smithsonian Institution, and Pat Willey, University of Tennessee. The Crow scaffold burial (Figure 1) was provided by the National Anthropological Archives, directed by Herman Viola. I express final appreciation to my wife, Maruja, for typing the initial draft and sacrificing many of our weekends and evenings during the writing of this manual.

Contents

Figures

Tables

APPENDIX

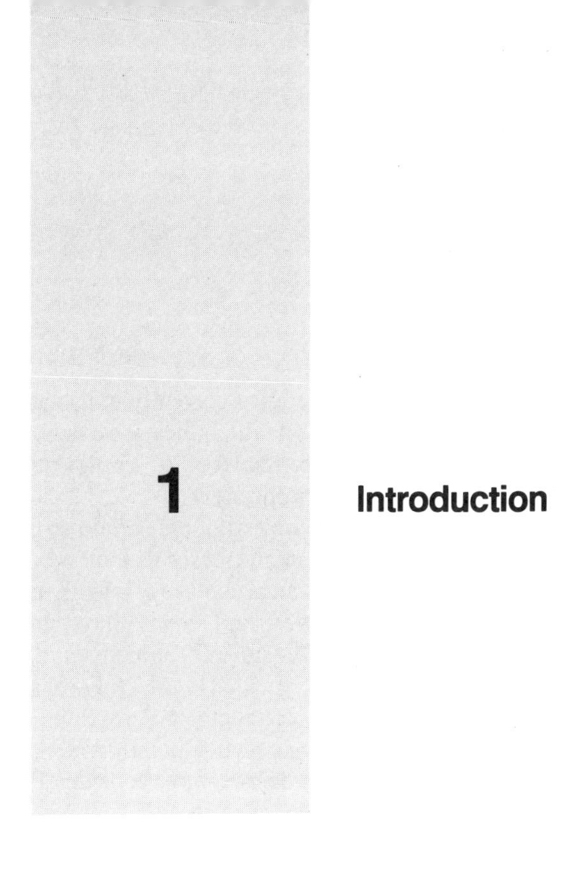

1 Introduction

Many non-anthropologists, and even some archeologists (such as Heizer 1974) have asked "Why excavate skeletons? What information can we gain to merit the disturbance of human interments?" This manual is designed to answer these questions. It attempts to demonstrate the range of data and interpretations potentially obtainable from human skeletal remains and to show how this information can contribute to the solution of various anthropological problems. Basic techniques of skeletal excavation and analysis are also described and evaluated.

For convenience, the book is divided into two sections. The first (Chapters 2–3) reviews the techniques and information needed for excavating and describing skeletal remains and for achieving reliable estimates of stature, sex, and age at death. These chapters should improve the capacity of non-specialists to undertake skeletal excavation and preliminary analysis. The second section (Chapters 4–6) discusses additional kinds of information that can be gleaned from suitable samples by experienced skeletal biologists.

Excavation and analysis are complicated by the heterogeneous circumstances in which human remains are encountered and the diversity of mortuary customs around the world. Early European visitors to the Americas reported direct interment in cemeteries, exposure on scaffolds, cleaning bones before burial, deposition in charnel houses, cremation, exposure in trees or caves, and placement in rivers or dwellings. Prehistoric remains recovered archeologically exhibit similar variability. Skeletons have been found in cemeteries, beneath house floors, in plazas, in cache and fire pits, in mounds and rock shelters, in pottery urns, and scattered in village refuse (to mention the most common practices). In some instances, methods of disposal left no tangible clues, interments were isolated and unmarked, or deterioration has been complete. Burial may be primary (complete articulated skeletons) or secondary (disarticulated artificial arrangement). Either type may occur in any of the contexts listed above, alone or in combination. Forensic samples range from a few bone chips to complete skeletons. Skeletal parts have been lodged in house walls, bottled in formalyne, buried under porches, and even sent through the mail. Many have been differentially affected by unique environmental and human action between death and discovery.

Two widely held beliefs have caused loss of much valuable data. One is that the skull is the only part worth saving; the other is that badly broken, incomplete or deteriorated bones are useless. It must be emphasized that every bone should be saved. An experienced skeletal biologist can often estimate age at death, sex, and even stature from poorly preserved remains. Small fragments may possess diagnostic features. For example, a scrap of skull from a burial urn at the mouth of the Amazon was sufficient to identify intentional deformation. A cranial bone from the scene of a crime in New England, identified through radiograph comparison, lead to the conviction of the assailant. Hence, anyone who discovers human bones should excavate them carefully, record position and other details in situ, and save everything, keeping in mind that continuing development of new methods for inferring information may permit using samples in the future that are inadequate for existing techniques.

Most of the procedures outlined in this volume

are applicable to the recovery and analysis of individual human remains in both traditional archeological and forensic contexts. Others require large samples, usually preserved in cemeteries. These provide a basis for inferences about diet, disease, behavior, and other aspects of prehistoric populations. A single skeleton, or a few individuals, cannot be assumed to be typical and consequently cannot be used to evaluate the frequency of disease, to reconstruct mortality curves, or to make other kinds of demographic analyses.

The number of individuals in a collection from a single context is not the only consideration; the sample must also be unselected. Skeletal collections in museums often contain mostly adults because it was previously believed that little could be learned from immature individuals. Even the skeletons of adults may not be representative because normal individuals were saved less often than those showing pathology. An investigator unaware of these criteria of selection might conclude, erroneously, that the population represented by a sample had little or no subadult mortality and a remarkably high frequency of bone disease. Samples can also be biased by natural factors, such as differential rates of decomposition in the ground, or by archeological procedures, such as incomplete excavation of a cemetery. The existence of numerous potential sources of distortion makes sampling procedures the most important factor underlying skeletal analysis and interpretation.

Properly collected samples of adequate size and composition can provide considerable amounts of information. Although many of the techniques are simple and objective, they must be employed flexibly and critically. A choice among them depends on the types of bones and their conditions, the nature of the prehistoric population, the experience of the researcher, and the equipment available. A regression formula that provides an accurate estimate of the living stature of White females may not be suitable for American Indian males. The microscopic method of establishing age at death from changes in the structure of cortical bone cannot be employed without access to slide-producing equipment and a microscope. My intention is to describe the principal procedures currently in use and the assumptions and samples on which they are based. It is the responsibility of the readers to employ the procedures cautiously and appropriately.

Archeologists and forensic investigators should strive to include a skeletal biologist in all phases of their work. Archeologically oriented physical anthropologists are employed at most universities and many have had extensive field experience. Forensic anthropologists may be contacted through the American Board of Forensic Anthropology, Inc., which conducts a program of certification, at <http//www.csuchico.edu/anth/ABFA>. Diplomates must hold a Ph.D. degree, have considerable forensic experience, and pass an examination.

The information on the following pages is introductory and many aspects have been treated in greater detail by others. Bass (1971) has published an excellent text on basic human osteology and methods for distinguishing left from right side. Steele and Bramblett (1988) and White (1991) also provide excellent texts on human skeletal anatomy. Krogman (1962) and Stewart (1968) have discussed techniques for inferring age, sex, stature, and other information from the skeleton. Auferheide and Rodriguez-Martin (1998), Morse (1969), Brothwell (1972), Ortner and Putschar (1981), and Steinbock (1976) provide data on bone pathology. Brothwell also elaborates techniques of excavation and other aspects of research. Additional references on special topics are included in the text.

I have avoided technical terminology where possible, but accurate recording and description cannot be accomplished without employing the names of individual bones and other skeletal landmarks. Terms most commonly needed for description are included in the Glossary. More detailed skeletal nomenclature has been provided by Bass (1971). Chapters 4 and 6 are the most technical. Diagnosing disease and reconstructing population dynamics require expert knowledge of skeletal biology and sophistication in using computers beyond what can be provided in a manual of this kind. Hopefully, the discussion is sufficiently clear to illustrate what can be done with well documented samples of human skeletal remains. If it stimulates readers to collect the data on which such inferences depend, my mission will have been accomplished.

2 Skeletal Recovery

Complete and accurate recovery of skeletal parts and information on their associations with one another and with other items constitutes the all-important first step in skeletal analysis. The only rule is to employ the techniques that maximize the quantity and quality of the data relevant to the goals of the investigation. These techniques vary according to the type of inhumation (primary, secondary, cremation, etc.), the time and amount of funds available, and the nature of the sample desired (total, random, etc.). The need for flexibility cannot be over-emphasized. Every situation is unique. A "cookbook" approach can cause unusual data to be overlooked and this, in turn, limits the opportunities for analysis. The following discussion consequently is not a definitive statement, but a resumé of methods that are easy to employ and that provide the information necessary for the principal types of analysis.

PREPARATIONS

If there is reason to believe that human remains have been encountered or if an archeological excavation is planned, appropriate authorities must be contacted. In most forensic situations, the police and/or local medical examiner's office should be notified immediately. These authorities are responsible for assessing the situation and deciding how to proceed. Many medical examiners are forensic pathologists, who usually have had some exposure to techniques of skeletal recovery and analysis. Alternatively, the medical examiner's office may employ a forensic anthropologist as a consultant or know a local person qualified to assist.

If the remains are clearly archeological, authorities still must be consulted. National, state, and local laws protect archeological resources and require official approval prior to excavation. In some cases, local groups may have strong cultural, ancestral, and religious involvements. In the United States, concern over the excavation of ancient human remains is especially intense among American Indians. Effective communication with such groups prior to excavation is essential. It is important to remember that human remains represent not only unique sources of valuable scientific information, but also unique individuals. As such, they command respect, as do the complex feelings and attitudes of others about them.

Prior to initiating archeological fieldwork, all the relevant data on mortuary customs and grave locations should be compiled to facilitate intelligent selection of methods of excavation. Sources of such information include descriptions of the population or related groups by early observers, published or unpublished data obtained by previous investigators at the site or other sites in the region, and examination of the surface of the ground.

In some forensic situations, reports of a missing person or suspicious activity may precipitate a search. In such cases, it is important to gather as much preliminary information as possible about the individual and the circumstances. The situation may benefit from including an archeologist and/or forensic anthropologist from the beginning.

Ethnohistorical Research

If the archeological site dates from shortly before or during historic times, ethnohistorical documents may provide information on the locations, depths, and positions of the burials, and alert the excavator to features and artifacts likely to be encountered during excavation. General summaries and discussions of the variability in mortuary customs have been provided by Yarrow (1880) and Bushnell (1920, 1927), but primary sources on the region should always be consulted. The following examples illustrate the utility of this procedure.

Hidatsa. On June 24, 1811, John Bradbury provided an eye-witness account of Hidatsa mortuary customs:

I passed through a small wood, where I discovered a stage constructed betwist four trees, standing very near each other, and to which the stage was attached, about ten feet from the ground. On this stage was laid the body of an Indian, wrapt in a buffalo robe. As the stage was very narrow, I could see all that was upon it without much trouble. It was the body of a man, and beside it there lay a bow and quiver with arrows, a tomahawk, and a scalping knife. There were a great number of stages erected about a quarter of a mile from the village . . ." (1817:147).

This procedure would probably leave little for an archeologist to find. Excavation might reveal the post holes of the scaffold supports, however, and perhaps remnants of human bone (Fig. 1).

Arikara. The Arikara, another tribe of the North American Plains, used a different method. Lewis Henry Morgan provided the following description in 1862:

Just out of the village is the burying ground. The Arickarees did not scaffold the dead but buried them in the ground. The most of the graves, and there are hundreds of them visible, are on the segment of a great circle. Others are grouped together. They wrapped up the body, dug a grave, and put it either in a sitting posture or doubled it up, I do not know which. I saw the size of some of the graves. They could not have extended the body and I could not tell whether it was an empty grave and timbered roof like the Omaha, or the earth was placed upon the body . . . (in White 1959:162).

This description alerts the archeologist to watch for impressions or other traces of wrappings in the soil; it also provides details that could permit correlating a cemetery with this historically described tribe.

Kansa. The fact that considerable variation may exist within the same general region is emphasized by W. J. Griffing's account of another Plains tribe, the Kansa:

The Kaws, while living at their old village near Manhattan, buried their dead in graves on the bottom land near the village, leaving no permanent markings of any kind which might lead to the identification of the spot. In later years, stones were heaped over the graves, to protect the bodies from wolves. Often a horse was killed over the spot, whose spirit was supposed to convey that of the departed to the happy hunting grounds (1904:134–5).

This description is corroborated by Morgan, who reported in 1860 that

The Kaws still bury in a sitting posture facing the west, arms crossed and knees flexed. A bow and arrow on the left side, a little brass or earth kettle between the legs or feet, containing corn or beans or dried buffalo meat, and their tobacco pouch and pipe. The hole in the ground is about one foot deep, the body is set up erect, and covered with bark, this is then covered with dirt lightly, after which stones are piled up around the body loose so as [to] cover the body fully about one or two feet over their head. This is to secure the body against wolves, etc. In the case of a distinguished man the Kaws saddle a horse, lead him up to the grave and shoot him at the grave, and leave him there unburied. Sometimes the saddle is buried in the grave (in White 1959:82–3).

Huron. Occasionally, ethnohistorical data are sufficiently explicit to permit archeological identification of an actual site. One of the most complete and vivid descriptions of native American mortuary customs is Jean de Brebeuf's account of the Feast of the Dead ceremony and subsequent ossuary burial held at the Huron village of Ihonatiria in 1636. He wrote as follows:

Let me describe the arrangement of this place. It was about the size of the place Royale at Paris. There was in the middle of it a great pit, about ten feet deep and five brasses wide. All around it was a scaffold, a sort of staging very well made, nine to ten brasses in width, and from nine to ten feet high; above this staging there were a number of poles laid across, and well arranged, with cross-poles to which these packages of souls were hung and bound. The whole bodies, as they were to be put in the bottom of the pit, had been the preceding day placed under the scaffold, stretched upon bark or mats fastened to stakes about the height of a man, on the borders of the pit. . . . About five or six o'clock, they lined the bottom and sides of the pit with fine large new robes, each of ten Beaver skins, . . . some went down to the bottom and brought up handfuls of sand. . . . They put in the very middle of the pit three large kettles which could only be of use for souls; one had a hole through it, another had no handle, and the third was of scarcely more value. I saw very few Porcelain collars: it is true, they put many on the bodies . . . (Thwaites 1896–1901: Vol. X, 293–7).

Fig. 1. Scaffold grave of a Crow Indian from the vicinity of Big Horn River. (Courtesy National Anthropological Archives, Smithsonian Institution)

So many features in a Huron ossuary excavated by Kidd (1953) coincide with the details provided by de Brebeuf, including dimensions of the pit, placement of complete skeletons on the floor of the ossuary, presence of sandy soil, and association of copper kettles and other specific artifacts, that it seems likely to be the one described.

Unfortunately, few prehistoric cemeteries can be assigned even a tribal affiliation, much less correlated with a specific ceremony. Most can be identified only with a cultural horizon or tradition that lasted several hundred years. Even in such situations, however, the investigation should begin with an exhaustive review of the literature on the area.

Historic Records

If there is reason to believe the interments are very recent, a survey of nearby known historic cemeteries may be called for. Most local officials and long-term residents know the locations and histories of local cemeteries and family burial grounds. Government and church records usually reveal considerable detail about the size of cemeteries and the individuals buried. Local police departments may have reports of vandalism that assist in identifying disturbed graves.

Data from Previous Excavations

Investigations previously conducted at related sites may provide enough information on the location and position of burials to eliminate the need for preliminary testing. This was the case when T. Dale Stewart and I were informed by a land owner of the discovery of human skeletal material on her farm in Maryland in the spring of 1971. The find was only 30 to 40 meters from an ossuary that Stewart had excavated 18 years earlier. Careful review of Stewart's unpublished field notes and the relevant ethnohistorical literature on ossuary burial permitted us to plan an excavation that not only preserved all the skeletal material with minimal damage, but also produced the data necessary to resolve ambiguities in the literature (Ubelaker 1974).

Similarly, museum skeletal collections from the vicinity may provide clues to the physical type of the aboriginal inhabitants and indications of the quality of bone preservation. This kind of information may prove critical for establishing the relative antiquity of the remains.

Types of Surface Indications

After the published information has been reviewed, the surface of the ground should be examined for indications of plowing, burrowing, or erosion, because these disturbances often reveal subsurface features. At the Mobridge site in northern central South Dakota, the location of a cemetery containing several hundred skeletons was revealed by glass trade beads and fragments of human bone observed in the backdirt around a rodent burrow during a Sunday afternoon stroll.

Clues may also be provided by construction activities or stream erosion that cut through the topsoil. We discovered another site in northern South Dakota by exploring the river banks from a boat. One burial had been cross sectioned by the water. The skeleton was covered with a slanting layer of wooden sticks, which were separated from the surface of the ground by about 30 cm. of sterile topsoil (Figs. 2–3). This information on the depth and construction of the grave covering allowed us to remove the topsoil rapidly with minimal damage to the archeological features.

Local residents are other sources of useful information. Farmers may have observed pottery, bone, or other kinds of evidence while working the soil. Neighbors may recall accounts of such discoveries and remember their exact locations. Prior to excavation, it is advisable to visit as many people as possible and to show them pictures or examples of the objects you are looking for. This procedure helps elicit interest and may increase the willingness of the owner of the property to grant permission for trespass or excavation.

Topography may also provide clues. A grave may be revealed by a small mound or a depression; a cemetery may have the form of a large earthwork. When interment was within the village, the distribution of potsherds and other refuse on the surface of the ground helps to delimit the general area for investigation. Ceremonial activities associated with burial sometimes produce telltale debris. For example, the Feast of the Dead conducted by the Huron brought together great numbers of people at the time of interment in the ossuary. Consequently, the amount of domestic refuse associated with such sites is considerably greater than in the vicinity of the burial areas of agricultural tribes in the Plains. Accumulation may be sufficient to alter the acidity and texture of the soil, which affects the vegetation. An aerial view of the habitation area of the Nordvold site in South Dakota shows this clearly (Fig. 4). The variation in the growth is so great that individual houses can be recognized. Such dramatic contrasts are seldom seen in cemetery sites, but subtle distinctions may exist.

Detecting Subsurface Features

Some circumstances warrant searching for human remains where no surface clues exist. In a forensic case, bones may be dispersed or the location of suspected disposal may be known only vaguely. In archeological sites, such as Middle Eastern tells, cemeteries presumably associated may be obscured by sedimentation. Since they can reveal considerable information on disease, diet, longevity, status, and other biological and cultural aspects of the populations, their discovery is desirable.

An early example of successful use of a linear electrode device for detecting subsurface bones is Helmut De Terra's discovery of Tepexpan Man in the Valley of Mexico. Exploration of an area of 1,360,000 square meters along the shore of a Pleistocene lake revealed a small sector with unusually high electrical resistance. Excavation brought to light one of the oldest known human skeletons in the New World (DeTerra 1949).

Fig. 2. Cross section of a burial at the Rygh site, South Dakota. A few bones projecting from the bank of the river led to discovery of this cemetery. Clearing the surrounding area showed that the grave was covered with a sloping layer of sticks, seen here in profile.

Fig. 3. View from above after excavation, showing the arrangement of the sticks used to cover the grave in Figure 2.

Fig. 4. Aerial view of the Nordvold site, a prehistoric village in South Dakota. Variation in the composition and humidity of the soil causes differences in the vegetation, revealing the locations of circular houses inside a large defensive palisade.

When metal objects are associated, as commonly occurs in forensic cases, historic burials, and some prehistoric cultures, metal detectors have been successfully used (Morse, Duncan, and Stoutamire 1983). Proton magnetometers have also revealed burials (YaKubik et al 1986). Other potentially useful techniques include ground-penetrating radar and soil resistivity (Thomas 1987).

Sampling Considerations

When not restricted by factors of an extra-scientific nature, such as imminent destruction, the choice of the site and the intensity of its excavation should be based on several kinds of considerations.

A long-range program directed toward reconstructing paleodemography should begin with a survey of the region to be studied. This permits selection of a site or sites for excavation based on cultural affiliation, relationship to village remains, degree of disturbance, state of preservation, and other kinds of factors.

After the site has been selected (or if the selection is dictated by the circumstances of discovery), the excavator must decide whether to uncover the entire area or to secure a sample. To a large extent, this choice depends on two factors: (1) the amount of funds and time available, and (2) the type of information desired.

b

Fig. 5. Using an elevated scraper to remove overburden at the Larson site, a prehistoric cemetery in South Dakota. **a,** Removing the sterile topsoil. **b,** Close-up of the clean surface, which facilitates the discovery of pits and other features.

The larger and more representative the sample, the greater the reliability of the interpretations; hence, an effort should be made to acquire as much data as possible. Some archeologists prefer to test part of a site systematically and leave the rest undisturbed for the future, when techniques may be improved. This approach may not be advisable for unprotected sites where excavation and publication would attract attention and enhance the risk of destruction by looting.

If the entire site cannot be excavated, the fact that great spatial variability in burial pattern often exists in prehistoric cemeteries must be taken into consideration in planning the fieldwork. To ascertain and compensate for possible spatial variability, burials should be obtained from as many parts of the cemetery as possible. One method is to divide the surface into numbered squares and select at random those to be excavated. If a specific hypothesis is being tested, this may determine which areas are included in the sample.

PRELIMINARY FIELD PROCEDURES

Use of Test Pits

Even if a selected sample is sought, it is advisable to define the limits of the cemetery as accurately as possible before excavation begins. The best way to do this is by systematic use of test pits. These should be large enough to allow the excavator to work comfortably and small enough to produce the desired information with minimal expenditure of time. Traditionally, pits five feet square have been employed in North America and one meter square in Latin America.

Test pits not only delimit the cemetery, but also provide data on stratigraphy, soil texture, and archeological conditions. When cultural refuse is plentiful, good samples of faunal remains, ceramics, and other types of artifacts may be obtained. The possibility that the stratigraphy was disturbed by intrusion of the burials or that the cemetery is later than and unrelated to the village must be kept in mind in evaluating the trends in pottery types.

Heavy Power Equipment

Heavy machinery can be used for large-scale, rapid excavation of cemeteries in certain situations. Bass employed an elevated scraper (Fig. 5a) to remove the sterile topsoil over large areas and reveal individual graves in cemeteries threatened with inundation in the vicinity of Mobridge, South Dakota. The elevated scraper has the advantage over a bulldozer or other earth-moving machines because it collects all the soil and leaves a smooth surface that may be examined for evidence of burial pits (Fig. 5b). Bass directed the scraper to deepen a trench about 3 inches at a time until a pit outline, wood covering, or (rarely) bone was encountered. This procedure is rapid and economical, and does minimal damage to the skeletons. It is best suited to excavation of cemeteries that (1) consist of graves scattered over a wide area (Fig. 6); (2) are covered with a layer of sterile soil, and (3) possess soil conditions that make pit outlines easy to distinguish. In some situations, heavy power equipment can be used to locate and define the limits of a cemetery, after which a grid system can be established for more exact spatial control.

Use of a Grid

Exact recording of the location of each burial is essential. This can be achieved in two ways: (1) by employing a transit or alidade to map each feature or (2) by superimposing a grid and plotting the coordinates of each grave within a square. The first method may be more convenient if the burials are widely spaced over a large area. In such situations, I prefer to use an

Fig. 6. Aerial view of the Larson site, South Dakota. The parallel rows left by the elevated scraper show clearly, as well as the pits that remained after the burials were excavated.

alidade because it permits all the measurements to be made from a single central point within the cemetery. If the cemetery is relatively small (30 by 30 meters or less), a grid system generally provides the easiest and best spatial control both for the burials and for artifacts and features around or between them. In some situations, the two approaches can be combined. A grid may be employed for recording the locations of the graves and an alidade used to map the relationship of the grid system to natural features in the surrounding area.

I used the grid technique during 1973 in the excavation of a late precolumbian cemetery on the south coast of Ecuador (Ubelaker 1981). The large pottery jars employed as "coffins" contained as many as 23 skeletons, most of them completely disarticulated. Although the site appeared to contain hundreds of these jars, limitations of time and labor required me to confine the initial excavation to an area 8 by 10 meters. This produced nearly 50 urns. Since they had been interred in a habitation site representing an earlier culture, I wanted to record not only the locations of the urns, but also all artifacts and faunal remains in the surrounding earth. I consequently divided the area into 2 by 2 meter squares and excavated each square in artificial levels 10 centimeters thick (Fig. 7). This procedure forced the workmen to expose the urns slowly and also permitted accurate recording of the depth and position of artifacts. Although the stratigraphy of the habitation refuse was disturbed in the immediate vicinity of the urns, there were gaps between urns that might not have been affected. This possibility could not be assessed until the pottery was analyzed, however. In such cases, the probability of obtaining useful stratigraphic results must be weighed against the desirability of acquiring a larger sample of skeletal remains.

For surface forensic discoveries, a grid provides a context for mapping. When ground cover is dense, the vegetation must be cleared to within three centimeters (one inch) of the ground surface (Fig. 8). The clearing should be extended well beyond the imme-

Fig. 7. A grid of two-meter squares superimposed over the portion of a cemetery to be excavated on the coast of Ecuador. The locations of all burials, pottery, and other artifacts were mapped with reference to these squares.

a

b

Fig. 8. Locus of a forensic discovery. **a,** Before clearing surface debris. **b,** After clearing. (After Morse et al 1983).

Fig. 9. A slightly darker color differentiating the fill of a grave from the undisturbed soil at the Mobridge site in South Dakota.

diate area of discovery, since foraging animals may have dislocated parts of the skeleton. Remember also that crania and other bones have a tendency to roll or be displaced downhill from their original location. All bones and other features of forensic interest should be recorded by measuring from the item to at least two points on the grid.

PRIMARY BURIALS

Initial Soil Removal

After the locations of the skeletons have been established within the confines of the cemetery, excavation can begin. The soil must be removed from above and around each skeleton and its associated artifacts to allow observations and measurements of important features. Unfortunately, this procedure is more complicated than it first appears. The initial step is to define the boundary of the pit at its uppermost level. If a contrast in soil color or texture led to its discovery, the complete outline can be exposed by scraping the surrounding area clean (Fig. 9). The dimensions, orientation, and shape of the pit should be recorded, as well as the depth of the rim below the surface of the ground. The latter measurement indicates the amount of earth or residue deposited since the pit was filled. The contour of the pit must be examined carefully for indications of disturbance by subsequent interments (Fig. 10), since this may explain the existence of an incomplete skeleton (Fig. 11) or of artifacts at lower depths than expected.

Exposing the Skeleton

After the outline of the pit has been defined and recorded, excavation should proceed slowly downward. This may be done either by (1) removing the soil inside the pit or (2) clearing a large area around the pit. The first method facilitates measurements and description, since it restores the original shape as closely as possible. If the pit is small, however, working within its confines may be awkward or impossible; photography and note taking are also difficult. Consequently, the second approach is usually more practical. Enough dirt should be removed to permit freedom of movement and observation. A sufficient number of measurements must be made of the outline at different depths to permit reconstruction of the shape and dimensions of the pit after it has been destroyed.

Fig. 10. Clearing the surface to expose a burial at the Mobridge site, South Dakota. The dramatic contrast in soil color identifies a pit recently dug into the aboriginal grave.

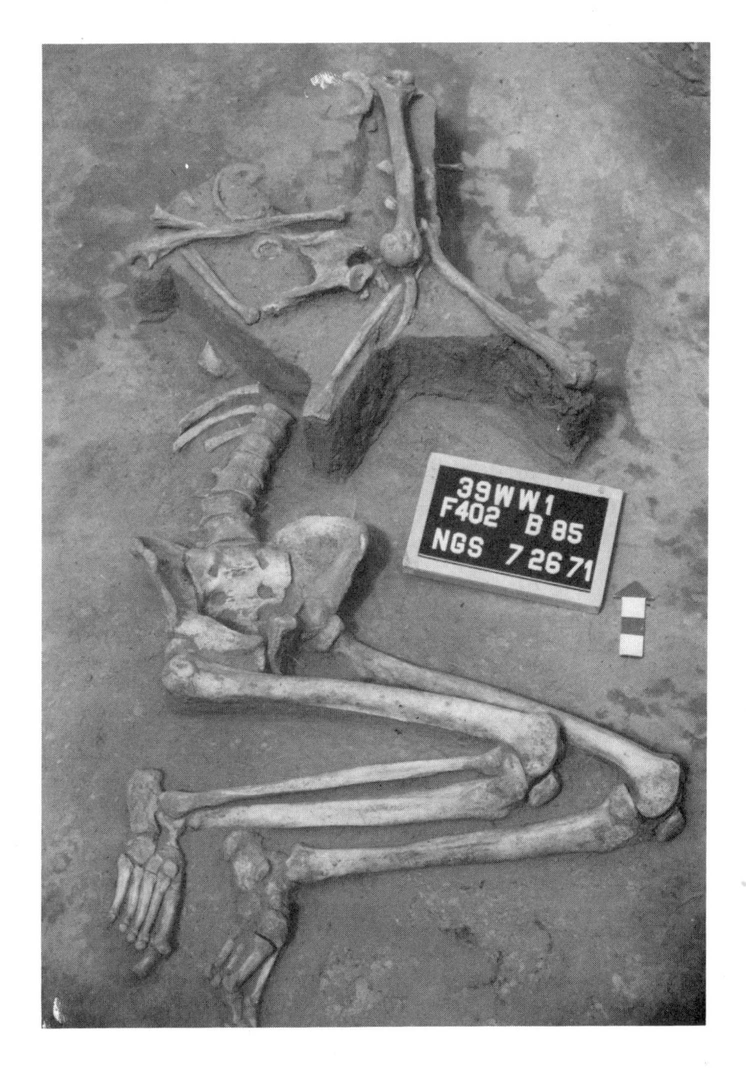

Fig. 11. A disturbed primary burial. The articulated spinal column, pelvis, and legs of the skeleton are visible in the lower two-thirds of the photograph. The upper part, represented by the disarticulated bones, was disturbed by digging an intersecting grave for another individual.

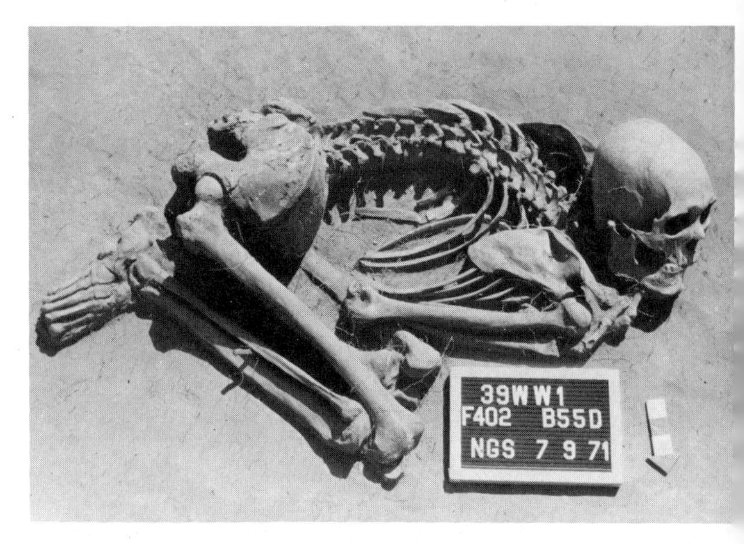

Fig. 12. An excavated skeleton, showing the kind of identifying label, scale, and arrow pointing north that should be included in photographs of burials.

Whichever method is chosen, the excavator should record all changes in the color and texture of the soil and any other details (such as roots.) While this must be done carefully, it should also proceed as rapidly as possible to minimize damage to the bones from exposure to sun and air. As the bones are encountered, they should be left in place and any earth adhering to them should be removed. This treatment applies to any artifacts associated.

Photography

The skeleton and associated artifacts should be photographed and described immediately after exposure is completed. Both black-and-white and color pictures should be taken to provide a permanent visual record of the over-all position and arrangement, and to preserve details that are difficult to describe or may be overlooked when the notes are taken. Whenever possible, a label with the feature number, site number, date, and other information relevant to future identification of the burial and its contents should be included in the photograph (Fig. 12). An arrow pointing to magnetic north should be included for orientation, along with a scale. Before photography, all extraneous features, such as footprints, lumps of dirt, and digging tools, should be removed. The soil around the skeleton should be level and smooth to provide a uniform background.

Shadows reduce clarity and should be minimized by waiting until the entire feature is in the shade, by using an electronic flash, or by reflecting sunlight into the part of the area in shadow. Electronic flash offers the most efficient control, but the apparatus is easily damaged in field situations. A reflector board composed of plywood covered with aluminum foil is an inexpensive alternative (Fig. 13). Excellent results can be obtained by shading the entire feature using a large cloth (bed sheet) and adjusting the exposure of the camera for the reduced intensity of the light.

Fig. 13. Using a reflector made from a board covered with aluminum foil to create more even illumination when photographing a burial.

Description

Recording should be as thorough and objective as possible, making use of sketches and diagrams to complement the narrative. The location, deposition, position, orientation, and depth of the skeleton must be recorded, along with complete measurements of the bones, artifacts, and pit. Terminology proposed by Griffin and Neumann (1942), Bass (1962), and Sprague (1968) has been reviewed in compiling the following recommendations.

Location. All bones and artifacts within a pit should be recorded in both horizontal and vertical dimensions. The distances from each bone or artifact to reference points on the margin of the pit or in the natural stratigraphy should be measured and incorporated in the notes or used to make scale drawings or both.

Deposition. This term refers to the over-all configuration of the body in the ground. Common varieties include lying on the back, on the face, or on the side (left or right); sitting, or standing. Intermediate positions may occur. Terms such as "prone," "supine," "reclining," "dorsal," "ventral," and "lateral" appear in the literature, but should be avoided because they are vague and subject to misinterpretation.

Position. Sprague (1968:481, after Anderson 1962:159) defines this as "the relationship of segments of the body to each other. Position concerns only the body and is not described in reference to the grave, axis of the earth, cardinal directions, or any other natural features; rather, it should be thought of as if the body were suspended in space." Unfortunately, satisfactory description of this relationship has proved difficult to judge from the picturesque terms in the literature. "Contracted," "fetal," "squatting," "hunched," "frog," "bunched," "folded," "closely flexed," "jitterbug," and "crouched" are a few examples. Difficulties are compounded by inconsistent definitions. Both Sprague (1968) and Bass (1962) employ "extended," "semiflexed," "flexed," and "tightly flexed," but define them differently.

The position of a skeleton is best described by reference to three anatomical components: (1) the legs, (2) the arms, and (3) the head. Sprague (1968) recommends using "extended," "semiflexed," "flexed," and "tightly flexed" to describe leg positions. *Extended* indicates the legs are straight, joining the trunk at an angle approaching 180 degrees (Fig. 14). *Semiflexed* applies when the angle between the axis of the trunk and the axis of the femur is between 90 and 180 degrees (Fig. 15). *Flexed* signifies an angle less than 90 degrees between the axes of the trunk and femur (Fig. 16). *Tightly flexed* means the angle approaches zero (Fig. 17). While this classification is an improvement over previous efforts, it is incomplete because it does not describe the position of the lower part of the leg. For greater accuracy, I recommend applying Sprague's terms separately to the upper and lower parts and providing estimates of both angles. Thus, the position shown in Figure 18 would be recorded as left upper leg semiflexed, 120 degrees; left lower leg flexed, 25 degrees; right upper leg semiflexed, 120 degrees; right lower leg tightly flexed, 10 degrees.

Arm position generally falls into one of four categories: (1) extended beside the body (Fig. 19), (2) crossed on the pelvis (Fig. 20), (3) folded over the chest (Fig. 21), and (4) raised toward the head (Fig. 22). Exceptions are common, however, and each arm may also be treated differently (Fig. 15). The exact locations

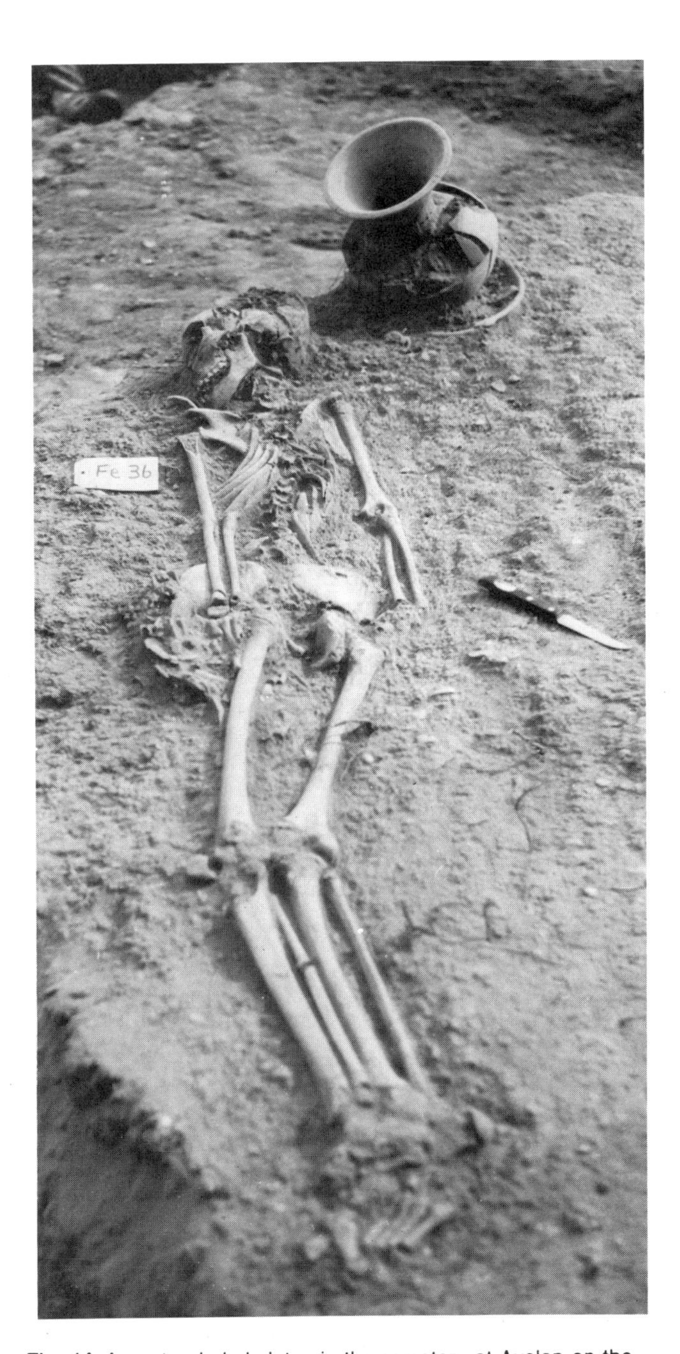

Fig. 14. An extended skeleton in the cemetery at Ayalan on the coast of Ecuador. The legs join the torso at an angle of about 180 degrees. The knife serves as an improvised scale.

of the arm and leg bones should be recorded, preferably using a sketch.

To describe skull position, Sprague (1968:482) advocates the term "looking" combined with a direction:

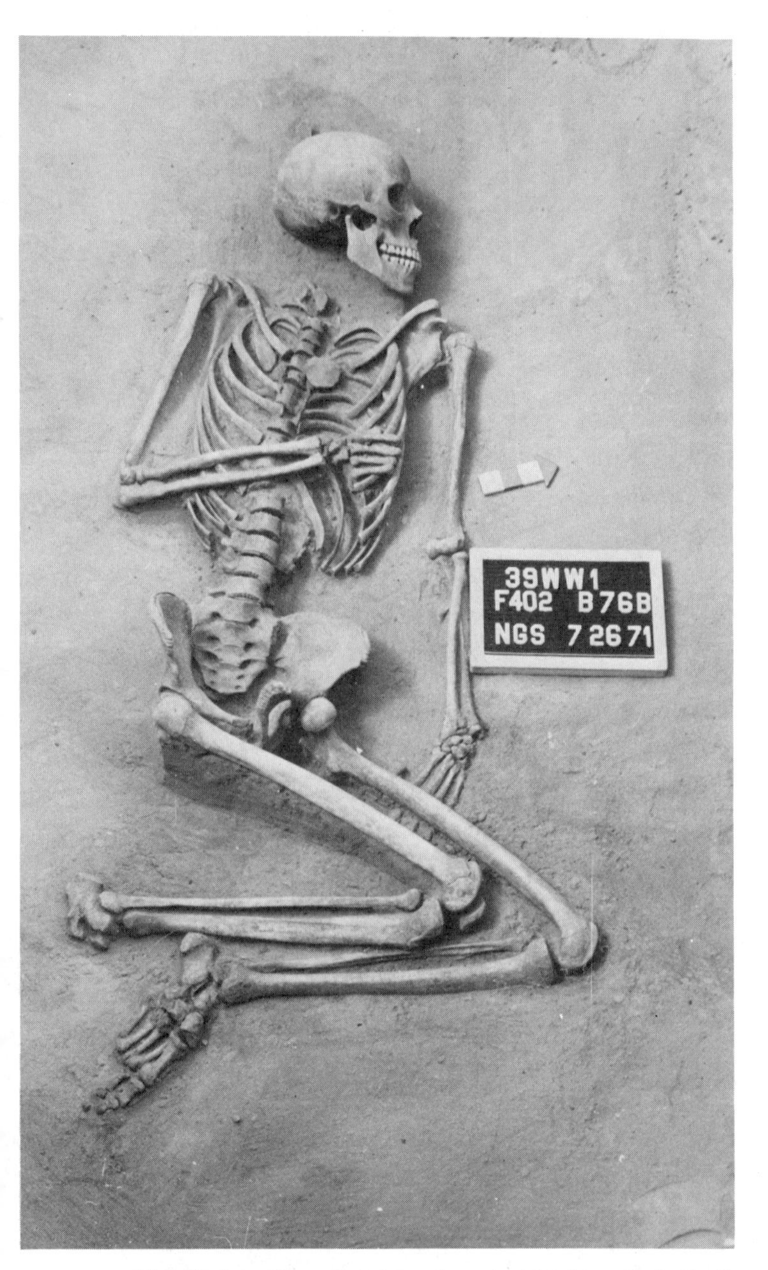

Fig. 15. A semiflexed skeleton from the Mobridge site in South Dakota. The upper legs join the torso at an angle greater than 90 degrees but less than 180 degrees.

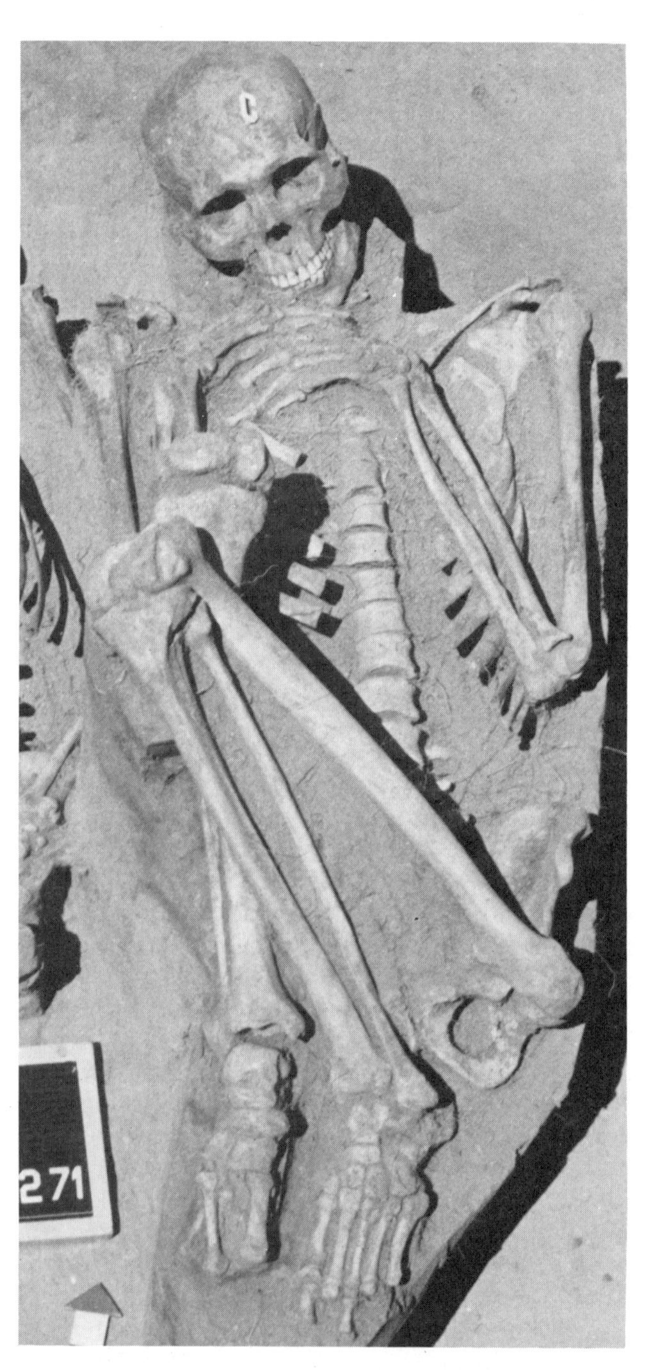

Fig. 16. A flexed skeleton from the Mobridge site, South Dakota. The upper legs form an angle less than 90 degrees with the axis of the torso.

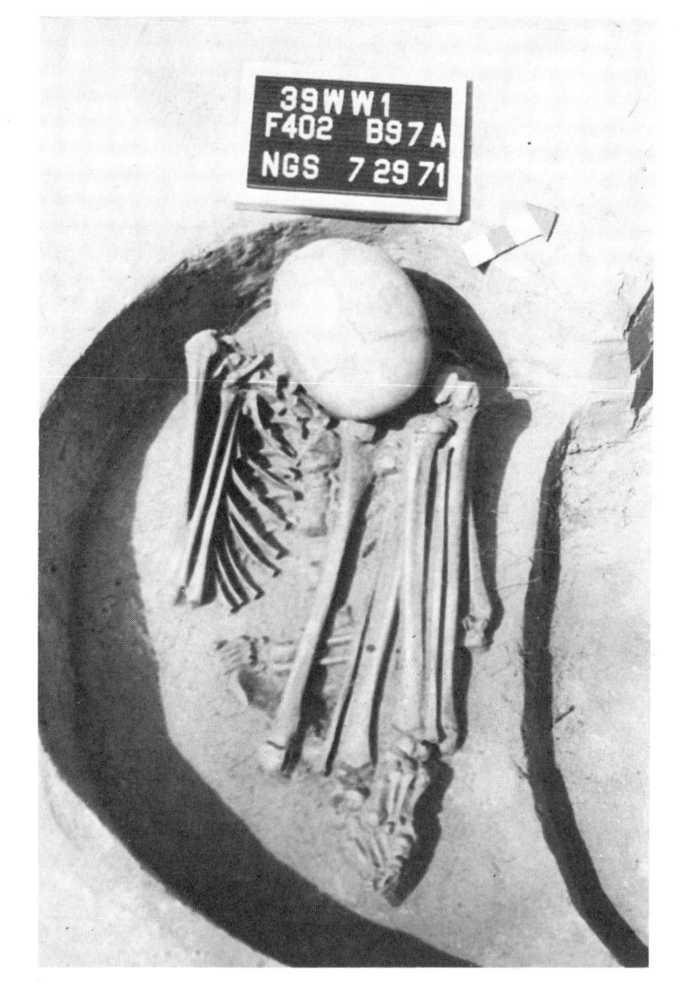

Fig. 17. A tightly flexed skeleton from the Mobridge site, South Dakota. The angle between the axis of the torso and the upper legs approaches zero.

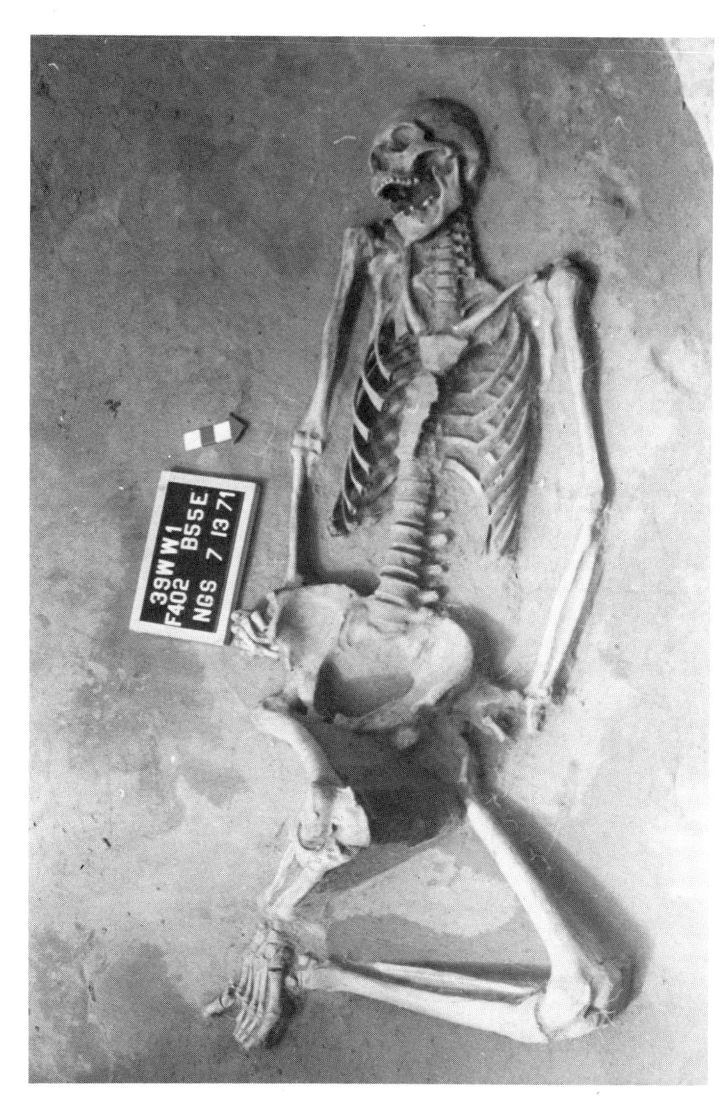

Fig. 19. A skeleton with the arms in an extended position.

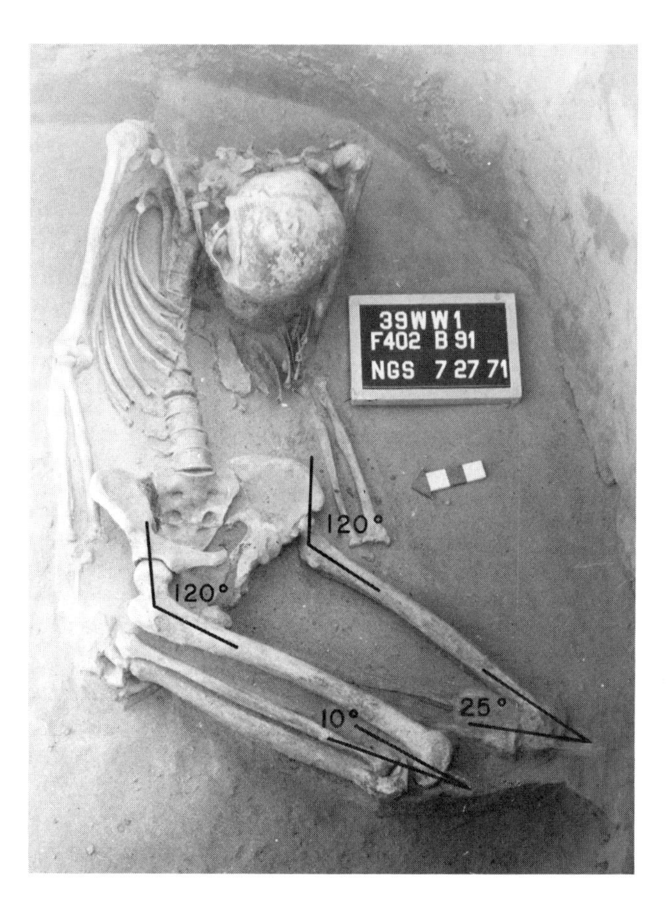

Fig. 18. A skeleton from the Mobridge site, South Dakota, illustrating the terms for describing the positions of the upper and lower parts of the leg. Both upper legs form an angle of about 120 degrees in relation to the axis of the torso, placing them in the semiflexed category. The lower left leg joins the femur at an angle of 25 degrees, categorizing it as flexed. The angle between the lower right leg and the femur is 10 degrees, classifying it as tightly flexed.

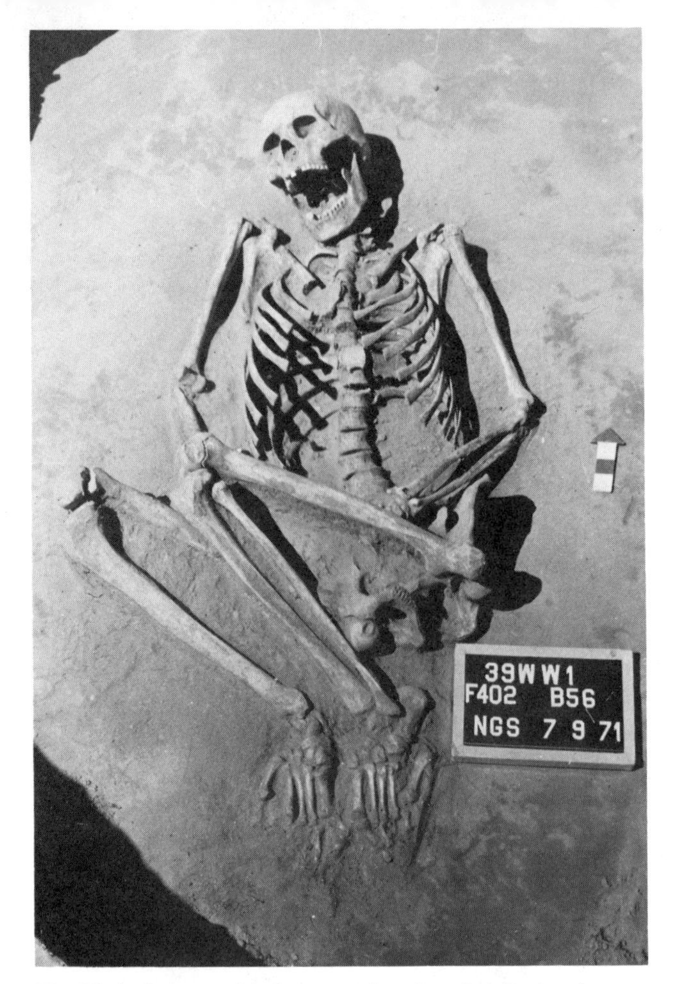

Fig. 20. A skeleton with the arms placed so that the hands cross on the pelvis.

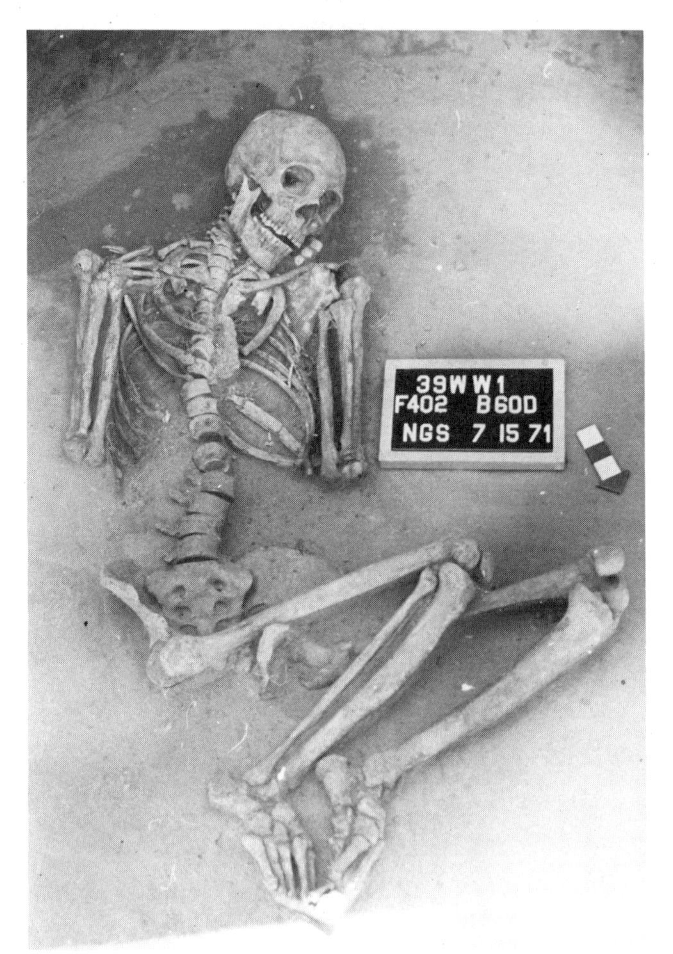

Fig. 22. A skeleton with the arms raised toward the head.

Fig. 21. A skeleton with the arms folded on the chest.

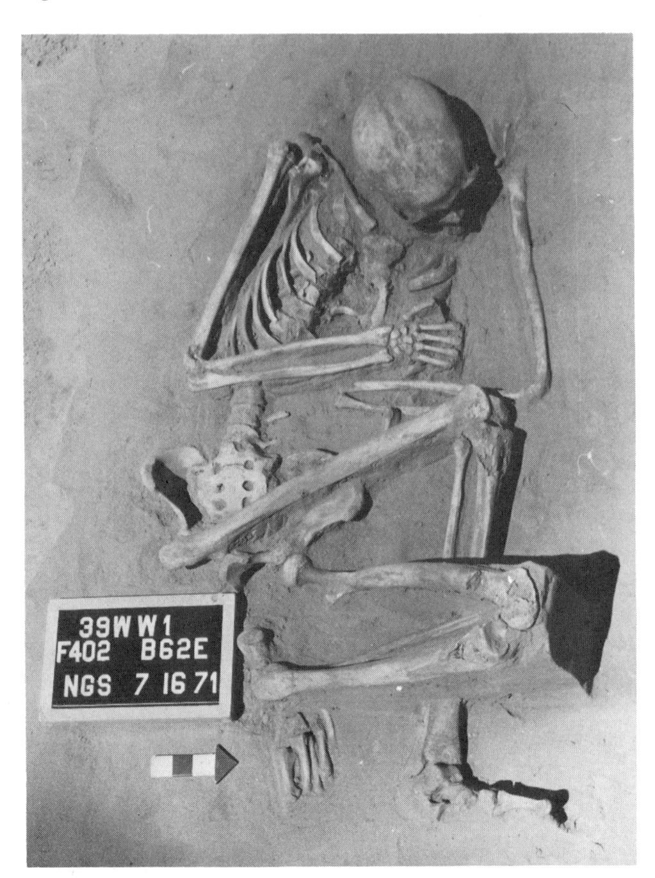

"looking left," "looking right," "looking straight ahead." Other variations include "chin compressed to the chest" and "head extended backward." I prefer a more anatomical description, such as "resting on the right parietal facing northeast" or "resting on the occipital facing upward." The more detailed the field description, the better. Analysis may reveal consistencies in burial position, but a field record that is overly generalized can never be elaborated after the bones have been removed.

Orientation. When a skeleton is lying on the back, on the face, or on the side, its orientation is "the direction in which the head lies in relation to a line between the skull and the center of the pelvis" (Heizer 1958:65). Orientation of a seated burial is the direction the body is facing (Sprague 1968:482). Orientation should be recorded in terms of compass directions, natural landmarks, cultural features, or preferably combinations of these reference points. A burial might be described as oriented "16 degrees east of south, toward Feature B and the village area." The direction of the head should be specified.

Fig. 23. Measuring the depth of a burial. A line has been placed across the excavation at the surface of the ground and secured in a horizontal position using a line level (nearly concealed beneath the measuring tape). The distance is being measured vertically from this line to the bones.

Depth. A line level (Fig. 23) and a metric tape or rod should be used to measure distances from the present surface of the ground and all significant stratigraphic features to the highest and lowest parts of the skeleton, all artifacts, and any bones whose position or significance warrants special attention. Care must be taken to keep the line taut because slack will cause errors in the measurements.

Measurement. Distances between the edge of the pit and items inside, as well as between items should be measured. Measurements of the skeleton should include maximum length and width, and any other dimensions that will aid in description or possible reconstruction. The positions of all artifacts, patches of discolored soil, and other details can be recorded using two reference points on the skeleton. For example, "the scraper is 0.6 meters west of the right greater trochanter and 0.3 meters south of the distal end of the right ulna" (see Figs. 85 and 161 for identification of parts of the skeleton). The exact location of an artifact may provide a clue to its function. This was the case in a cemetery attributed to the Arikara Indians; examination of the positions of blue glass pendants indicated that they were not only worn on forelocks in the hair, but also on other parts of the body (Ubelaker 1966, Ubelaker and Bass 1970, Howard 1972).

Observations

Pathological Conditions. Any abnormality in the bone should be described in detail, since some of the evidence may be destroyed by disarticulation and removal of the bones. This is especially likely if the disease or the composition of the soil have accelerated deterioration. Pathological lesions, healed fractures, and other unusual conditions should be photographed in situ. Descriptive terms applicable to bone pathology are provided by Aufderheide and Rodriguez-Martin (1998), Brothwell (1972), Morse (1969), Anderson (1962), and Ortner and Putschar (1981).

Perishable Materials. Hair, skin, cloth, leather, and other perishable types of evidence may survive even in humid contexts because of unusual circumstances. Great care should be taken to describe such materials before their removal and to preserve them for technical study.

Soil Samples. Samples of earth should be collected from beneath and around skeletons. Analysis of the chemical composition and acidity of the soil may explain preservation, erosion or other conditions of the bones. These samples may contain important particles, such as kidney stones or pollen grains, too small to be detected during excavation. All soil removed during excavation of a pit should be sifted through a fine screen to recover small artifacts and fragments of bone. Particular attention should be given to the recognition of infant bones, some of which are so small they are not likely to be discovered without screening.

Non-cultural Items. Natural inclusions, such as insect remains, may provide important information about the season when the burial took place or details about the procedure. An example is provided by great quantities of chitinous exuvia of Calliphorid or Sarcophogid fly pupae with skeletons in archeological cemeteries in South Dakota (Gilbert and Bass 1967). Metamorphosis of these flies (commonly known as blue bottle, green bottle, and flesh flies) involves two stages between the egg and the adult: (1) a larval or maggot stage and (2) a pupal or resting stage. The exuvia represent the pupal stage. Since these flies emerge in late March and disappear during mid-October, the presence of exuvia indicates that burial occurred during this interval. Ubelaker and Willey (1978) noticed that the fly pupal cases were open, implying the adults had emerged, yet no adult parts were found near the skeletons. This suggested that the skeletons were exposed above ground for at least three weeks prior to burial, probably as a regular part of the funerary procedure. Close inspection also revealed parts of a genus of beetle *(Trox),* which is usually one of the last visitors to decomposing carrion. Since the beetle cannot dig, its presence implies the bodies were above ground much longer than previously suspected. Although these data are not conclusive, they show how seemingly irrelevant phenomena can offer leads for interpretation.

SECONDARY BURIALS

Definition

Secondary inhumations consist of non-articulated collections of bones. They represent a complicated method of treatment of the dead involving two or more stages. The first is removal of the flesh, which may be accomplished with tools or by allowing decomposition to proceed naturally above or below ground. The second stage is collection or disinterment of the bones, which may be kept briefly or for many years. The third stage is reburial individually or in a mass grave. Some of the variations employed in North America were described by early explorers.

Choctaw. One of the most vivid descriptions of intentional defleshing is the account of Choctaw practices written in 1775 by Bernard Romans. After death, the body was placed upon a stage above ground and left for three or four months. According to Romans,

> A certain set of venerable old Gentlemen who wear very long nails as a distinguishing badge of the thumb, fore and middle finger of each hand, constantly travel through the nation. . . . the day being come, the friends and relations assemble near the stage, a fire is made, and the respectable operator, after the body is taken down, with his nails tears the remaining flesh off the bones, and throws it with the entrails into the fire, where it is consumed; then he scrapes the bones and burns the scrapings likewise; the head being painted red with vermillion is with the rest of the bones put into neatly made chest . . . and deposited in the loft of a hut built for that purpose, and called bone house; each town has one of these; after remaining here one year or thereabouts, if he be a man of any note, they take the chest down, and in an assembly of relatives and friends they weep once more over him, refresh the colour of the head, paint the box red, and then deposit him to lasting oblivion (Romans 1961: 61–2).

Such an interment, if found archeologically, would consist of a collection of non-articulated bones, perhaps with some red pigment remaining on the skull. Since the flesh was removed with the finger-nails, there would be no cut marks, although the scraping may have altered the surface of the bone.

Huron. Most secondary burials represent interment of the bones after natural decomposition of the flesh. A well known example is the Feast of the Dead ceremony and subsequent ossuary burial practiced by speakers of Iroquoian and Algonquian languages in the Great Lakes area of the United States and Canada. The ceremony was described in detail by a Jesuit missionary named Jean de Brebeuf, who observed it at the Huron village of Ihonatiria on July 16, 1636. It was held every 10 to 12 years, at which time the Huron gathered the bones of all individuals who had died since the previous ceremony. After feasting, they buried the remains in a single large pit. The majority of the individuals had been dead long enough for the skeleton to be clean of flesh and totally disarticulated. Since all those who had died since the last ceremony were included, however, some were less completely decomposed. Concerning these, de Brebeuf wrote:

> The flesh of some is quite gone, and there is only parchment on their bones; in other cases, the bodies look as if they had been dried and smoked, and show scarcely any signs of putrefaction; and in still other cases they are still swarming with worms. When the friends have gazed upon the bodies to their satisfaction, they cover them with handsome Beaver robes quite new: finally, after some time they strip them of their flesh, taking off skin and flesh which they throw into the fire along with the robes and mats in which the bodies were wrapped. As regards the bodies of those recently dead, they leave these in the state in which they are, and content themselves by simply covering them with new robes. Of the latter they handled only one Old Man, . . . who died this Autumn on his return from fishing: this swollen corpse had only begun to decay during the last month, on the occasion of the first heat of Spring: the worms were swarming all over it, and the corruption that oozed out of it gave forth an almost intolerable stench; and yet they had the courage to take away the robe in which it was enveloped, cleaned it as well as they could, taking the matter off by handfuls and put the body into a fresh mat and robe, and all this without showing any horror at the corruption (Thwaites 1896–1901: Vol. X, 283–5).

This type of secondary burial differs from that practiced by the Choctaw in the range of variation in disarticulation. Recently deceased individuals, such as the one described above, were interred before decomposition was advanced and evidence of articulation will survive in the ossuary. More than 200 ossuaries have been recorded in the Canadian province of Ontario alone (Anderson 1964), although few have been excavated professionally or remain undisturbed (Kidd 1952:73). Ossuaries reflecting similar mortuary procedures have been reported in the mid-Atlantic region (Ubelaker 1974) and the Central Plains (Strong 1935). Smaller secondary deposits and isolated urns are relatively common throughout the New World.

In forensic cases, the discovery of non-articulated remains or isolated articulated skeletal parts indicates a complex sequence of events prior to deposition at the site of discovery. Secondary interments may represent burial of parts of a body severed through criminal activity or, more likely, reburial of isolated bones or parts after the soft tissue was destroyed through natural decomposition.

Excavation Approaches

After secondary burials have been encountered, an appropriate excavation approach can be selected. This choice will depend mainly on two factors: (1) the size of the burial deposit and (2) the type of data desired. Size is an important factor, since the material should be removed as rapidly as feasible after exposure to avoid damage. Examples of the procedures employed in excavating two kinds of secondary burials will illustrate some of the problems and solutions more concretely than a general discussion.

Large Ossuaries. During the summers of 1971 and 1972, T. Dale Stewart and I excavated a Late

Fig. 24. Map showing the locations of Ossuaries I and II at the Juhle site in Maryland in relation to topography and modern structures. Note the inclusion of a scale and an arrow designating north.

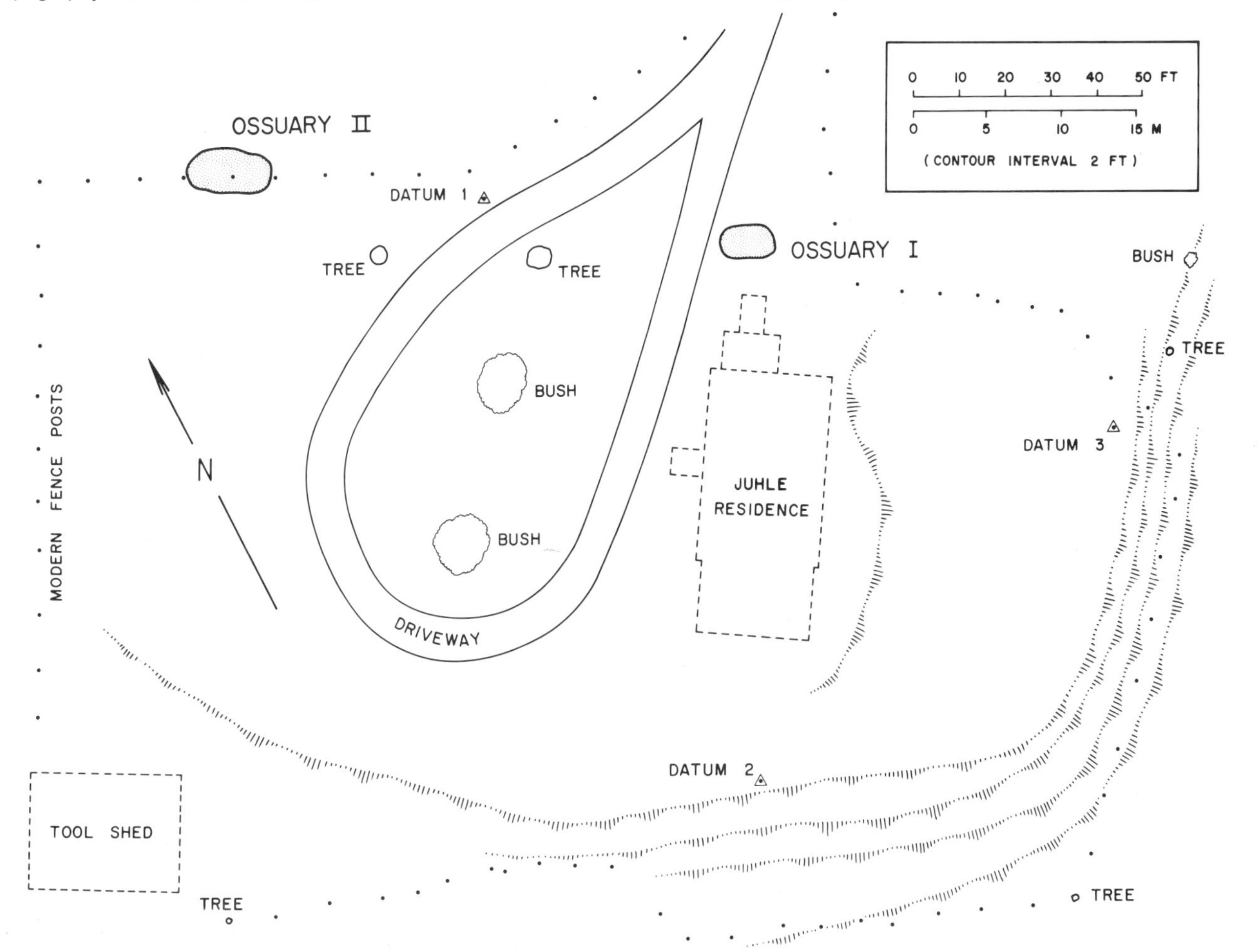

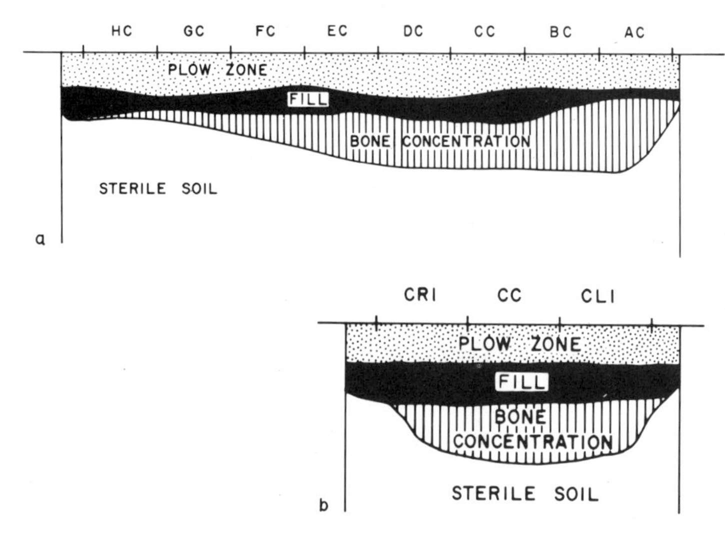

Fig. 25. The stratigraphy of Ossuary II at the Juhle site, Maryland. **a,** Longitudinal view, showing the sloping floor of the pit **b,** Cross section.

Woodland ossuary at the Juhle site in southwestern Maryland (Fig 24). Preliminary testing revealed that the pit was in the form of a rectangular trench with rounded corners and was 5.2 meters long by about 2.1 meters wide. Depth tapered from 0.5 meters at one end to 0.9 meters at the other (Fig. 25). The bones represented the largely disarticulated remains of at least 188 individuals.

The large size of the ossuary created problems of excavation. To permit detailed observation of the arrangement of the bones, we wished to expose, photograph, and describe the skeletal material completely before removing anything. Since the area was so extensive, however, the length of time needed for clearing would require subjecting some of the bones to prolonged exposure, with possible damage. In addition, the density of the material created logistic problems. Where could we sit while digging? How could we remove the soil? How could we excavate in the center of the ossuary without damaging bones along the sides? Another complication was a three-strand, barbed-wire fence that accidentally bisected the ossuary longitudinally. The land owner preferred that the fence be left intact, so we had to work around it.

The above considerations led us to divide the ossuary into three longitudinal sections: one third to the left of the fence (as one faces northwest), one third to the right of the fence, and the middle third below the fence. Each section was about 0.6 meters wide by 4.8 meters long, and could easily be excavated, recorded, photographed, and cleared with minimal damage to the bones. If the notes, drawings, and photographs were accurate, a composite picture of the complete ossuary could be assembled. We first excavated and removed the contents of the left section (Fig. 26), then the right section (Fig. 27), and finally the center section (Fig. 28). In each section, the soil was removed without disturbing the bones so that observations and photographs could be taken of the material in situ. Figure 29 shows the northern corner before excavation; the outline of the pit is clearly defined by the darker color of the fill and the bone concentration on the floor of the ossuary is exposed in the profile. Figure 30 shows the same area after the soil has been removed.

Initially, we planned to preserve the original pit by removing the soil inside because this would permit observation of the relationships of the bones to the walls. This proved impractical because the excavator was forced to lean down (Figs. 31–32). Later, we placed boards across the excavation, enabling the excavator to lie above the bone concentration while working (Fig. 33). The best results were obtained, however, by removing the soil immediately outside the bone concentration to the depth of the pit (Fig. 34). Of course, as the soil was removed, exact records were kept of the shape of the pit.

Small Bone Concentrations. The excavation approach can be simplified when the concentrations of secondary burials are smaller. A cemetery of the late Integration Period on the southern coast of Ecuador contained large pottery jars buried upright and covered by a similar jar inverted over the mouth (Fig. 35). Inside were the largely disarticulated remains of as many as 23 persons. After the urns had been measured, described, and photographed, the upper and middle portions were removed. The fact that most urns were fractured facilitated this procedure (Fig. 36). All soil inside was carefully cleared away to reveal the arrangement of the bones. Excavation usually proceeded from top to bottom, and notes and photographs were taken as the bones were exposed. In some instances, we were able to leave the entire contents intact by approaching the excavation from the side (Fig. 37; Ubelaker 1981).

Fig. 26. Southwestern third of Ossuary II after exposure of the bones.

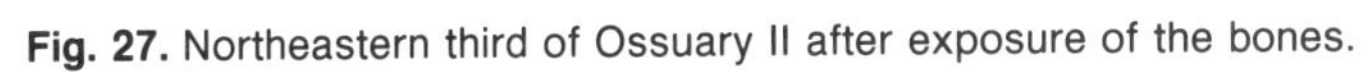

Fig. 27. Northeastern third of Ossuary II after exposure of the bones.

Fig. 28. Central third of Ossuary II after exposure of the bones. The upper three horizontal lines are part of the barbed-wire fence that bisected the feature. The lower two lines are part of the grid employed for mapping the bones.

Fig. 29. Northern corner of Ossuary II. The pit is clearly defined by the darker color of the soil. The bone concentration can be seen in the vertical wall of the cut.

Fig. 30. Northern corner of Ossuary II after the earth shown in Figure 29 was removed to expose the bones. The soil surrounding the pit has also been cut away to facilitate access and photography.

Fig. 31. Excavating Ossuary II. The workers are attempting to preserve the outline of the pit while clearing the earth from the bones. To do so, they are forced into awkward and uncomfortable positions.

Fig. 32. Close-up of an excavator working within the confines of the ossuary pit.

Fig. 33. Excavating Ossuary II. A board has been laid across the bone concentration so that the excavator can lie on it and reach the bones more easily.

Fig. 34. Using a small knife to loosen the soil around the bones in Ossuary II. Note the ease of excavation when the soil surrounding the bone concentration has been removed to the level of the skeletons.

Fig. 35. Excavating a cemetery at Ayalan on the south coast of Ecuador. Large pottery urns were used as "coffins." A grid of squares measuring 2 by 2 meters has been placed over the area to facilitate mapping.

Fig. 36. Cleaning the interior of a burial urn at Ayalan. The lid and rim of the jar have been removed. The urn was badly broken, permitting its piecemeal removal.

Fig. 37. The contents of a burial urn from the Ayalan cemetery, revealed by removing one side of the broken vessel. This jar contained the remains of 15 individuals.

Documentation of Bone Position

An important and difficult problem in excavating concentrations of secondary burials is documenting the positions of the bones. Their locations and relationships may shed light on the manner and sequence of deposition or provide other significant information on the mortuary procedure. When there are few individuals, as in the Ecuadorian urns, the best procedure is to record the exact position of each bone using diagrams, descriptions, measurements, and photographs. Accomplishing this is more difficult in large concentrations.

Ossuaries. In excavating the Maryland ossuary, Stewart and I wished to collect the data in a manner that would permit recognizing the bones belonging to the same individual and establishing whether individuals were distributed randomly throughout the pit. To achieve maximum precision in mapping, a grid of 0.6 meter squares was superimposed over the bone mass (Fig. 38). We chose this magnitude for three reasons: (1) it coincided approximately with the widths of the three longitudinal sections into which the pit was divided, (2) it was large enough to include a considerable quantity of bones, and (3) it was small enough to facilitate the detection of variations in distribution and the identification of scattered bones belonging to the same skeleton. Since the grid system was arbitrarily imposed over the bone concentration, many of the long bones extended into two or more squares. In these cases, the bone was assigned to the square containing more than 50 percent of its length.

In the laboratory, the bones from each square were analyzed independently. The total number of bones of each type (femora, tibiae, etc.) was listed, and the sex and age of adults were estimated. The contents of the squares were compared to see whether the spatial distribution of all variables was random. This analysis revealed that three groups of bones were not randomly distributed. These were: (1) large bones of both adults and subadults, (2) small bones of adults, and (3) small bones of subadults. This patterning suggested that when the time came for the ossuary burial, the Indians travelled to the places of initial deposition (probably scaffolds or death houses), gathered together all the large bones, and placed them in the ossuary. They may have made a second trip to collect the smaller bones left behind and deposited them separately. The fact that the smaller bones of adults were in a different part of the pit from those of subadults might imply that the scaffolds or death houses were segregated according to age and that this segregation was maintained during transfer of the bones. The existence of ethnohistorical information that distinctions between adults and subadults were important in many kinds of activities is in keeping with this interpretation. Although data must be obtained from other ossuaries to evaluate these inferences, they illustrate how employing spatial control during excavation of a mass of bones can lead to hypotheses about the mortuary activities.

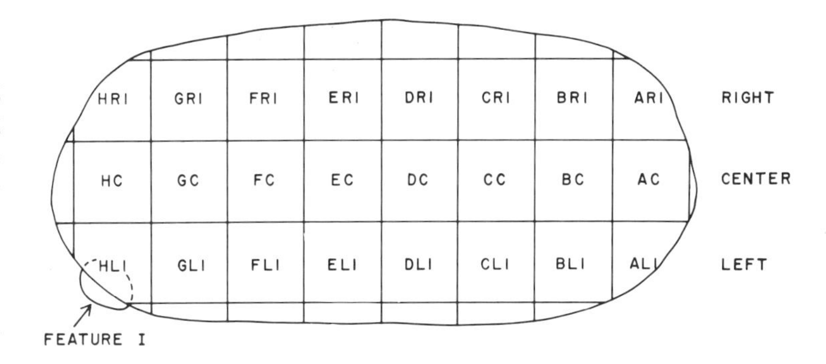

Fig. 38. The grid employed for excavating Ossuary II in Maryland. The bone concentration was divided lengthwise into three sections (right, center, left), which were crossed at regular intervals to produce squares measuring 0.6 by 0.6 meters.

Urns. Even when the concentrations are smaller, information on the position of the bones can be informative. An example is provided by Feature 30 in the Ecuadorian cemetery described earlier (Figs. 35–37). This consisted of a large pottery urn covered with the inverted lower part of a similar urn (Fig. 39). After the initial photographs, measurements, and descriptions were completed, we removed the cover, most of the lower urn, and the soil to reveal the incomplete remains of at least 13 individuals, as well as artifacts (Fig. 40). As we removed the cover, we saw small bones from several individuals between the exterior of the jar and the interior of the lid (Fig. 41). Since they were leaning against the side of the lid, they could not have been introduced after it was in place. A possible explanation is that the lid was used to transport the remains to the cemetery. When it was inverted, a few bones fell onto the shoulder of the urn and remained trapped there.

Fig. 39. Fragmentary burial urn from Ayalan, southern coastal Ecuador. Although part of the jar is missing, the lid remained in place.

Fig. 40. Contents of the urn shown in Figure 39. At least 13 individuals were encountered, along with several artifacts. The knife serves as an improvised scale.

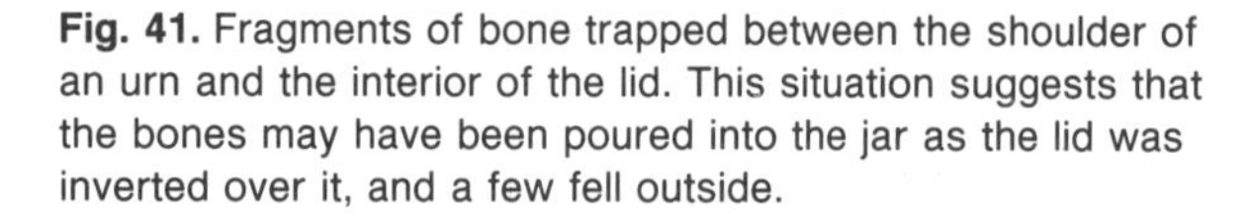

Fig. 41. Fragments of bone trapped between the shoulder of an urn and the interior of the lid. This situation suggests that the bones may have been poured into the jar as the lid was inverted over it, and a few fell outside.

Articulation

In excavating secondary burials, it is important to observe and record any evidence of articulation because arrangement of the bones in correct anatomical position indicates that flesh, ligaments, or other perishable tissue were holding them together at the time of burial. Articulation can be of two types: (1) complete, in which all bones of the skeleton are in correct position, or (2) partial, in which some bones are in correct position. Bass (1962:43) calls partial articulation "semi-articuloses," and stresses the importance as well as the difficulty of observing it during excavation. Partially articulated remains not only indicate a lapse of time between death and burial, but can provide a basis for estimating its duration.

We encountered all three of the principal states in the Juhle ossuary in southwestern Maryland: (1) completely articulated, (2) disarticulated, and (3) partially articulated. Only three completely articulated skeletons were found. One was a young female lying on her back in an extended position on top of the bone concentration (Fig. 42). The others were adults laid tightly flexed on the ossuary floor (Figs. 43–44). These skeletons represent individuals deposited soon after death, when decomposition of the body was minimal. Most of the bones in the ossuary were disarticulated, implying sufficient passage of time between death and final interment for total decomposition of the soft parts. At least 23 adults showed partial articulation or semi-articuloses (Table 1). Although bones in any part of the body may remain attached, articulation is maintained the longest with the bones of the foot (Fig. 45), lower leg, and thoracic portion of the vertebral column (Fig. 46). This condition can only be detected if the bones are left in situ and the soil is removed carefully until all the articular surfaces and the surrounding area are exposed. Recognition of articulation requires familiarity with skeletal anatomy and if the excavator does not have this expertise, he should consult a good field guide to osteology (such as Bass 1971 or Anderson 1962).

Table 1. Frequencies of partially articulated bones in Ossuary II, Charles County, Maryland.

Articulated Bones	Number of Occurences	Minimum No. of Individuals
Bones of foot	46	23
Tibia-fibula	40	20
Thoracic vertebrae	59 (238 vert.)	20
3–7 cervical vertebrae	21 (62 vert.)	13
Skull-mandible	12	12
1–2 cervical vertebrae	11	11
Lumbar vertebrae	19 (5 vert.)	11
Occipital-1st cervical vertebra	9	9
Sacrum-lumbar vertebrae	88	8
Radius-ulna	14	7
Tibia, fibula-bones of foot	12	6
Bones of hand	7	4
Sacrum-pelvis	3	3
Femur-innominate	2	1
Radius, ulna-bones of hand	2	1
Humerus-scapula	1	1
Humerus-radius, ulna	1	1
Femur-tibia, fibula	1	1
Femur-patella	1	1
Tibia, fibula-patella	1	1

The presence of bones in various stages of articulation in a mass burial provides a basis for estimating the length of time between the earliest and most recent deaths. Assuming a relatively constant death rate and rate of decomposition, the ratio of individuals exhibiting each state of articulation can be used to estimate the length of time the bodies were stored prior to interment. If an ossuary contained the remains of those who died during the month preceding interment, we would expect all the skeletons to be completely articulated. If the length of time between ossuary deposits was slightly longer, for example three months, we would expect a combination of completely articulated and partially articulated remains. If the interval was six months or longer, we should find a similar number of completely and partially articulated skeletons as in the three-month accumulation, but disarticulated remains should also be represented. The longer the interval between ossuary deposits, the higher the proportion of disarticulated remains should be. In the Juhle ossuary, about 20 percent of the adults over the age of 20 retained some articulation, implying that they were not completely defleshed at the time of burial. The remaining 80 percent had been dead long enough for total disarticulation. If we assume that bodies exposed on scaffolds or in death houses would be defleshed by eight months after death (allowing for slower decomposition during the winter), the existence of a 20 percent ratio of articulation can be converted into an interval of three to four years since the previous mass burial.

The rationale for this inference, the variability in accuracy, and the need for more exact data on decomposition rates are illustrated by Table 2. If two years were required for complete decomposition and disar-

Fig. 42. An articulated extended skeleton lying on top of the central section of Ossuary II. The correct anatomical arrangement of these bones contrasts with the confusion of the disarticulated remains.

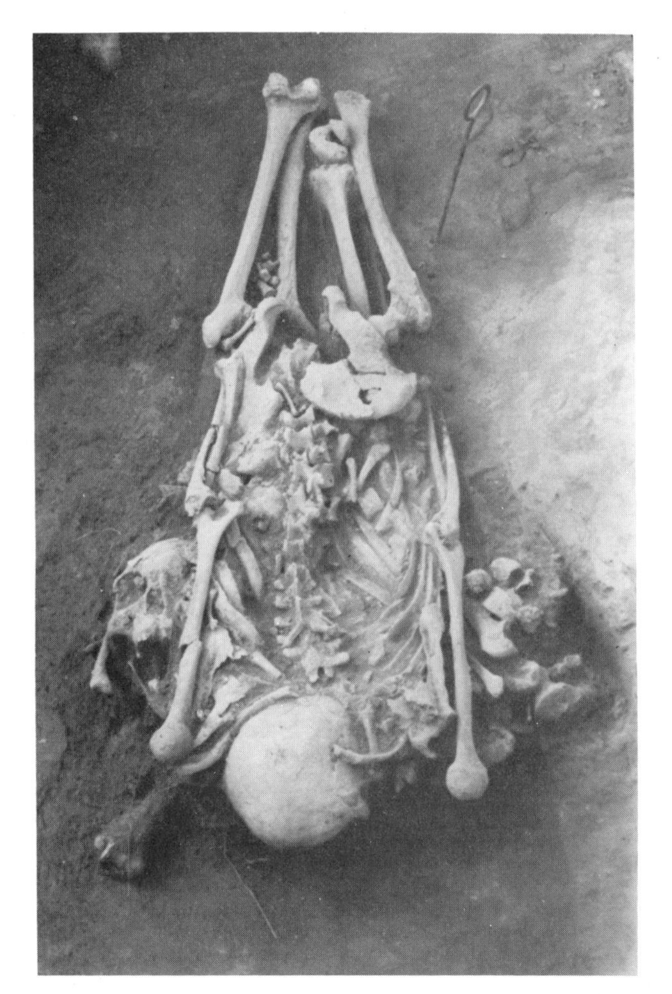

Fig. 43. A tightly flexed, fully articulated skeleton encountered on the floor of Ossuary II.

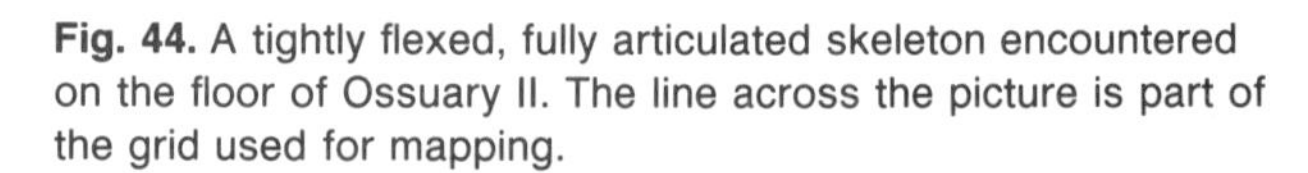

Fig. 44. A tightly flexed, fully articulated skeleton encountered on the floor of Ossuary II. The line across the picture is part of the grid used for mapping.

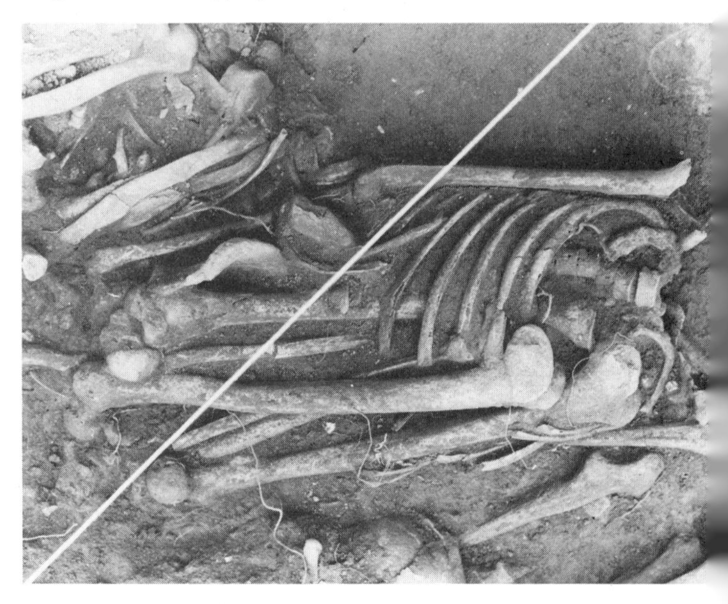

Fig. 45. Thoracic vertebrae and metatarsals (foot bones) encountered in correct anatomical relationship, indicating that the flesh had not completely decomposed when this individual was placed in the ossuary.

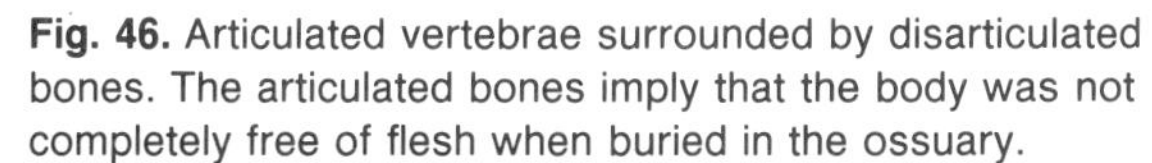

Fig. 46. Articulated vertebrae surrounded by disarticulated bones. The articulated bones imply that the body was not completely free of flesh when buried in the ossuary.

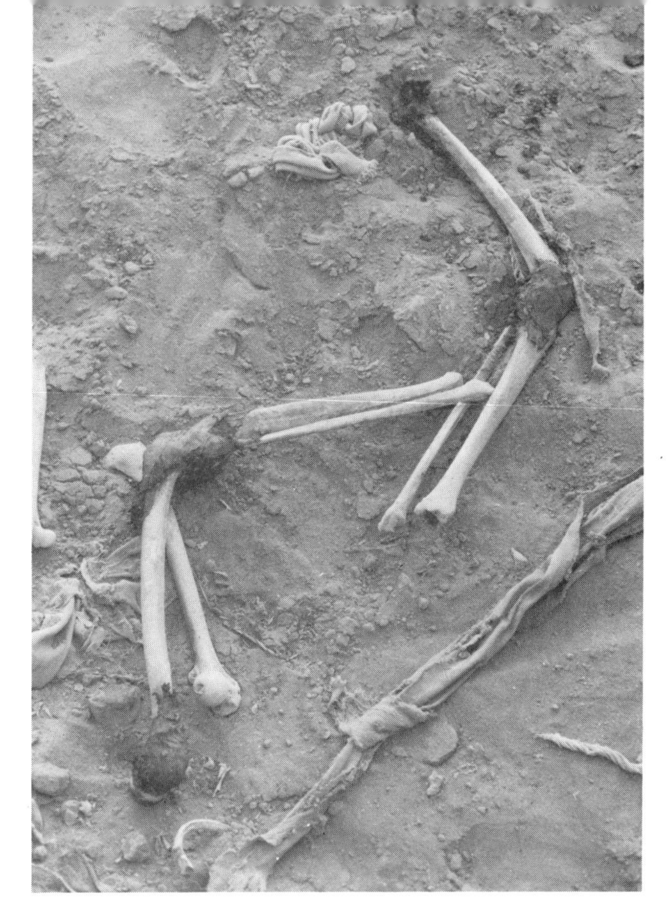

Fig. 47. Articulated bones of the upper and lower leg lying on the surface of a prehistoric cemetery on the coast of Peru. With exceptional conditions, such as aridity, articulation can be preserved for centuries after burial.

Fig. 48. The articulated pelvis and lower limbs of an adult discovered at the bottom of a burial urn from the cemetery of Ayalan on the coast of Ecuador, after several disarticulated secondary burials had been removed. This unexpected situation illustrates the need for careful observation before removing secondary burials.

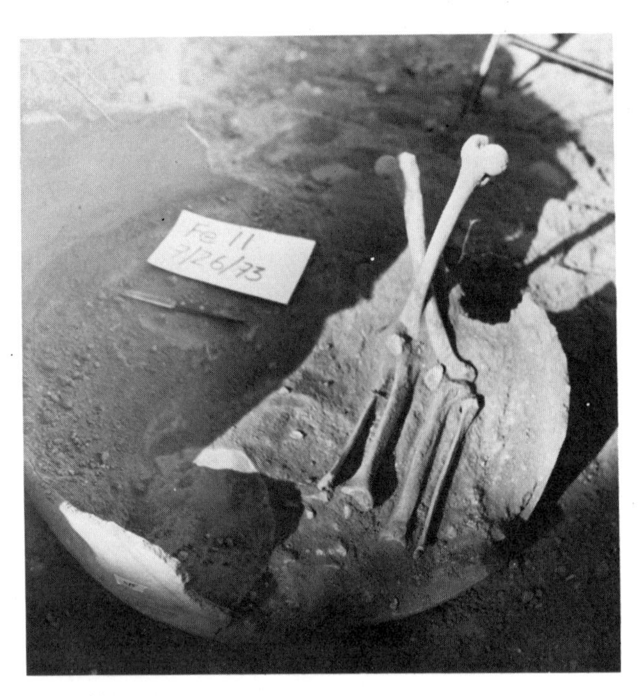

Fig. 49. Another example of articulated leg bones encountered beneath several completely disarticulated individuals in a burial urn from Ayalan on the coast of Ecuador.

Table 2. Time intervals represented by Ossuary II, calculated from different rates of decomposition.

Decomposition Rate (t) (months required for disarticulation of lower leg bones)	Time Interval (T) (years represented by the ossuary)
1	0.4
2	0.8
3	1.3
4	1.7
5	2.1
6	2.5
7	2.9
8	3.4
9	3.8
10	4.2
11	4.6
12	5.0
18	7.6
24	10.0

ticulation, then the ossuary would represent an accumulation over a period of 10 years. If only one month was required for decomposition, however, the ossuary would represent the deaths during a period of only about five months. Unfortunately, rates of decomposition vary widely depending on the presence and intensity of such variables as cultural treatments of the body prior to burial, moisture, temperature, bacterial activity, and flesh-eating animals. Under desert conditions, such as prevail on the coast of Peru, hair, flesh, and other dried tissue often preserve articulation for centuries after burial (Fig. 47). The opposite extreme was recorded by Che Guevara. On March 23, 1967, seven of his men were killed in ambush in the Bolivian jungle. When he returned to the spot only 10 days later, he found the bodies stripped to the bones.

Another example of the importance of observing articulation is provided by the Ecuadorian urns. Many contained articulated leg bones of one adult, usually a male (Figs. 48–49). Absence of articulation in the skeletons of other individuals in the same urn implied that they had been dead longer. Differences in age at death and other data were compatible with the interpretation that the contents of an urn corresponded to an extended family. Combining these two clues leads to the hypothesis that final interment followed the death of the head of a household, at which time all who had died since the last such event were placed in the same urn. In this case, there are no ethnohistorical accounts of the kind available from parts of North America to add credibility to the hypothesis. The example illustrates, however, how data on articulation can permit recognizing the order of death of individuals within a skeletal deposit and how this information can be used to reconstruct aspects of the mortuary practice.

In forensic contexts, data on bone articulation offer direct evidence of the degree of soft-tissue preservation at the time of interment and help reconstructing events since death and/or criminal activities. For example, some years ago a largely articulated skeleton was found in an unused cistern near Omaha, Nebraska (Ubelaker and Sperber 1988). The individual was identified through dental radiographs as a 19-year-old white woman from the local area, who had been missing for nine years. Some testimony suggested that after killing her, the assailant had deposited a very caustic substance (acid or lye) on her face to thwart identification. The face bones and teeth showed alterations consistent with this treatment (Fig. 50). The question was, could these physical changes have been produced by caustic substances in the cistern? Careful examination of the mouth area showed erosion extending from the maxilla to the mandible in a pattern

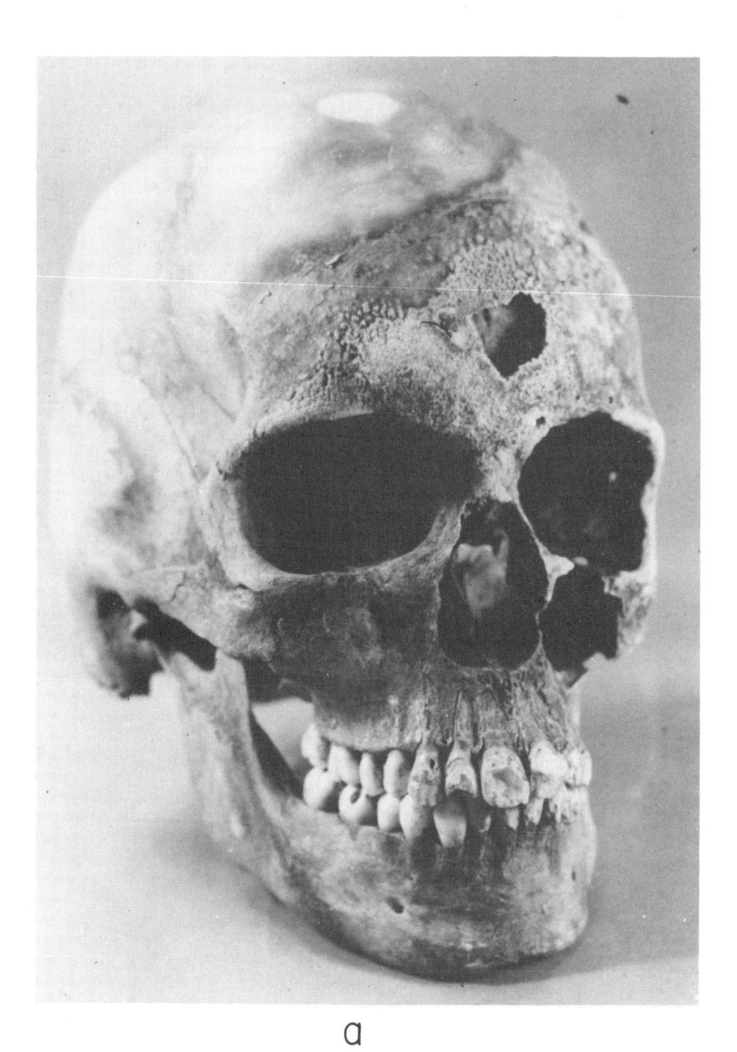

a

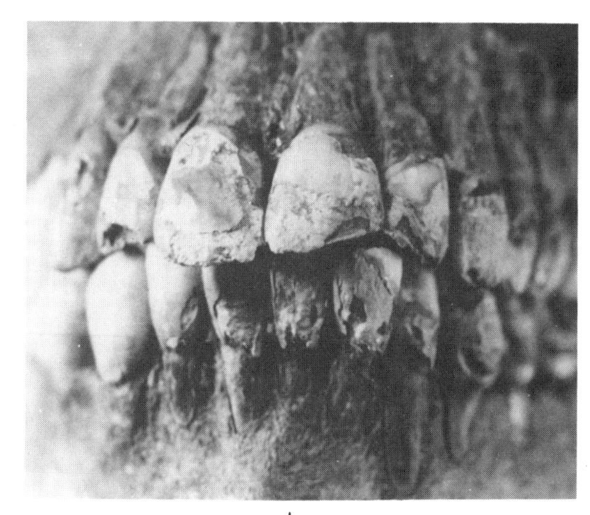

b

Fig. 50. Erosion produced by a caustic substance.
a, Extension over face and mouth.
b, Detail of damage to the maxillary and mandibular teeth.

possible only if they were articulated when the substance was applied. Yet, photographs taken at the time of discovery showed the mandible at a considerable distance from the maxilla on the floor of the cistern, where it might have been in contact with caustic debris. Observations on the positions of the bones at the time of discovery eliminated post-depositional events as the cause of the disfigurement.

CREMATIONS

Cremations are collections of human bones that have been burned. If incineration is nearly total, the remains will be very small and are often considered unworthy of preservation. Much can be learned, however, if the data are properly collected. The principal goals of excavation are: (1) to identify and remove all fragments of bone, (2) to record the position of every fragment, (3) to establish whether the remains were burned on the spot or were burned elsewhere and redeposited, and (4) to observe details relevant to reconstructing the firing procedure.

Identification

The identification of cremated remains is a problem both because the bones are usually very fragmentary and because the firing process can produce changes in their sizes and shapes. The amount of shrinkage observed during experimental firings ranged from 1 percent to 25 percent, depending on the density of the bone and the temperature and duration of the fire (Van Vark 1970:102). No shrinkage occurred until the temperature reached 700 degrees C. There was a progression between 700 and 900 degrees C, but higher temperatures produced no further shrinkage. Extensive shrinkage is accompanied by changes in color, first to black or gray and then to white. Thus, anyone identifying white fragments must allow for shrinkage up to 25 percent. This consideration is especially important if the sizes of the bones are used for inferring sex and age.

Position

The positions of all identifiable fragments should be recorded before they are removed. This may be accomplished using diagrams, photographs, measurements, and descriptions. If the quantity of bones is large, the concentration should be photographed, measured, and described as a whole. Only the positions of the larger bones and others useful for reconstructing the condition of the body prior to cremation need be plotted. Position data are important for distinguishing secondary deposits of bones burned elsewhere from cremations at the site of discovery. In the former case, the remains should be scrambled and none of the bones should have distributions that approximate their anatomical relationships. Also, no evidence of fire should be detectable in the surrounding soil. If the body was cremated on the spot, the soil around the bones should show indications of fire in the form of ash, charcoal, or staining. If cremation followed soon after death, the distribution of the bones may correspond to their anatomical positions in the body or at least reveal some concentrations of foot bones, skull bones, etc. Even if deterioration of the fragments is advanced, the distributional pattern may provide information on the process.

Fig. 51. Cremated bones exhibiting minimal evidence of exposure to fire in the form of smoking and slight scorching.

Color

Numerous studies have shown that regular changes in color occur with increasing temperature during the crematory process. Minimal heat causes slight scorching or smoking, but otherwise the appearance of the bone is normal (Fig. 51). Combustion of the organic material remains incomplete below 800 degrees C, and bones in this condition have been called "incompletely incinerated or smoked" (Baby 1954:2). Increased exposure to temperatures above 800 degrees C results in calcined bone that ranges from bluish-gray to white. White color indicates longer exposure to higher temperatures than blue and gray.

Fracture Patterns

Patterns of fracture are another source of useful information. Experiments by Baby (1954) and Binford (1963) suggest that cremation of dried bones produces different patterns than cremation of bones with flesh around them. Burning dry bones causes cracking or "checking" on the surface and longitudinal splitting, but no warping or twisting (Fig. 52). Burning of "green" or flesh-covered bone creates curved transverse fracture lines, irregular longitudinal splitting, and marked warping (Fig. 53). Thus, if the bones have undergone extreme combustion, observation of their fracture patterns can reveal whether or not the individual was cremated in the flesh.

Recording the locations of bones showing different fracture patterns and colors may permit reconstructing the crematory procedure. For example, a group of Archaic Period cremations from New Jersey that I examined exhibited transverse, curved fracturing and severe warping, strongly suggesting that the individuals had been cremated soon after death. Most of the bones from the center of the body were brittle, white, and extremely warped. By contrast, many bones

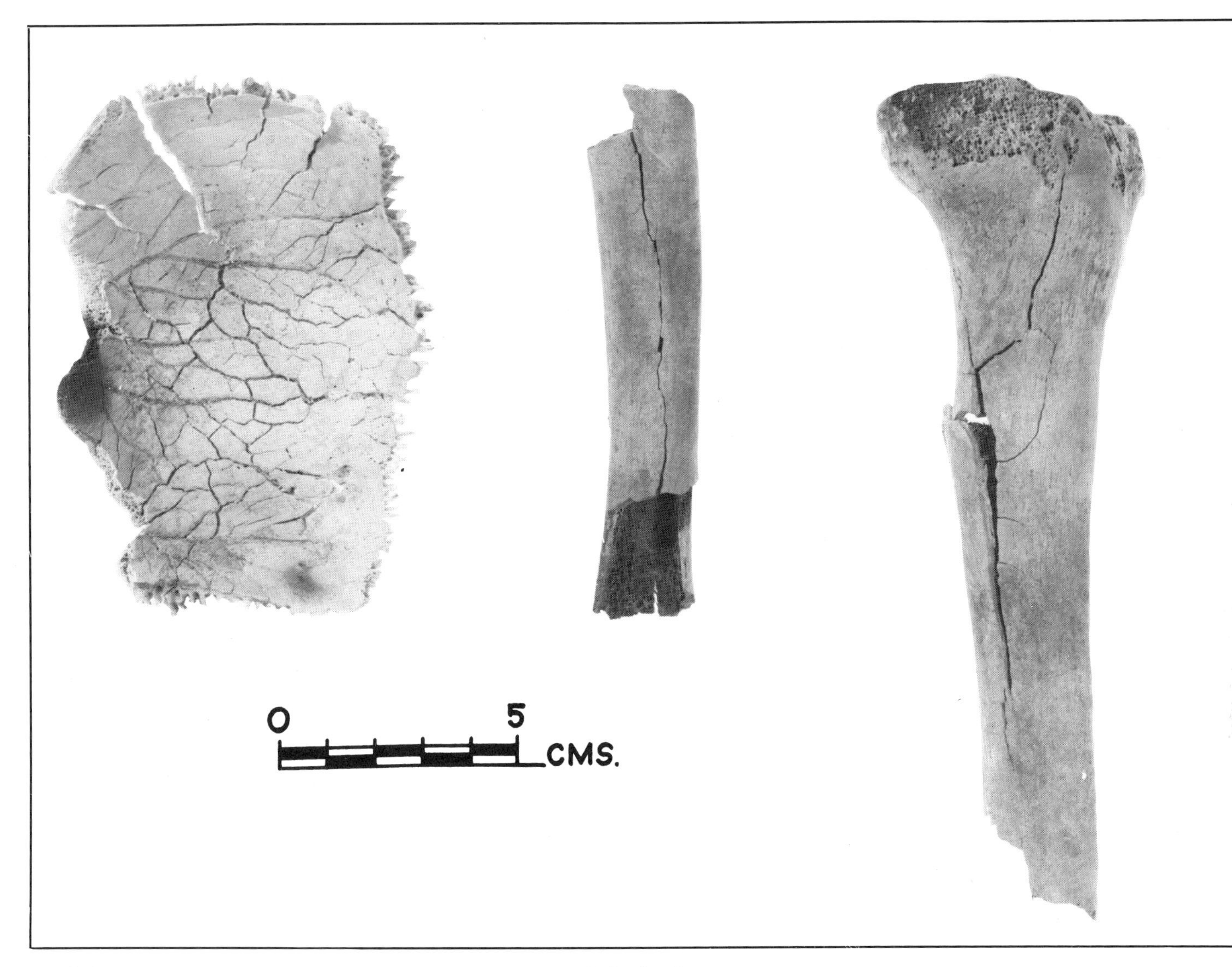

Fig. 52. Cremated bones showing cracking and longitudinal splitting, indicating that they were dry when burned.

of the feet and parts of the skull were merely scorched and blackened, indicating they had been subjected to less intense heat. In addition, calcination (and thus combustion) was usually more complete on the dorsal (back) surfaces of the bones than on the ventral (front) surfaces. These data suggest that the bodies were cremated in the flesh, in an extended position, either lying on the back on top of the fire or on the stomach beneath the fire.

Observations on exposure of bone to fire frequently provide important evidence of criminal activity. Figure 54 illustrates remains of a four-year-old child recovered from a crime scene in New England, where fire was used in an attempt to destroy the body. Note the incompleteness of the skeletal parts and their fragmentary nature. Note also the variability in extent of oxidation; some parts are completely calcined, others are charred, and some are hardly affected. Such variation is typical of the damage done by small fires of relatively short duration. In contrast, a major house fire will reduce a body to small calcined fragments.

Bones may reveal evidence that fire or intense

Fig. 53. Cremated bones with transverse fracture lines, irregular lengthwise splitting, and marked warping, indicating that cremation was done when the bones were "green" or covered with flesh.

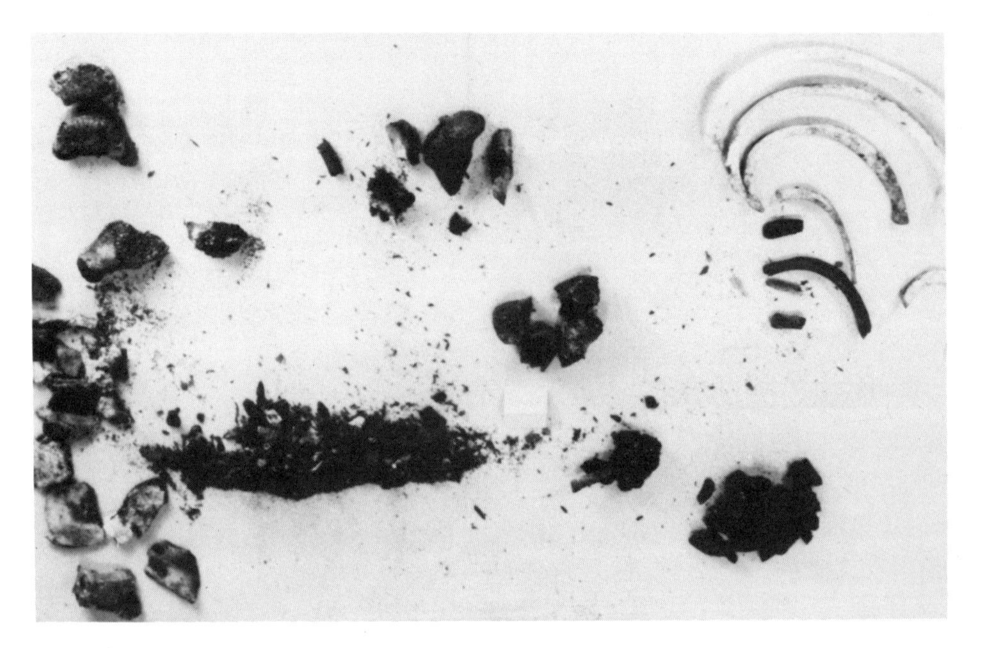

Fig. 54. Cremated remains of a four-year-old child.

heat has been applied in an attempt to disfigure a murder victim and thwart identification. In such cases, the head and face are usually targeted because of their assumed value in identification. Such behavior may have occurred in a case from the Pacific island of Palmara. In 1974, a couple visiting the island was reported missing and their stolen boat was later recovered. In 1981, human bones found on the beach were identified as the female of the couple. The cranium revealed a large, irregular, white, calcined area over most of the left frontal and parietal (Fig. 55) produced by exposure to intense heat. The sharply defined border separating the affected from the unaffected areas implied the heat was applied while flesh protected the skull. Bones without such protection show a gradient of change.

REMOVAL AND TREATMENT OF BONES

Field Data Forms

The variability of mortuary features that may be encountered in the field requires similar variability in techniques of excavation. This situation prevents compilation of a simple, general form for recording the data. If a form or data sheet is employed, it should

Fig. 55. Calcined area on the left frontal and parietal of female. The sharp border indicates that heat was applied while flesh protected the bone surface.

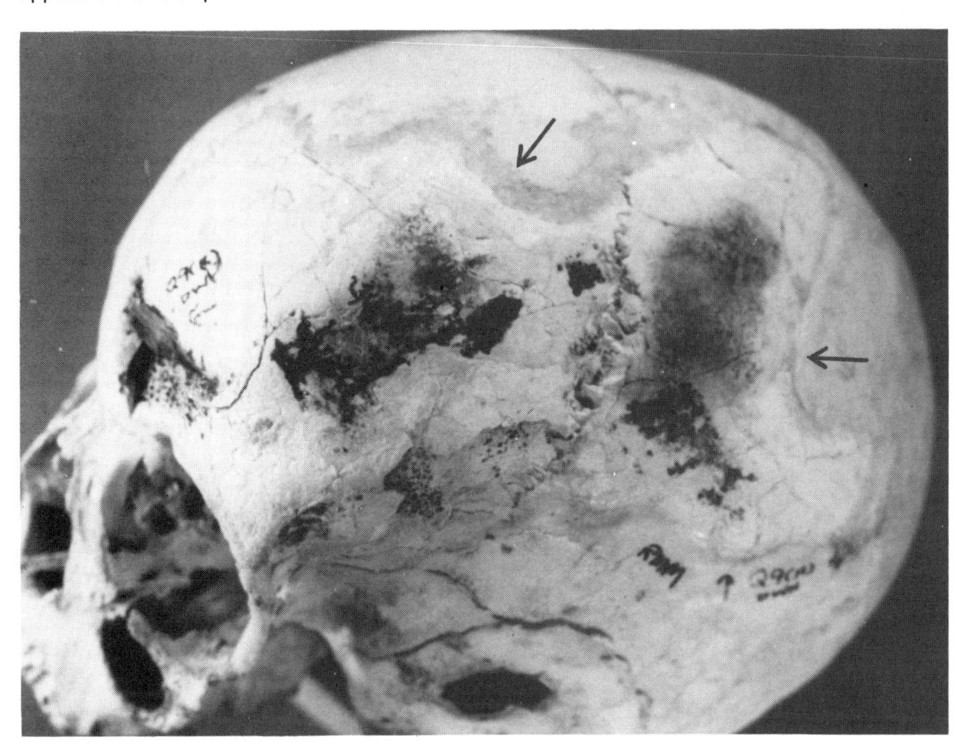

be designed after preliminary investigations have revealed the types and ranges of data that can be recorded. A data sheet that has proved useful for documenting primary inhumations in the Plains area of the United States is reproduced as Figure 56.

Removal of the Bones

After the skeletal material has been cleaned, photographed, described, and mapped, it should be removed as carefully and as rapidly as possible. Under no circumstances should a skeleton remain exposed and unprotected because of the risk of irreparable damage from desiccation, rain, accident, or other causes. This requirement applies regardless of how well the remains appear to be preserved. Any soil holding the bones in place should be removed or loosened. The excavator should never attempt to pull out a bone that is securely imbedded in the earth.

Under no circumstances should any bone be discarded. Even fragments of ribs and small bones of the hands and feet may offer important information. Isolated complete or incomplete bones found in cache pits, house floors, or the general refuse of a site should also be recorded and saved.

Preservatives

Under normal conditions, preservatives should not be applied in the field. If a bone appears delicate, it should be measured before removal. Broken bones can usually be restored in the laboratory. Preservatives applied in the field usually fail to prevent extremely fragile specimens from shattering: furthermore, they also complicate cleaning and may interfere with certain kinds of chemical and microscopic studies. If a skeleton is destined for exhibit, field preservation may be justifiable. In such cases, an acetone-soluble adhesive (such as PVA) should be employed.

Removal of a Burial Intact

Occasionally, it may be desirable to remove intact all or part of a burial or concentration of human remains for later analysis or possible display. If the bones and matrix are sufficiently durable, it may suffice to excavate deeply around the entire feature until it can be undercut, removed intact, and packed for shipment.

An alternative approach is to isolate the feature as described above, but then enclose it in a plaster jacket before removal. The process begins by covering the skeleton (Fig. 57) with a five-centimeter layer of

Smithsonian Institution
RIVER BASIN SURVEYS
BURIAL FORM

Feature No.________________ Site No.________________

Burial No.________________ State ________________

Reservoir ________________ County ________________

1. LOCATION
 a. Square ________________
 b. Horizontal________________
 ____
 c. Depth from surface ________
 d. Depth from datum ________

2. BURIAL TYPE
 a. Extended ____d. Reburial ______
 b. Flexed ______e. Cremation _____
 c. Semiflexed ___f. Part crem._____
 g. Other ________________

3. BURIAL DIMENSIONS
 a. Max. length _____ Dir. ________
 b. Max. width _____ Dir. ________
 c. Thickness ______

4. DEPOSITION
 a. Position ________________
 b. Head to ________________

5. GRAVE TYPE
 a. Surface ________c. Cist.______
 b. Pit ___________d. Other ______
 e. Shape ________________

6. GRAVE DIMENSIONS
 a. Max. length______Dir. ________
 b. Max. width ______Dir. ________
 c. Depth ________________

7. STRATIFICATION
 a. Inclusive________c.Precedent_____
 b. Intrusive________d.Disturbed_____
 e. ________________

8. ASSOCIATIONS
 a. Features ________________
 ____
 b. Specimens ________________
 ____

9. PRESERVATION
 a. Poor ____Fair____Good______

10. COMPLETENESS
 ____

11. SEX
 a. M_____F_____Indeterminate_____

12. AGE
 a.Infant_______d. Adult ________
 b.Child________c. Mature _______
 c.Adolescent___f. Senile _______

13. NEG. Nos. ________________

14. REMARKS ________________
 ____

Recorded by ________________ Date ________________

Fig. 56. Burial form of the type used for recording information on primary inhumations in the Plains area of the United States.

wet tissue paper or cloth (Fig. 58). A layer of plaster, also about five centimeters thick, should be applied to the entire surface, covering the wet tissue or cloth (Fig. 59). A trench should then be excavated around the area to be removed (Fig. 60). After the plaster is dry, the casing can be freed by undercutting at the inner edge of the trench. It should be turned over so that a layer of plaster can be applied to the back, totally enclosing the feature (Fig. 61).

Neither of these approaches is completely successful in preventing disturbance. Accordingly, it is important to document all aspects of the feature thoroughly prior to its removal.

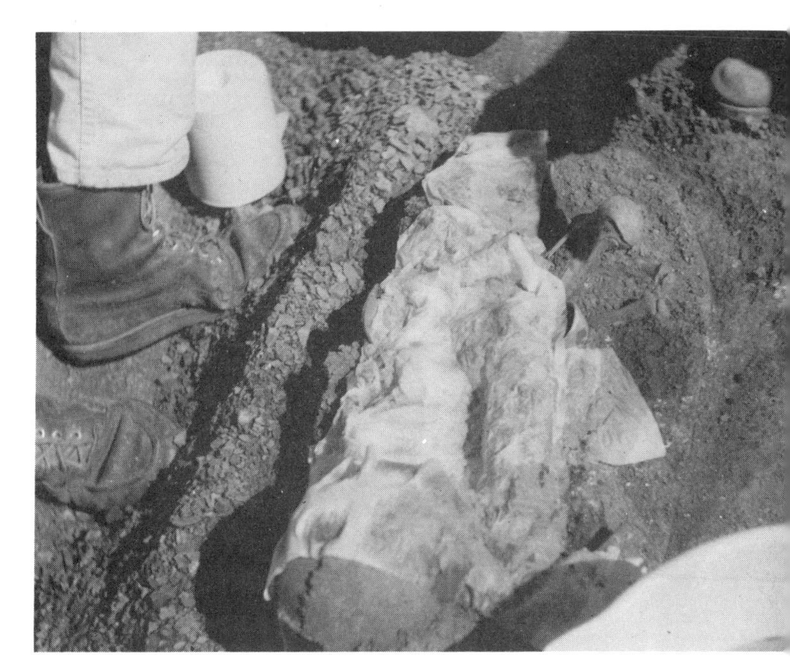

Fig. 58. The burial in Figure 57 after being covered with a layer of wet tissue paper, the first step in encasing it in plaster for transport intact. Part of the pistol remains visible in the upper right.

Fig. 57. Primary burial of an infant accompanied by a flintlock pistol (along the right side), from the Leavenworth site in South Dakota.

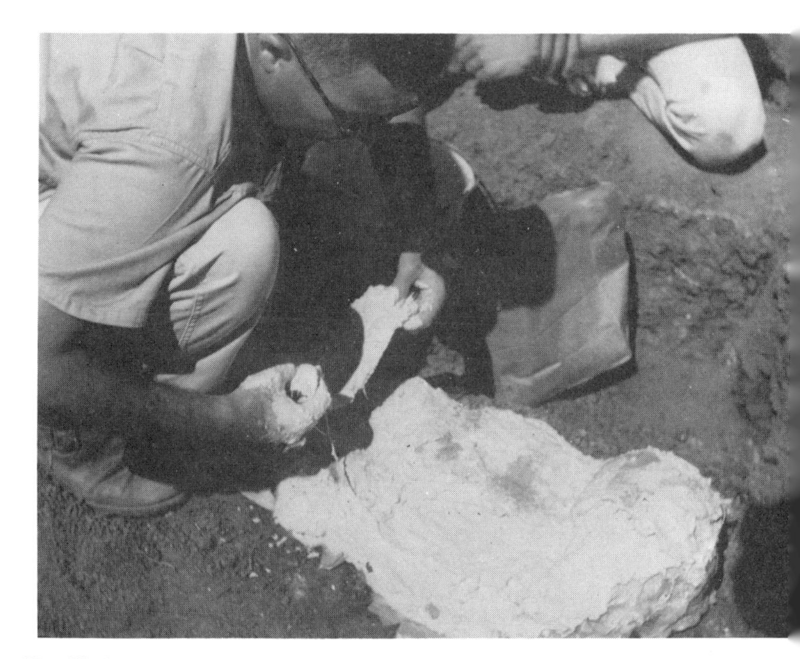

Fig. 59. Applying a layer of plaster-impregnated cloth over the wet tissue paper shown in Figure 58.

Fig. 60. The plaster-encased burial after the surrounding area has been deepened to permit undercutting and lifting out the stabilized remains.

Fig. 61. Placing layers of wet tissue paper and plaster over the bottom, exposed by turning the cast upside down after separation from the soil matrix.

Removal of Adhering Soil

Most soil adhering to bones should be removed immediately after exposure, both because it will crack or even disintegrate the bone as it dries and because it will be difficult to extract from cranial orbits or the interior of the vault after it is hard. If the soil is washed off, the bones should be allowed to dry in the shade before they are packed. Moisture retained will condense inside the transporting containers; breakage of containers and loss of identifying marks may occur if proper precautions are not taken.

After all visible bones have been removed, the earth in the pit should be sifted through a fine screen to recover any remaining fragments. All fragments should be saved for use in reconstruction and analysis.

Placement in Containers

Once cleaned and dried, the bones should be placed in containers (paper or cloth bags) that have been labeled appropriately. Paper bags are cheaper, but deteriorate when exposed to moisture. The container should be labeled in a clearly visible area using waterproof, smearproof ink. The label should include the site name, burial or feature number, date of excavation, name of the excavator, and the nature of the contents (Fig. 62).

Bones of similar type and size should be packed together to facilitate processing and minimize damage. A single articulated skeleton is best packed in at least three bags or containers, one for the cranium, one for the long bones, and the third for the other bones. Fragmentary remains should be wrapped individually in tissue, cloth or newspaper. Fragile bones and fragments should never be placed in the same container with large heavy bones. Packages should be stored in a safe and dry place until shipment or processing (if the latter is done in the field).

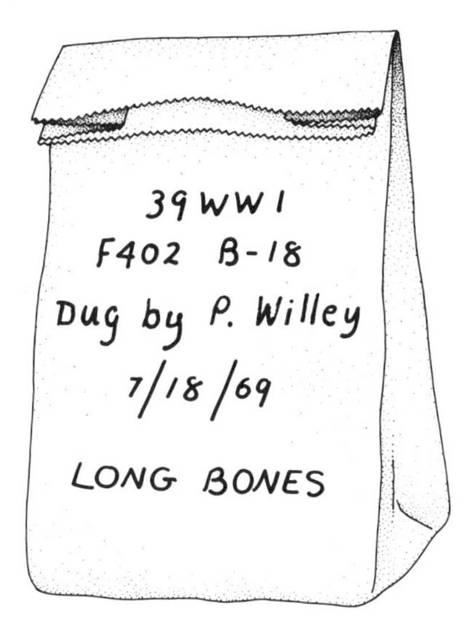

Fig. 62. Paper bag labeled to facilitate identification of the contents.

Cleaning and Cataloguing

In the laboratory, the material should be cleaned and catalogued before analysis. Cleaning should be done with water and soft brushes. Brushes and other tools should be selected according to the state of preservation of the bone. Care should be taken not to damage the outer surface, since this can affect determination of age and disease. Extremely fragmentary material should not be washed; instead, it should be carefully brushed while dry. If washing is done in a sink, a fine screen should be placed over the drain to prevent inadvertent loss of teeth and small bones. To prevent accidental mixing of two burials, wash only one at a time and be certain that the field identification information is always kept with the bones.

As soon as the bones are clean and dry, they should be labeled clearly with waterproof ink. The museum catalogue number or site information should be written on each bone in a place where it does not obscure anatomical landmarks or obvious pathological lesions. Broken bones should be glued together with Duco cement or a similar kind of adhesive. If the glued pieces are set in a box of fine sand for support until firm, care should be taken to avoid imbedding the glued areas so deeply that the sand adheres. Modeling clay, wooden splints or light wire may be used for additional support. If reconstruction is incomplete, the broken areas or missing fragments should never be supplemented or altered by adding artificial fillers, such as plaster or cement. Also, bones should never be coated with oil, paint, glue, varnish, wax, or other materials, since these inhibit observation and future chemical and microscropic research. If specimens are deteriorating, they may require a preservative. In such cases, after the bone has been labeled it can be dipped in a transparent solution containing Alvar, Duco, Ambroid, or Gelva. In all situations, bones should be handled carefully and stored where they are protected from disturbance.

Under special circumstances, soft tissue may be preserved and may conceal evidence of trauma, disease, and other phenomena. Therefore, it should be carefully examined and then removed so that all bone surfaces can be observed. Small quantities of dessicated tissue can be dislodged using tweezers or other small blunt instruments. Scalpels and knives should be avoided since they can leave misleading marks. Larger amounts of tissue can be eliminated using chemicals (Krogman and Iscan 1986:40–43) or by exposing the remains to colonies of flesh-eating dermestid beetles. The latter method is ideal for preserving details of bone surfaces, but may require weeks or months to complete.

3 Sex, Stature, and Age

Estimates of sex, stature, and age from skeletal remains employ information obtained from living or recently deceased individuals whose age, sex, and relevant medical history are known. These studies indicate that considerable variability exists between males and females; between individuals of the same sex, age, and ancestral affiliation, and between those exposed to different environments and diets—to mention only the primary causes. Applying standards derived from such studies to skeletons of unknown individuals and populations incorporates two important kinds of uncertainty, one resulting from the variability within the documented populations and the other from differences of unknown magnitude between the documented populations and those being studied. Although estimates of sex, stature, and age can never be exact, errors can be minimized by intelligent selection of techniques and careful interpretation of data.

Because of the number and variety of judgments required to achieve reasonably accurate estimates of sex, stature, and age, a physical anthropologist with expertise in skeletal biology should be consulted. This chapter will make clear the reasons for this warning. Terms not included in the accompanying figures are explained in the Glossary.

HUMAN OR NOT?

Identifying isolated or fragmentary bones may be difficult even for a specialist. In archeological sites, human remains are often mixed with non-human fauna and may have been altered sufficiently by weathering, cultural practices or exposure to high temperature that anatomical landmarks are not easily discerned. Ten to fifteen percent of samples submitted for forensic analysis are not human. One of the first problems, therefore, is to differentiate human from non-human bones.

In my experience, the most commonly misidentified bones are from deer, bear, pig, dog, cow, and horse, although many other animals (and even objects such as rocks, garden hoses, plastic containers, etc.) are occasionally confused. The best basis for recognizing human bones is detailed knowledge of human anatomical particulars. In general, it is important to remember that long bones of most other adult animals of comparable size have different bone architecture and usually have thicker, more compact cortices.

Figure 63 compares adult human bones with those most frequently mistaken for human. In each illustration, the bone on the left is from an adult human female. From left to right, the animals are black bear (*Ursus americanus*), large dog (*Canis familiaris*), hog (*Sus scrofa*), deer (*Odocoileus virginianus venatorius*), domestic sheep (*Ovis aries*), and small dog (*Canis familiaris*). Figure 64 compares immature human and animal bones. In each, the human bone on the left is from a ten to twelve-year-old child from the Mobridge site in South Dakota and the one to the immediate right is from a newborn human from the Smithsonian fetal collection. From left to right, the animals are black bear (*Ursus americanus*), deer (*Odocoileus virginianus venatorius*), dog (*Canis familiaris*), adult gray squirrel (*Sciurus carolinensis*), pig (*Sus scrofa*), and very young dog (*Canis familiaris*).

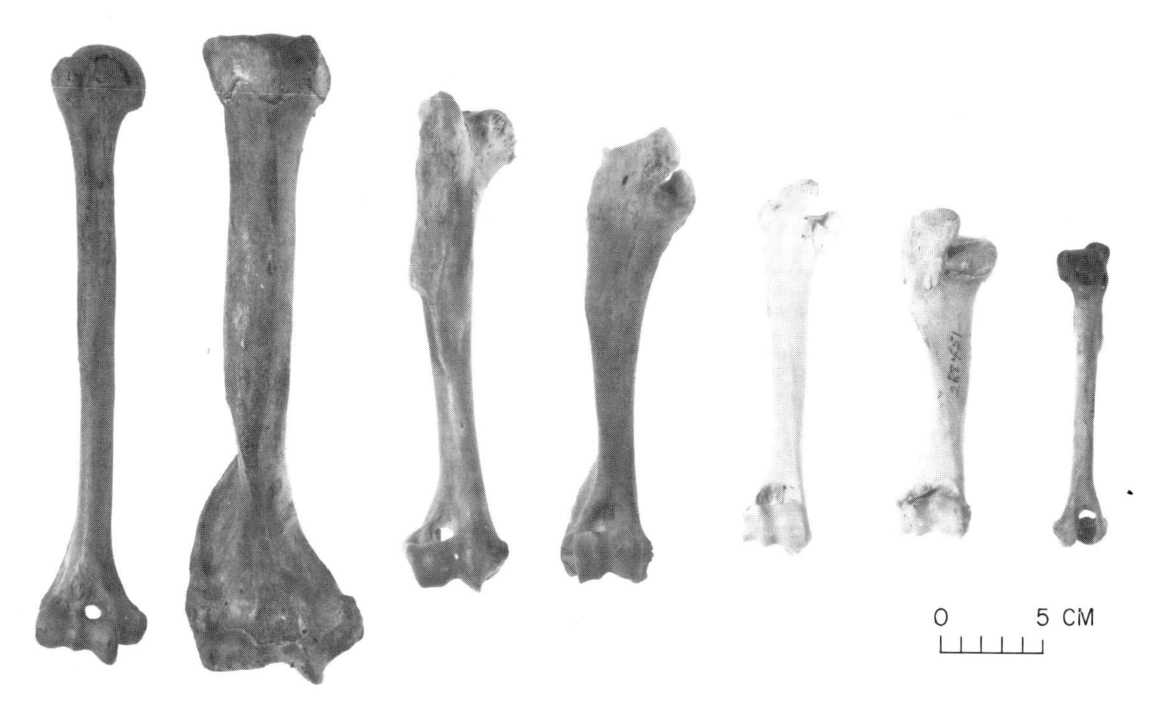

a

Fig. 63. Bones of an adult human female (left) compared with black bear, large dog, hog, deer, domestic sheep, and small dog (left to right). **a,** Humerus. **b,** Radius and ulna.

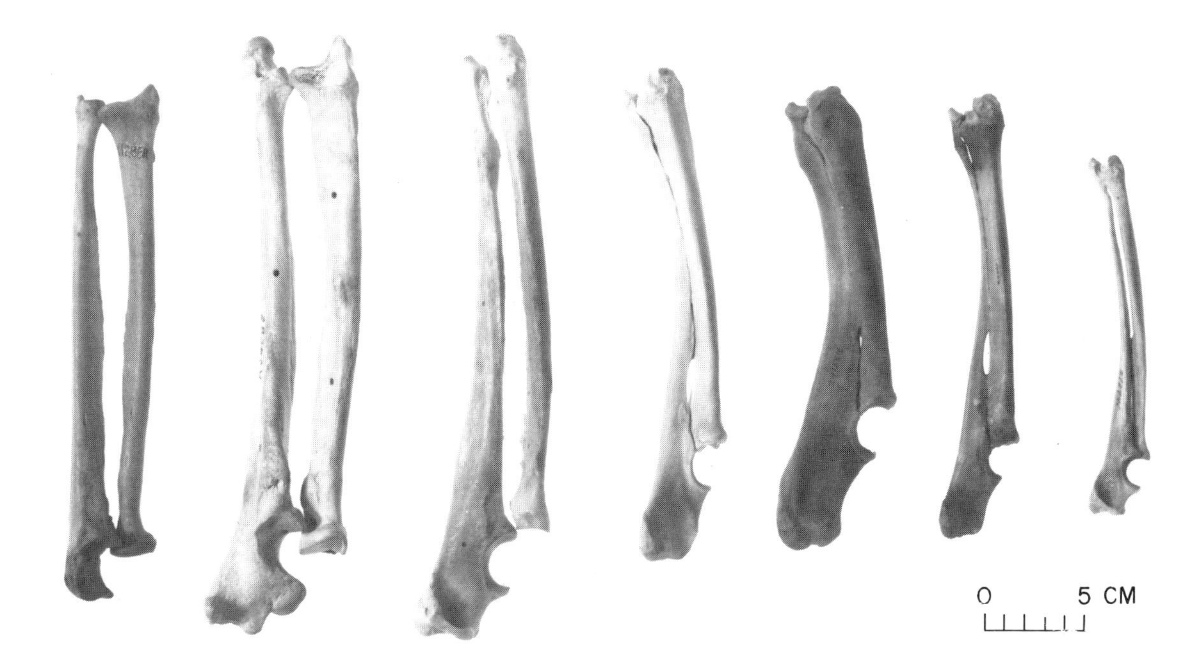

b

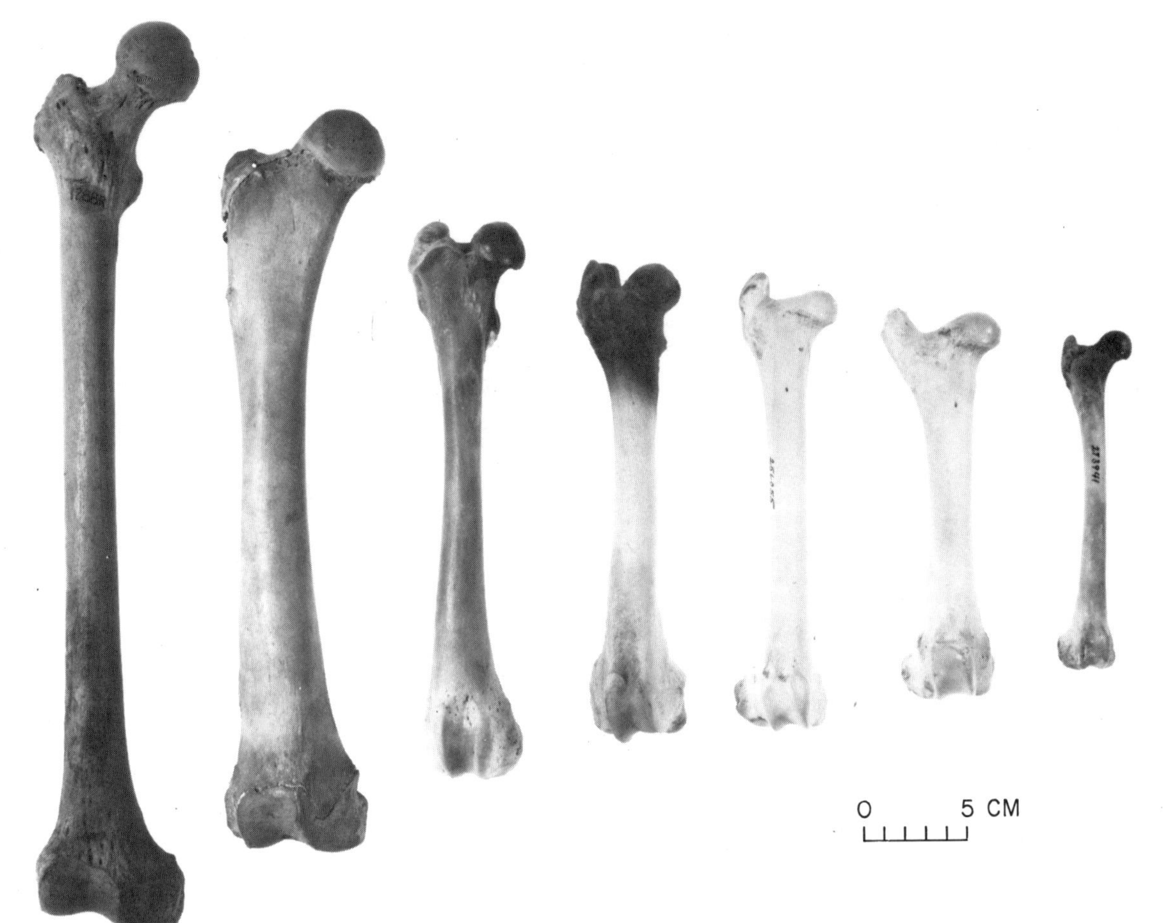

c

Fig. 63, cont. Bones of an adult human female (left) compared with black bear, large dog, hog, deer, domestic sheep, and small dog (left to right). **c,** Femur. **d,** Tibia. **e,** Scapula. **f,** Calcaneus.

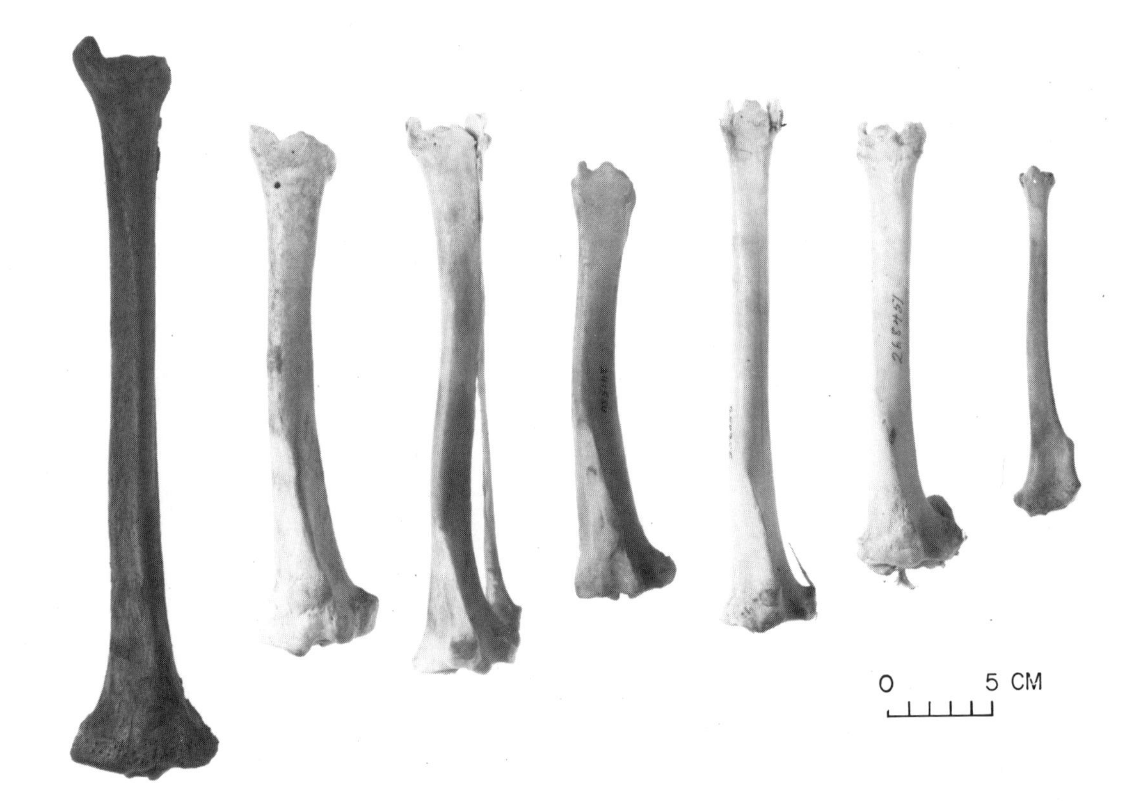

d

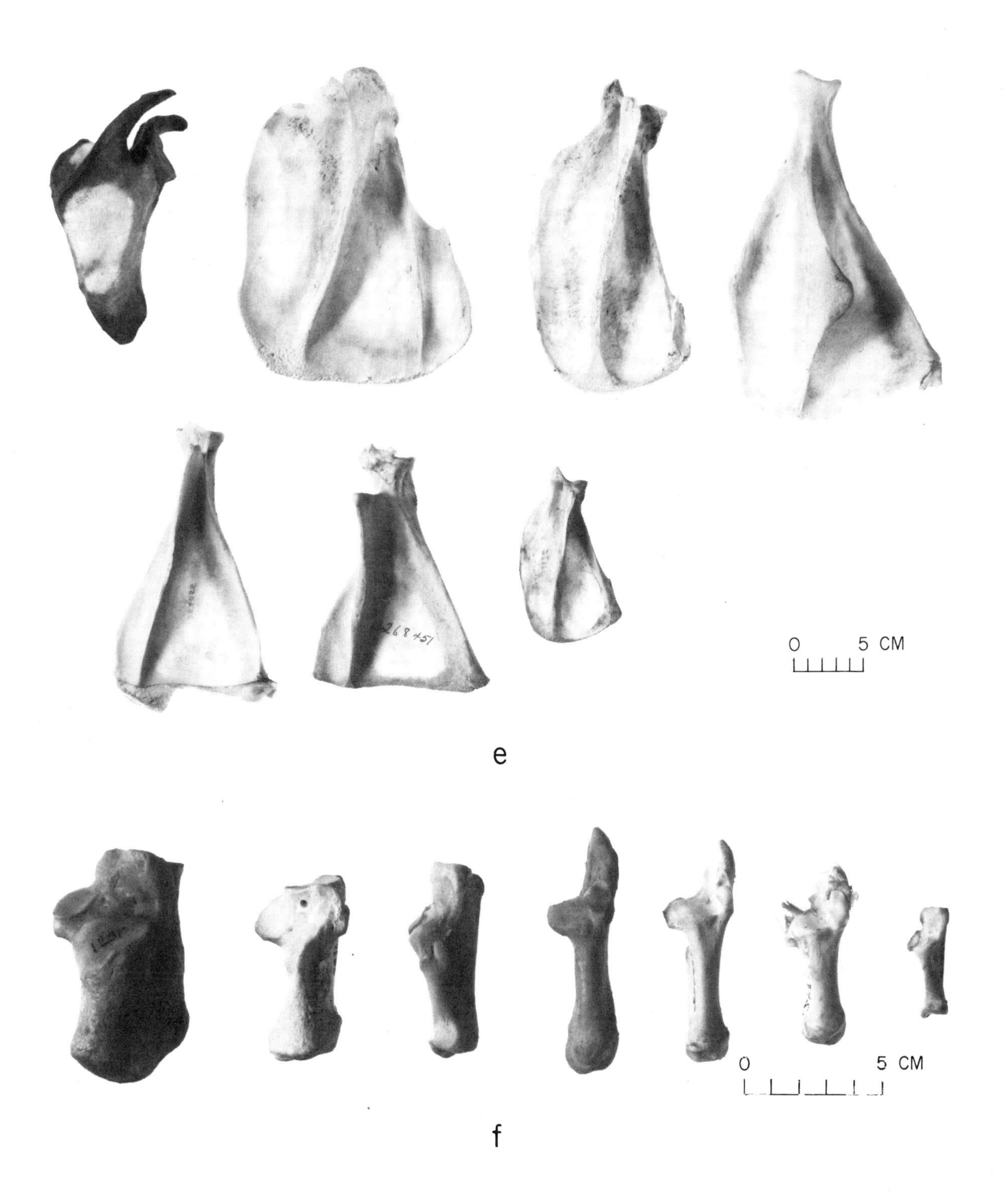
0 5 CM
e
0 5 CM
f

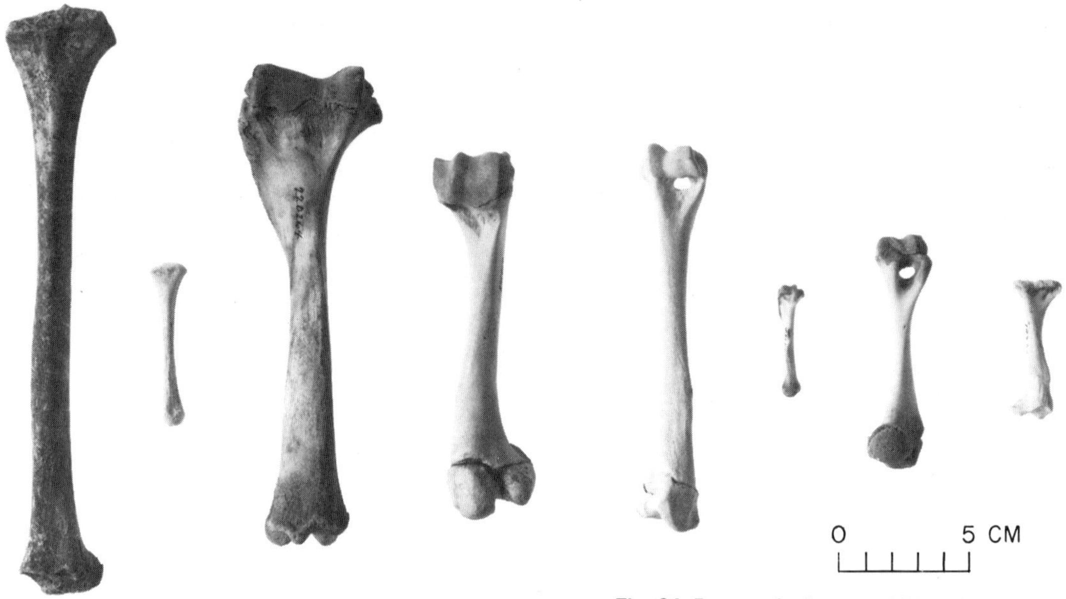

a

Fig. 64. Bones of a human child and a fetus (left) compared with black bear, deer, dog, adult gray squirrel, pig, and very young dog (left to right). **a,** Humerus. **b,** Radius. **c,** Ulna. **d,** Tibia.

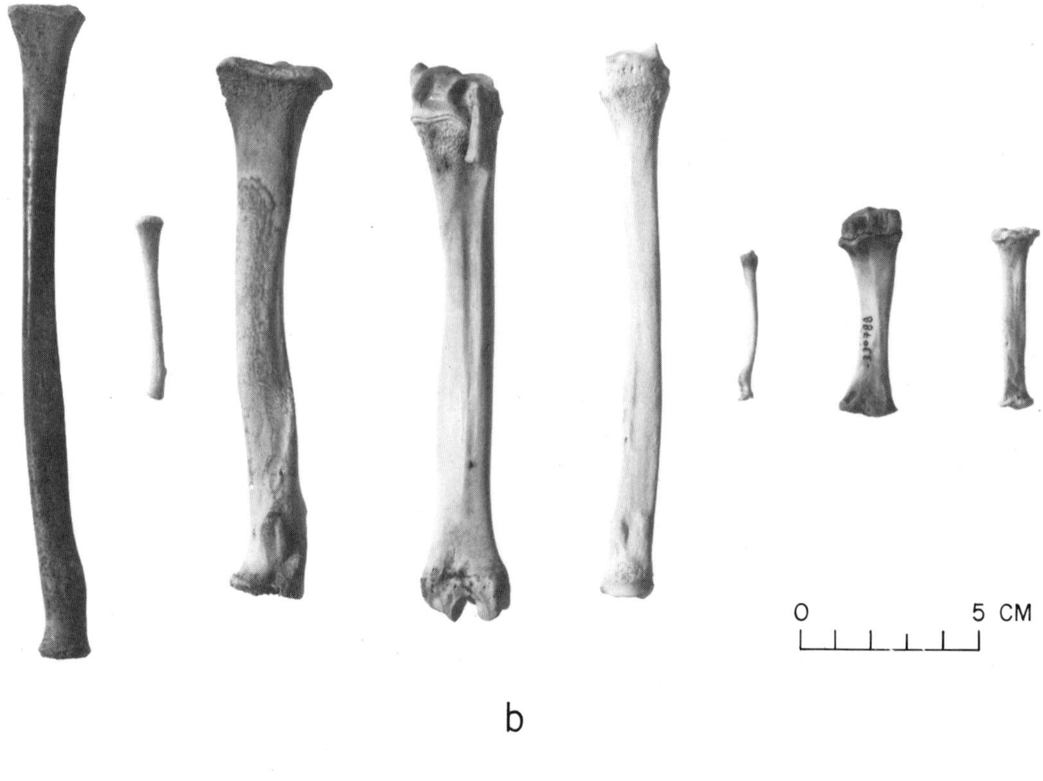

b

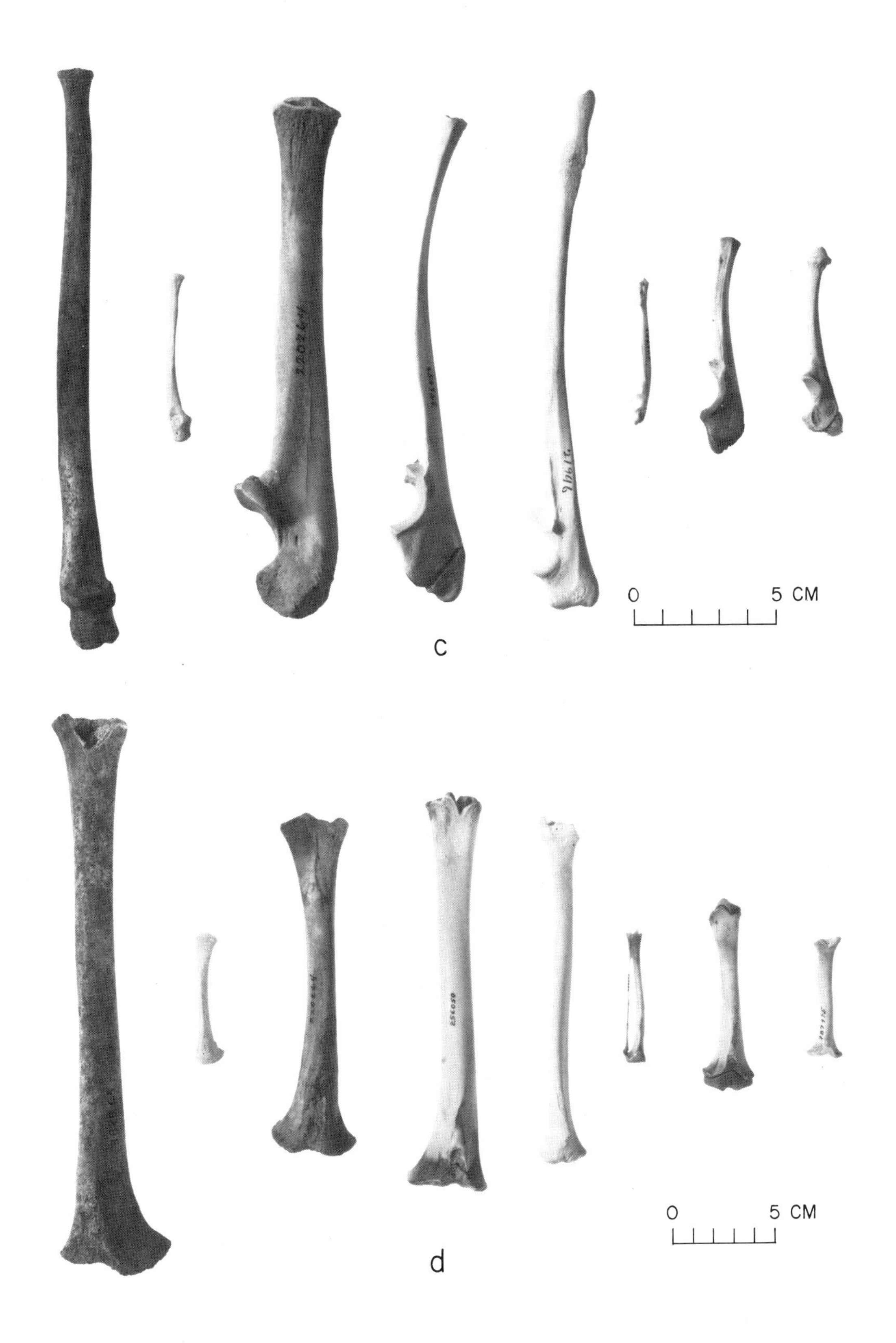
0
5 CM
c
0
5 CM
d

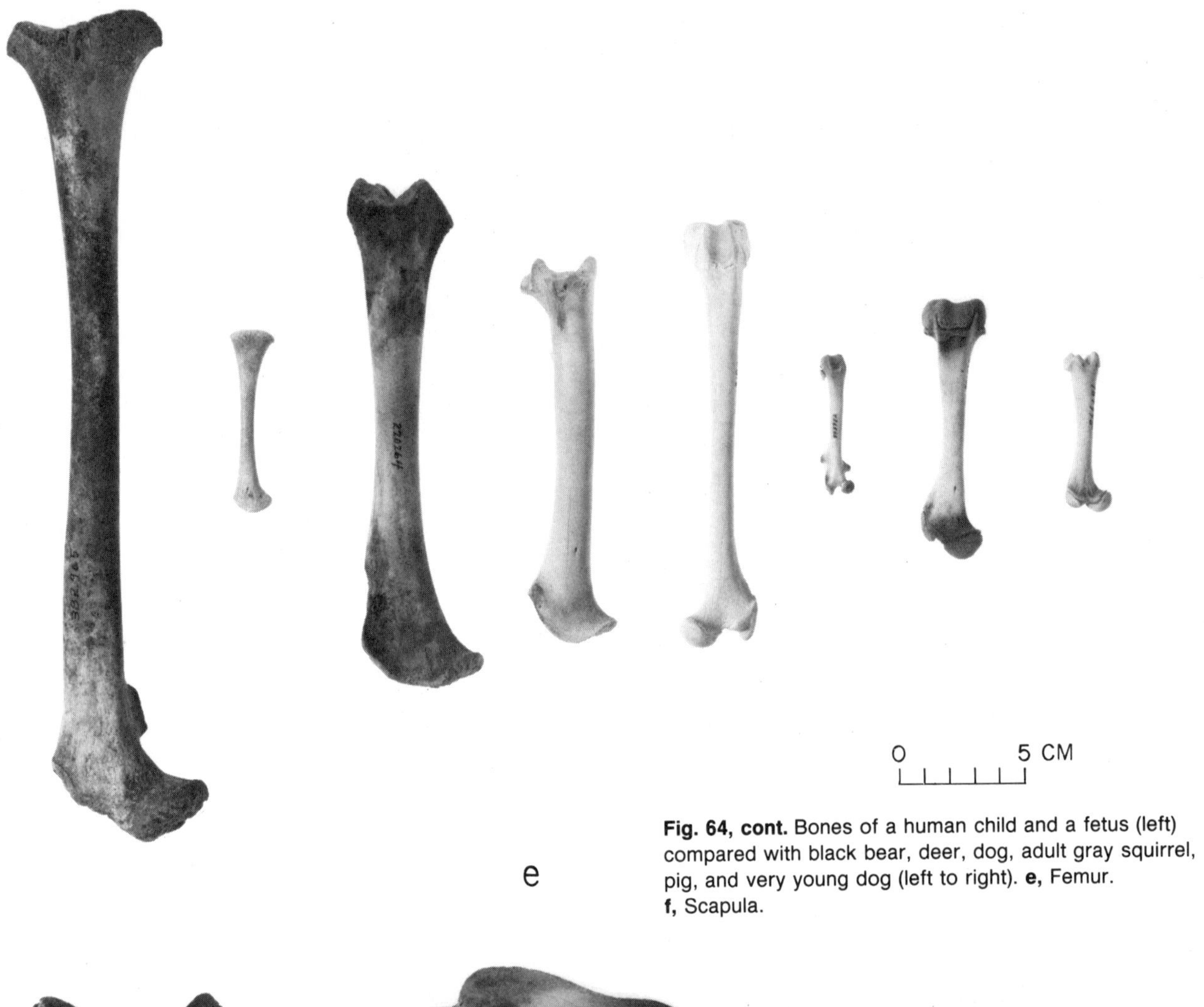

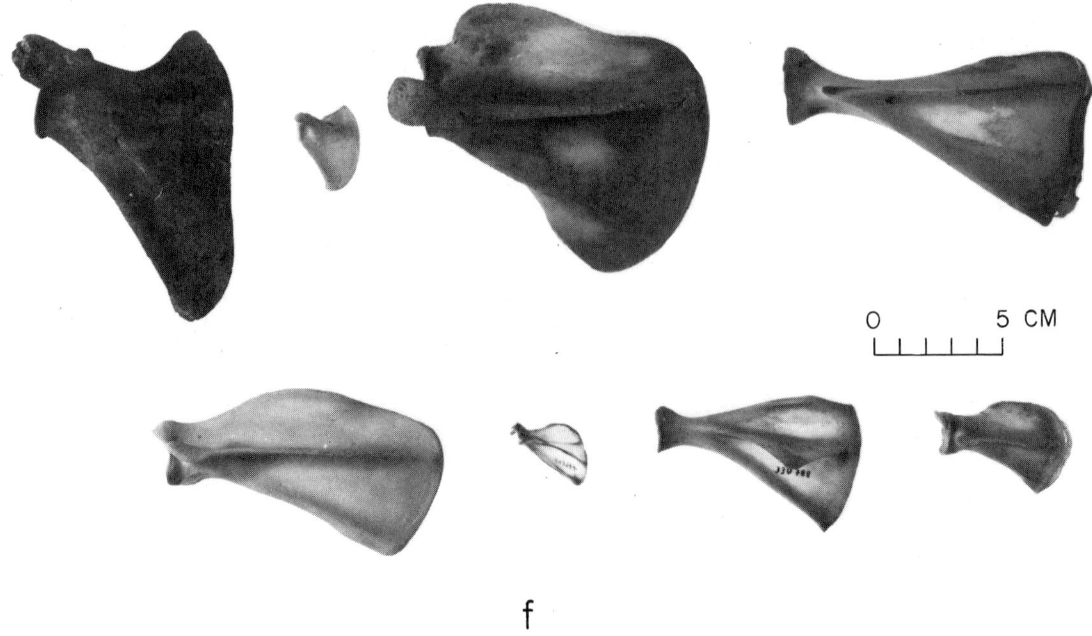

Fig. 64, cont. Bones of a human child and a fetus (left) compared with black bear, deer, dog, adult gray squirrel, pig, and very young dog (left to right). **e,** Femur. **f,** Scapula.

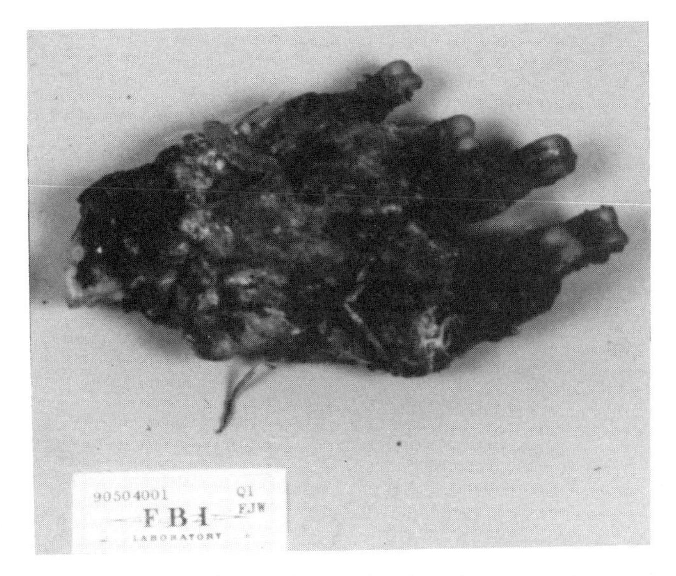

Fig. 65. Bear paw, frequently mistaken for a human hand.

Bear paws are especially prone to misidentification as human when hair and claws are absent (Fig. 65). Close comparison reveals that the bear carpal and tarsal bone are bigger and have much larger and more prominent articular surfaces.

Sorting fragmentary remains is more difficult, especially when normal features have been altered by diseases, weathering or surgery. In such cases, the microscopic appearance of the cortical bone may be diagnostic. In humans, the osteons are scattered and evenly spaced within circumferential lamellar bone, which has a layered appearance (Figs. 66a. 104), whereas in many non-human animals they tend to align in rows termed osteon banding (Fig. 66b) or form rectanguloid structures called plexiform bone (Fig. 66c). Although a plexiform pattern may rule out humans, the great variation among other species and even among bones of the same animal makes scattered osteon distribution inconclusive.

Two forensic cases illustrate application of these concepts. In the New England area of the United States, a few small bone fragments were found in a pool of human blood. Circumstantial evidence suggested a man had been shot in the head, but his body was not recovered. Although the microscopic structure was compatible with human cranial bone, the existence of non-patterned osteon distributions in other species made identification inconclusive.

In the second case, a bone fragment from a remote part of Alaska displayed an unhealed fracture (pseudoarthrosis) that had been bridged surgically by a metal plate (Fig. 67a). Considerable bone growth over the plate indicated the operation occurred long before death. In the absence of the potentially diagnostic epiphyses, the evidence for surgery led local authorities to identify the bone as human. After efforts to locate the surgeon failed, the bone was sent to the Smithsonian Institution, where a microscopic section revealed a non-human osteon banding closely matching that of a large dog (Fig. 67b). The surgeon could not be found because the work was done by a veterinarian!

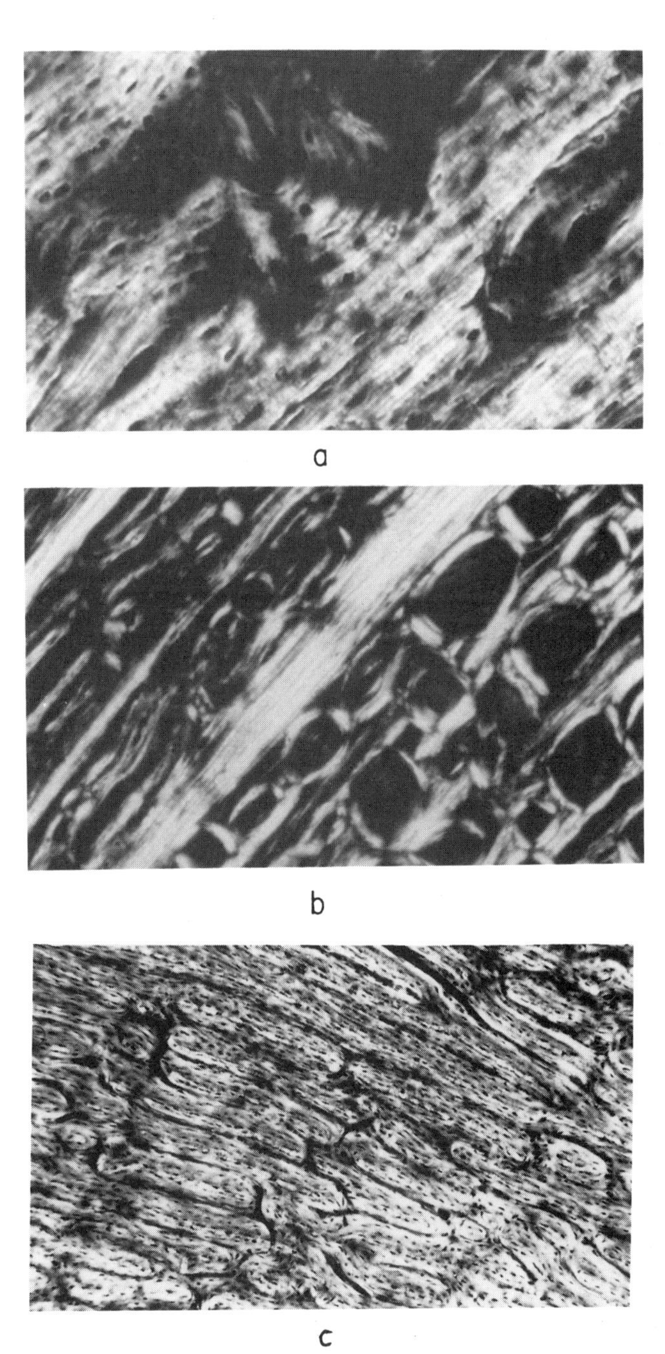

Fig. 66. Distinctive patterning of osteons in circumferential bone. **a**, Human. **b**, Non-human, osteon banding. **c**, Plexiform bone (sheep).

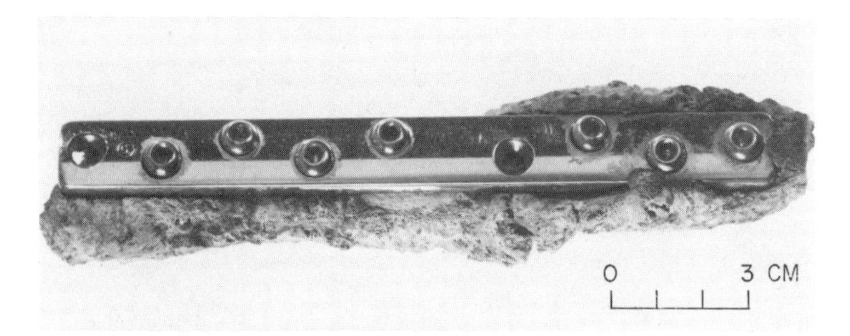

a

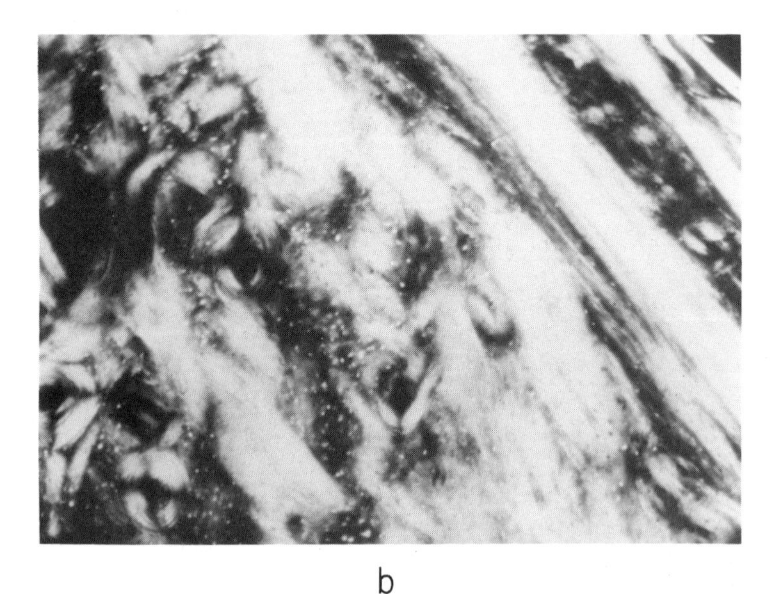

b

Fig. 67. Metal plate bridging an unhealed fracture. **a,** Condition of the specimen. **b,** Non-human osteon pattern, probably a large dog.

DETERMINATION OF SEX

Sexual differences begin to develop in the skeleton before birth (Boucher 1955, Thomson 1899). The width of the sciatic notch of the pelvis, which is one of the most distinctive features in adults (Fig. 68), increases faster in females during fetal growth (Boucher 1957). Through infancy, childhood, and into adolescence, sexual dimorphism becomes more marked and methods of recognizing sex from skeletal remains become more accurate. The fact that females grow faster and mature earlier than males makes it necessary to consider age when estimating sex in subadults.

For adults, the researcher should consider the morphology of the entire skeleton, but rely most heavily on details of the pelvis. Nearly 100 percent of sex estimates based on careful examination of a well preserved pelvis should be accurate. If the pelvis is missing or fragmentary, other criteria can be used, but the result will be less reliable. Because the criteria and judgments involved in estimating the sex of subadults and adults are different, they will be described separately.

Subadults

The sex of subadults may be estimated by comparing the stage of calcification of the teeth with the stage of maturation of the post-cranial skeleton (Hunt and Gleser 1955). The method is based on the fact that the post-cranial skeletons mature more slowly in boys than in girls, whereas the rate of calcification of the teeth is about the same. Thus, sex can be inferred by comparing dental development with post-cranial development. This is accomplished by aging independently the dentition and post-cranial skeleton of an unknown subadult using the standards established for males. If the two estimates agree, the skeleton is probably male. If the results diverge widely, the unknown individual is probably female.

When Hunt and Gleser tested their method on a sample of dental and skeletal radiographs of living children, they estimated sex correctly in 73 percent of the two-year-olds, 76 percent of the five-year-olds, and 81 percent of the eight-year-olds. The method is more difficult to apply to archeological specimens because the assessments of skeletal maturity are usually hampered by the incomplete and fragmentary condition of the bones. It should not be attempted unless complete skeletons are available and the researcher has expert knowledge of the processes of post-cranial and dental maturation. In most cases, it is advisable to limit identification of sex to mature skeletons where the sources of error are significantly less.

Weaver (1980) recorded five measurements and one observation on fetal and infant ilia in the Smithsonian collection. Although he found no significant correlations with sex among the measurements, the non-metric variation in the elevation of the auricular surface was promising. His criterion was as follows: "if the sacro-iliac surface was elevated from the ilium along its entire length and along both the anterior and posterior edges of the sacro-iliac surface, the auricular surface was considered elevated and was so scored. . . . If the surface was not elevated . . . , it was so scored"

(1980:192). This distinction correctly identified male ilia in about 92 percent of fetal skeletons, 73 percent of newborns, and 91 percent of six-month-old infants. It was less successful with females, correctly sexing only 75 percent of fetal ilia, 54 percent of newborns, and 44 percent of six-month-old infants. The method should be used with great caution since accuracy seems to diminish with age and may be markedly affected by population variation.

Adults

Above the age of about 18 years, sex differences are well defined on the skeleton and distinctions can often be made with confidence. The significant differences are of two types: (1) size and (2) function-related shape. In general, the bones of males are longer, more robust, and display more rugged features than those of females. Structural differences also exist and are especially pronounced in the pelvis, reflecting the birth process in females. Supplementary information can be obtained from the cranium and the long bones.

Pelvis. The pelvis provides the most abundant and accurate data for sex estimation. Differences between males and females are well marked and have been confirmed by considerable research. The landmarks described below are readily observable and suffice for preliminary identification. Summaries by Krogman and Iscan (1986) and Stewart (1968) should be consulted for more detailed discussions.

Size. The female pelvis is broader, even though the male pelvis may be heavier and more robust.

Sciatic Notch. The sciatic notch (Fig. 68a) is located at the junction between the ilium (upper, flat

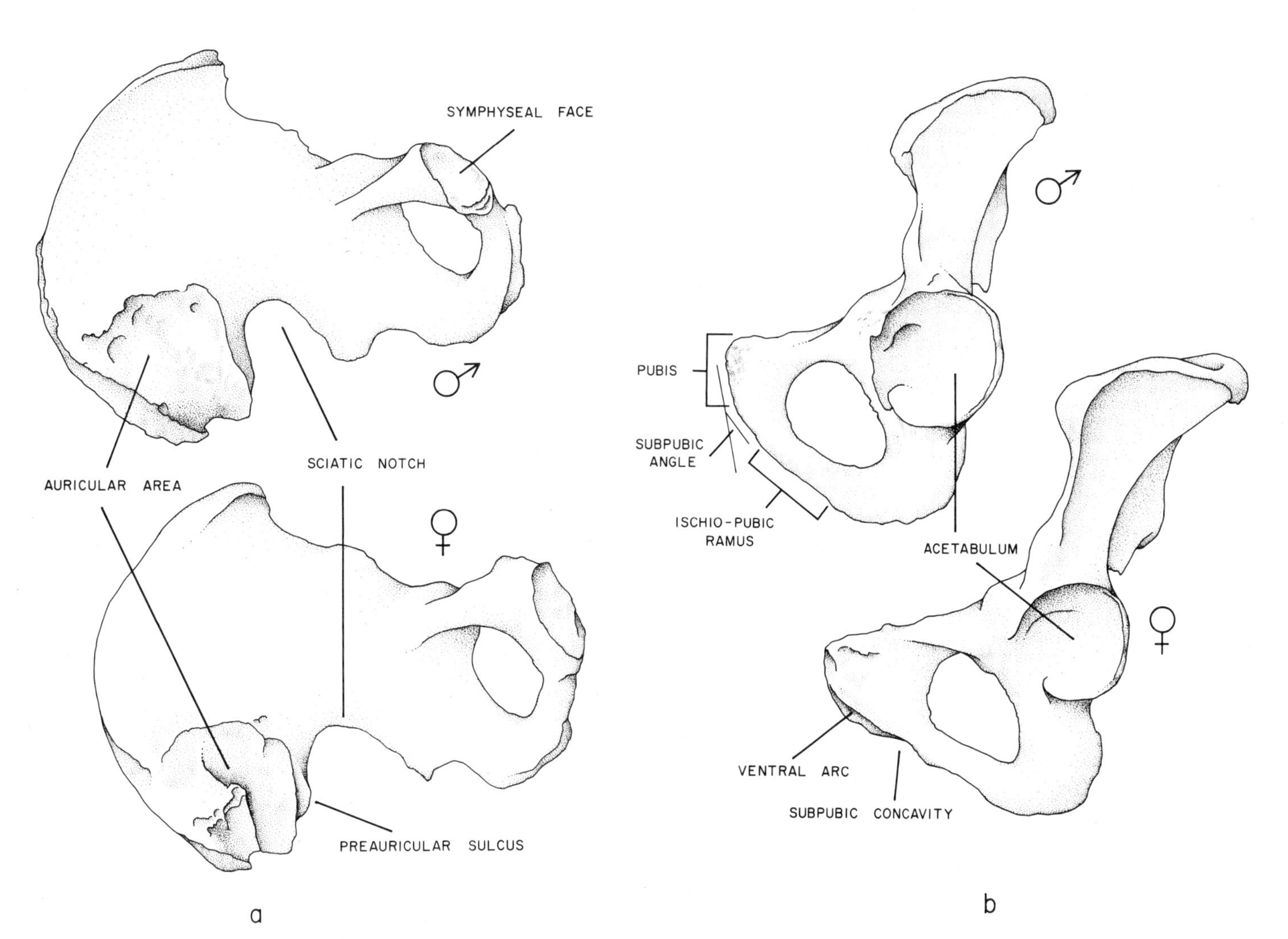

Fig. 68. The innominate bone of the pelvis, showing the features useful for differentiating males from females. **a,** Anterior view. **b,** Posterior view.

portion of the pelvis) and the ischium (lower portion of the pelvis). In females, the notch is wide, usually forming an angle of about 60 degrees. In males, it typically is narrower and forms an angle of about 30 degrees.

Auricular Area. The auricular area is the medial portion of the ilium that articulates with the sacrum (Fig. 68a, 92). It tends to be flatter in males than in females.

Pre-auricular Sulcus. This feature is a groove between the auricular area and the sciatic notch (Fig. 68a, 92). It nearly always occurs in females, but seldom in males. If present in males, it is shallower than in females.

Acetabulum. The acetabulum, a socket-like depression on the lateral side of the innominate that holds the head of the femur, is larger in males than in females (Fig. 68b).

Pubis. The pubis is longer in females and the subpubic angle is wider. Both features are correlated with the birth process (Fig. 68b). Three other characteristics are sufficiently distinctive to permit correct identification in 96 percent of the modern specimens examined (Phenice 1969a). These are the ventral arc, the subpubic cavity, and the contour of the medial part of the ischio-pubic ramus.

The ventral arc is a ridge on the anterior surface of the pubis that begins near the middle and extends downward and inward (Fig. 68b). Its lower margin is separated from the symphyseal face by several millimeters of bone. In males, the ridge is typically absent; a line can occasionally be observed corresponding to the site of the ridge in females, but its lower margin is never separated from the symphyseal face as it is in females.

The subpubic concavity is a depression in the medial border of the ischio-pubic ramus, just below the symphysis (Fig. 68b). In females, this concavity is large and obvious, whereas in males it is usually absent or, if it occurs, is very slight.

The medial portion of the ischio-pubic ramus, just below the symphysis, is usually flat in males (Fig. 69). In females, a ridge occurs at this site. Because variation is greater, this feature is less useful for estimating sex than the ventral arc or the subpubic cavity.

Cranium. Sex estimates made from the cranium are not as accurate as those based on the pelvis, but they may be useful in the absence of pelvic data. In general, males display larger mastoid processes, larger supra-orbital crests, and larger and more rugged muscle markings, especially on the occipital bone (see Figs. 162–163). Because sex differences in cranial morphology vary slightly among populations, a trained physical anthropologist should be consulted. Such a

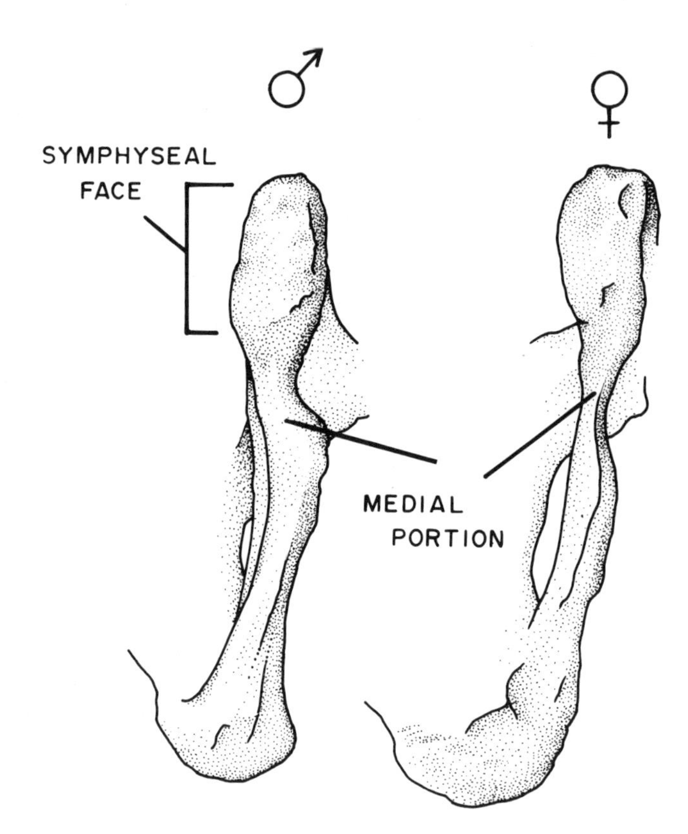

Fig. 69. The ischio-pubic ramus, showing the details useful for differentiating males from females.

specialist can identify sex with an accuracy of 80 to 90 percent. Kalmey and Rathbun (1996) provide a method employing only the petrous portion of the temporal bone

Long Bones. Sexual differences in the post-cranial skeleton are well documented, but less consistent than those on the pelvis and cranium. Although the bones of males tend to be larger and more rugged, the accuracy of sex identification is reduced by the overlap between the ranges in males and females in the same population and by the variation among different populations. For example, study of a documented sample may indicate that 95 percent of the femora displaying certain characteristics are from males. This would imply a 95 percent probability that a prehistoric femur with the same characteristics also belonged to a male if sexual differentiation or dimorphism had the same range of variation in the prehistoric and modern populations. This estimate thus includes two potential sources of error: (1) a five percent error resulting from the knowledge that five percent of the modern individuals possessing the characteristic were female, and (2) an error of unknown magnitude introduced by applying data derived from one population to another.

Maximum diameters of the heads of the femur, humerus, and radius are fairly good indicators of sex in adults when they fall outside the zone of overlap. Femoral heads with diameters exceeding 48 millimeters are usually male, whereas those measuring less than 43 millimeters are usually female. The maximum vertical diameter of the head of the humerus is usually more than 47 millimeters in males and less than 43 millimeters in females (Dwight 1904–1905, Stewart 1979: 100–101). When the minimum diameter of the head of the radius is 23 millimeters or more, sex is male. When the maximum diameter is 21 millimeters or less, sex is female (Berrizbeitia 1987).

Sex differences have also been documented in the bones of the arms (Holman and Bennett 1991), hands (Falsetti 1995, Lazenby 1995, Scheuer and Elkington 1993, Smith 1996), legs (Black 1978, Iscan and Miller-Shaivitz 1984), and left feet (Introna et al 1997, Robling and Ubelaker 1997, Smith 1996, Steele 1976). Applications of these techniques to documented samples predicted sex with an accuracy from 58 to 100 percent.

Equations. Discriminant function equations may be used to "calculate" sex. These employ measurements of individuals of known sex to predict the sex of unknown individuals. Because most equations are population specific, care should be taken to select the one that best matches the characteristics of the unknown individual.

Giles (1970) offers a useful summary of discriminant functions for various aspects of the skeleton. The measurements are described in Table 3 and the functions are presented in Tables 4–8. The left column in each table specifies by number the measurements used and the remaining columns give their multiplication factor components. Within each function, the value of the measurement is multiplied by the corresponding factor. The results are then added (if no sign precedes the factor) or subtracted (if a minus precedes the factor). If their sum is larger than the sectioning point at the bottom of the column used, the sex is male. If the sum is smaller, it is female. The bottom line shows the percentage of correct results obtained from applying the function to individuals of known sex.

A modern forensic case exemplifies use of these tables (Giles 1970:101). A cranium of apparent White affiliation produced the following measurement in millimeters:

1—168	4—94	7—72
2—140	5—125	9—29
3—128	6—93	

The presumed ancestry and locus of the measurements make the multiplication factors provided for Function 2 on Table 4 the most appropriate. These produce the following calculation:

$$(3.400 \times 168) - (3.833 \times 140) + (5.433 \times 128) - (0.167 \times 94) + (12.200 \times 125) - (0.100 \times 93) + (2.200 \times 72) + (5.367 \times 29) = 2544.05$$

This value falls below the sectioning point of 2592.32, indicating sex is female. Additional discussion and formulas are provided in Krogman and Iscan (1986).

Discriminant function approaches to estimating sex have elaborated considerably since the formation of a forensic data bank. Standard observations and measurements defined using identified forensic cases and augmented with comparable data from museum collections were entered into a central data bank. A computer program, Fordisc 2.0 (Ousley and Jantz 1996), provides a flexible system that enables the user to employ whatever standard measurements are available to create custom discriminant function calculations that classify unknowns. The advantages for forensic purposes are: (1) the flexibility that allows even minimal measurements to be used, (2) the forensic origin of the sample producing the data base, and (3) the presentation of results in a statistical format.

Parturition

Given the importance of estimating fecundity in forensic and paleodemographic studies, it is not surprising that attempts have been made to document the effects of childbirth on bone. Following Stewart's (1957) study of Eskimo skeletons and Putschar's (1931) study of German remains, Angel (1969) offered the following rationale for estimating the number of births from bony details on the female pelvis:

> These changes are consistently clearest around the pubic symphysis where pregnancy stresses the muscle and tendon attachments of the central belly wall (Rectus abdominis and both tubercle and pectineal attachments of the inguinal ligament or tendon of the Oliquus externus abdominis); where also during the birth process the arcuate and interpubic ligaments are stretched and torn and where cysts and knots of fibrocartilages follow the tears and small haemorrhages ("bruises") which occur under the ligaments, especially on the inner surface separated from the birth canal only by the bladder walls. On the anterior surface of the pubic symphysis, therefore, exostoses develop (not unlike those seen in older arthritic skeletons of either sex) and generally also a spiral fossa below the pubic tubercle begins to develop even after one or two births. Posteriorly, next to lip exostoses a series of small fossae from haemorrhages and cysts after a sufficient number of births (perhaps 4–8) may coalesce into a deep groove next to the exaggerated lip at the back edge of the symphyseal face. A clearcut development of these changes occurs after more than three births.

Additional data and opinions have been provided by Bergfelder and Herrmann (1978), Holt (1978), Nemeskéri (1972), Stewart (1970, 1972), Suchey et al (1979), and Ullrich (1975). The following changes have been detected: pitting on the ventral margin of the

Table 3. Measurements used in discriminant functions for inferring sex (after Giles 1970, Appendix).

1. Maximum length of the skull, from the most anterior point of the frontal, in the midline, to the most distant point on the occiput, in the midline.
2. The greatest breadth of the cranium perpendicular to the median sagittal plane, and avoiding the supramastoid crests.
3. Cranial height measured from basion (midpoint on the anterior border of the foramen magnum) to bregma (intersection of the coronal and sagittal sutures).
4. From basion (see *3*) to nasion (midpoint of the naso-frontal suture).
5. Maximum width between the lateral surfaces of the zygomatic arches measured perpendicular to the median sagittal plane.
6. From basion (see *3*) to the most anterior point on the maxilla in the median sagittal plane.
7. Lowest point on the alveolar border between the central incisors to nasion (see *4*).
8. Maximum breadth of the palate taken on the outside of the alveolar borders.
9. The length of the mastoid measured perpendicular to the plane determined by the lower borders of the orbits and the upper borders of the auditory meatuses (= Frankfort plane). The upper arm of the sliding calipers is aligned with the upper border of the auditory meatus, and the distance (perpendicular to the Frankfort plane) to the tip of the mastoid is measured.
10. Height from the lowest median point on the jaw (menton) to the lower alveolar point (bony process between the central incisors). If the menton is in a notch, then the measurement is taken from a line tangent to the lowest points on the margins lateral to the notch.
11. Mandibular body height as measured between the first and second molars.
12. From the most anterior point on the mandibular symphysis to an imaginary point formed by the posterior margin of the ramus and the antero-posterior axis of the body, and measured parallel to the axis.
13. The thickness of the mandibular body measured at the level of the second molar parallel to the vertical axis of the body.
14. The smallest antero-posterior diameter of the ramus of the jaw.
15. The distance between the most anterior point on the manidbular ramus and the line connecting the most posterior point on the condyle and the angle of the jaw.
16. Height measured from the uppermost point on the condyle to the middle of the inferior border of the body parallel to the vertical axis of the ramus. (The middle of the ramus on the inferior margin is not a distinct point but can be easily estimated.) For Japanese mandibles, measure to gonion (see *17*).
17. Maximum diameter, externally, on the angles of the jaw (gonion).
18. Femur length taken maximally, but perpendicular to a line defined by the distal-most points of the two distal condyles (so-called oblique or standing length).
19. Greatest diameter of femur head.
20. Least transverse diameter of shaft of femur.
21. Width of the distal end of the femur (epicondylar breadth).
22. Ischial length measured from where the long axis of the ischium crosses the ischial tuberosity to a point in the acetabulum that is defined as the intersection of the long axes of the pubis and the ischium.
23. Pubic length measured from the point in the acetabulum defined in *22* to the upper extremity of the symphyseal articular facet of the pubis.
24. Height of the sciatic notch, taken as a perpendicular dropped from the point on the posterior inferior iliac spine, where the upper border of the notch meets the auricular surface, to the anterior border of the notch itself.
25. Acetabulo-sciatic breadth, taken from the median point on the anterior border of the sciatic notch (half way between the ischial spine and the apex of the notch) to the acetabular border, and perpendicular, as far as possible, to both borders.
26. Taken from the most projecting point on the pubic portion of the acetabular border perpendicular to the innominate line, and thus to the plane of the obturator foramen.
27. The distance from the anterior iliac spine to the nearest point on the auricular surface, and subtracted from the distance from the anterior iliac spine to the nearest point on the border of the sciatic notch.
28. Maximum length of the humerus.
29. Maximum epicondylar width of the humerus.
30. Maximum length of the clavicle.
31. Minimum circumference of the shaft of the humerus.
32. Maximum length of the radius.

33. Circumference of the radius at the midpoint of the shaft.
34. Circumference of the head of the radius.
35. Maximum medio-lateral breadth of the distal epiphysis of the radius.
36. Maximum length of the ulna.
37. Transverse diameter of the shaft of the ulna taken at the point of greatest development of the crista.
38. Maximum diameter of the capitulum of the ulna.
39. Maximum length of the tibia from the lateral condyle to the malleolus.
40. Maximum antero-posterior diameter of the shaft of the tibia at its midpoint.
41. Least circumference of the shaft of the tibia.
42. Maximum breadth of the proximal epiphysis of the tibia.
43. Anatomical breadth of the scapula (maximum distance between the medial and the inferior angles).
44. Anatomical length of the scapula (the distance between the center of the glenoid cavity and the point where the spine, or its projection, intersects the vertebral border).
45. Projective length of the spine of the scapula (distance from the most projecting point on the acromion to the point where the spine, or its projection, intersects the vertebral border).
46. Length of the glenoid cavity of the scapula—from the most cephalic point to the most caudal.
47. Breadth of the glenoid cavity of the scapula taken perpendicular to *46*.
48. Ischium-pubis index formed by dividing the pubic length, measured from the upper end of the pubic symphysis to the nearest point on the border of the acetabulum, by the ischium length, measured from the point where the axis of the ischium crosses the ischial tuberosity to the most distant point on the border of the acetabulum.
49. Total breadth of the atlas measured between the apexes of the transverse processes.

Table 4. Discriminant function sexing by cranial measurements (after Giles 1970: Table 51).

Measurement**	American White						American Black						Japanese	
	1	2	3	4	5	6	7	8	9	10	11	12	13	14
1	3.107	3.400	1.800		1.236	9.875	9.222	3.895	3.533		2.111	2.867	1.000	1.000
2	−4.643	−3.833	−1.783		−1.000		7.000	3.632	1.667		1.000		−0.062	0.221
3	5.786	5.433	2.767				1.000	1.000	0.867				1.865	
4		−0.167	−0.100	10.714		7.062		−2.053	0.100	1.000		−0.100		
5	14.821	12.200	6.300	16.381	3.291	19.062	31.111	12.947	8.700	19.389	4.963	12.367	1.257	1.095
6	1.000	−0.100		−1.000		−1.000	5.889	1.368		2.778		−0.233		
7	2.714	2.200		4.333		4.375	20.222	8.158		11.778		6.900		0.504
8	−5.179			−6.571			−30.556			−14.333				
9	6.071	5.367	2.833	14.810	1.528		47.111	19.947	14.367	23.667	8.037			
Sectioning point	2676.39	2592.32	1296.20	3348.27	536.93	5066.69	8171.53	4079.12	2515.91	3461.46	1387.72	2568.97	579.96	380.84
Percent correct	86.6	86.4	86.4	84.5	85.5	84.9	87.6	86.6	86.5	87.5	85.3	85.0	86.4	83.1

Table 5. Discriminant function sexing by post-cranial measurements (after Giles 1970: Table 54).

	American White				American Black				Japanese			
Measurement**	1^R	2^L	3	4	5	6	7	8	9^R	10^L	11^R	12^L
18	1.000	1.000			0.070	1.000	1.000	1.980			1.000	1.000
19	30.234	30.716			58.140	31.400	16.530				9.854	9.351
20	−3.535	−12.643									11.988	8.369
21	20.004	17.565									4.127	3.575
22				0.607	16.250	11.120	6.100	1.000				
23				−0.054	−63.640	−34.470	−13.800	−1.390				
24			−0.115	−0.099								
25			−0.182	−0.134								
26			0.828	0.451								
27			0.517	0.325								
28					2.680	2.450			1.000	1.000		
29					27.680	16.240			8.726	6.198		
30					16.090							
31									7.394	3.221		
Sectioning point	3040.32	2656.51	9.20	7.00	4099.00	1953.00	665.00	68.00	1189.51	804.28	1431.82	1277.83
Percent correct	94.4	94.3	93.1	96.5	98.5	97.5	96.9	93.5	92.9	93.6	96.2	95.9

	Japanese											
Measurement**	13^R	14^L	15^R	16^L	17^R	18^L	19^R	20^R	21^R	22^L	23^L	24^L
32	1.000	1.000										
33	1.917	1.273										
34	2.991	3.163										
35	9.126	7.711										
36			1.000	1.000								
37			8.068	6.501								
38			5.551	2.881								
39					1.000	1.000						
40					4.264	2.954						
41					7.544	5.605						
42					12.213	10.212						
43							1.000	1.000	1.000	1.000	1.000	1.000
44									1.350			1.494
45							6.335	1.899		1.929	1.846	
46							12.664	11.922	10.940	6.949	7.107	6.800
47							10.991			2.120		
Sectioning point	763.92	696.97	441.54	370.25	1802.10	1494.54	1660.16	782.10	634.75	669.79	611.03	508.35
Percent correct	**96.7**	**97.0**	**88.9**	**90.5**	**95.7**	**95.3**	**96.8**	**96.0**	**95.6**	**94.8**	**94.7**	**94.1**

Table 6. Discriminant function sexing by combined cranial and post-cranial measurements (after Giles 1970: Table 55).

Measure-ment**	Japanese						
	1	2	3	4	5	6	7
1	1.000			1.000	1.000	1.000	1.000
3		1.000	1.000				
18	0.107	0.031	0.176	0.138		0.220	
46	6.644	4.390		8.117	8.035		4.757
48	−5.050	—2.654	—3.281	−5.156	—5.586	—3.816	
49	2.678		2.090		2.152	2.491	2.124
Section-ing point	299.18	117.11	142.12	157.76	233.09	194.55	494.36
Percent correct	99.0	98.8	96.4	98.6	98.8	97.4	92.5

Table 7. Discriminant function sexing by mandibular measurements (after Giles 1970: Table 52).

Measure-ment**	American White			American Black			Japanese
	1	2	3	4	5	6	7
10	1.390	22.206	2.862	1.065	2.020	3.892	2.235
11		−30.265			—2.292		
12		1.000	2.540		2.606	10.568	
13			−1.000			—9.027	
14			—5.954			—3.270	1.673
15			1.483			1.000	
16	2.304	19.708	5.172	2.105	3.076	10.486	2.949
17	1.000	7.360		1.000	1.000		1.000
Section-ing point	287.43	1960.05	524.79	265.74	549.82	1628.79	388.53
Percent correct	83.2	85.9	84.1	84.8	86.9	86.5	85.6

Table 8. Discriminant function sexing by combined cranial and mandibular measurements (after Giles 1970: Table 53).

	Japanese					American Black
Measure-ment**	1	2	3	4	5	6
1	1.000	1.000	1.000	1.000	1.000	1.289
3	2.614	2.519		2.560	2.271	
5	0.996	0.586	0.785	1.084	1.391	−0.100
7						1.489
9						4.289
10	2.364				2.708	−0.978
12						−0.544
16	2.055	2.713	1.981	2.604		3.478
17		0.661	0.404			1.400
Sectioning point	850.66	807.40	428.05	809.72	748.34	718.23
Percent correct	89.7	89.4	86.4	88.9	88.8	88.3

pubis, alterations on the superior aspect of the dorsal margin of the pubis, depressions associated with the pre-auricular sulcus of the ilium, and grooves on the auricular margin of the sacrum. These features are illustrated by Ullrich (1975).

Suchey et al (1979) found a weak correlation between the number of full-term pregnancies and the extent of dorsal pitting on the pubis, but some nulliparous females had medium to large pitting whereas others with several full-term pregnancies lacked bony changes. Alterations increased with age at death independently of the number of pregnancies, however, making them a strong indicator of female sex.

ESTIMATING STATURE

Using All Contributing Bones

The simplest way to estimate stature is to measure the lengths of the relevant bones and add a factor for the non-bone contribution. As early as 1894, Dwight introduced a rather cumbersome procedure that basically rearticulated the skeleton on a metrically graduated table (see Dwight 1894 and Stewart 1979:217 for details).

Apparently unaware of Dwight's efforts, Fully (1956) advocated using bone measurements and employing a single soft-tissue correction. His method combines the following measurements: basion-bregma height of the cranium; maximum height of the corpus of each vertebra from the second cervical to the fifth lumbar; anterior height of the first sacral segment; oblique length of the femur (both condyles in contact with the osteometric board); maximum length of the tibia without the spine; and the distance between the superior part of the tibio-talar articular surface and the most inferior part of the bearing surface of the calcaneus, measured with the talus and calcaneus articulated on a standard osteometric board. The following correction factors must be added to the sum of these measurements to calculate living stature:

10.0 cm, if the sum is 153.5 cm or less
10.5 cm, if the sum is between 153.6 and 165.4 cm
11.5 cm, if the sum is 165.5 cm or more

Using Representative Bones

Having observed that the sizes of many bones, especially vertebrae, are strongly correlated, Fully and Pineau (1960) concluded it was necessary to measure only a few representative ones. They provide two formulas:

$$\text{stature} = 2.09 \times (\text{femur length} + \text{sum of body heights of 5 lumbar vertebrae}) + 42.67$$

$$\text{stature} = 2.32 \times (\text{tibia length} + \text{sum of body heights of 5 lumbar vertebrae}) + 48.63$$

They note these formulas may be more practical since many skeletons are incomplete or have damaged bones.

Using Limb Bones

All other methods of calculating living stature are based on the correlation between body height and limb length. The considerable variation among different populations in the ratio of long-bone length to stature makes it necessary to generate population-specific formulas. Trotter (1970) has provided a useful summary of studies conducted to generate such equations.

Table 9 presents equations developed by Trotter (1970) after studying approximately 850 documented male and female skeletons in the Terry Collection in the National Museum of Natural History and about 4100 males killed during the Korean war and World War II. All measurements are the maximum length in centimeters and should be obtained using an osteometric board (Fig. 70). Because errors appear to have been detected in the measurements provided by Trotter for the tibia, this bone should be avoided. If the tibia must be used, Jantz et al 1995 should be consulted for specific instructions.

To use the formulas, one must measure the length of the bone in the correct manner, multiply the result by the factor in the formula, and add the amount specified. For example, if a femur is thought to belong to a Mexican male, one would measure the maximum length of the femur in centimeters, multiply that measurement by 2.44, and add 58.67. Each formula is accompanied by a plus-or-minus error reflecting the fact

Table 9. Equations to estimate living stature (cm) from long bones of males and females between ages 18 and 30 (after Trotter 1970: Table 28).

White Males		Black Males	
3.08 Hum + 70.45	± 4.05	3.26 Hum + 62.10	± 4.43
3.78 Rad + 79.01	± 4.32	3.42 Rad + 81.56	± 4.30
3.70 Ulna + 74.05	± 4.32	3.26 Ulna + 79.29	± 4.42
2.38 Fem + 61.41	± 3.27	2.11 Fem + 70.35	± 3.94
2.52 Tib + 78.62	± 3.37	2.19 Tib + 86.02	± 3.78
2.68 Fib + 71.78	± 3.29	2.19 Fib + 85.65	± 4.08
1.30 (Fem + Tib) + 63.29	± 2.99	1.15 (Fem + Tib) + 71.04	± 3.53
White Females		**Black Females**	
3.36 Hum + 57.97	± 4.45	3.08 Hum + 64.67	± 4.25
4.74 Rad + 54.93	± 4.24	2.75 Rad + 94.51	± 5.05
4.27 Ulna + 57.76	± 4.30	3.31 Ulna + 75.38	± 4.83
2.47 Fem + 54.10	± 3.72	2.28 Fem + 59.76	± 3.41
2.90 Tib + 61.53	± 3.66	2.45 Tib + 72.65	± 3.70
2.93 Fib + 59.61	± 3.57	2.49 Fib + 70.90	± 3.80
1.39 (Fem + Tib) + 53.20	± 3.55	1.26 (Fem + Tib) + 59.72	± 3.28
Asian Males		**Mexican Males**	
2.68 Hum + 83.19	± 4.25	2.92 Hum + 73.94	± 4.24
3.54 Rad + 82.00	± 4.60	3.55 Rad + 80.71	± 4.04
3.48 Ulna + 77.45	± 4.66	3.56 Ulna + 74.56	± 4.05
2.15 Fem + 72.57	± 3.80	2.44 Fem + 58.67	± 2.99
2.39 Tib + 81.45	± 3.27	2.36 Tib + 80.62	± 3.73
2.40 Fib + 80.56	± 3.24	2.50 Fib + 75.44	± 3.52
1.22 (Fem + Tib) + 70.37	± 3.24		

*To estimate stature of older individuals subtract 0.06 (age in years — 30) cm; to estimate cadaver stature, add 2.5 cm.

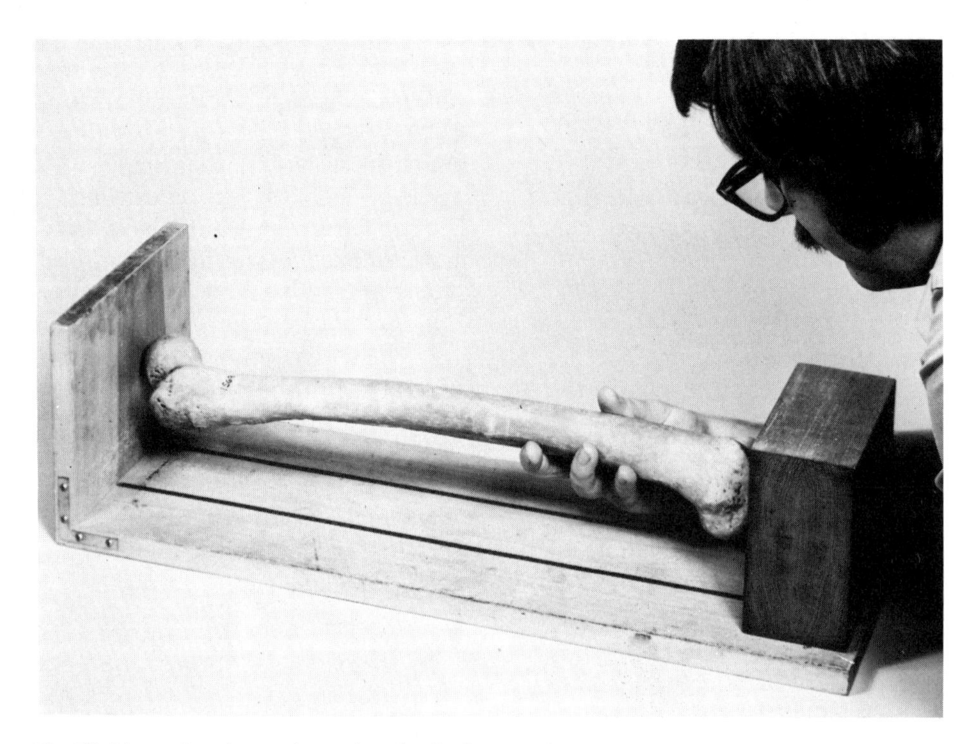

Fig. 70. Measuring the maximum length of a femur using an osteometric board. The black lines are a scale; the vertical board at the left is secured permanently, while that at the right is moved to establish the length of the bone.

that the ratio of long-bone length to stature varies within populations. The standard error of 2.99 means that only about two-thirds of the estimates will be within 2.99 centimeters taller or shorter than the actual stature. One-third will more than 2.99 centimeters taller or shorter than the estimate. The error is likely to be greater when the formulas are applied to different populations. Other corrections must be made for individuals over age 30 and for cadavers. Tables provided by Trotter for the expected maximum statures of U.S. Black and White males and females are reproduced in Appendix 1, Tables 1–4.

Table 10. Regression equations for estimating stature of prehistoric Mesoamericans. Subtract 2.5 centimeters to obtain the living stature (after Genovés 1967:76).

Males:		
Femur:	Stature = 2.26 Fem + 66.379 ± 3.417	
Tibia:	Stature = 1.96 Tib + 93.752 ± 2.812	
Females:		
Femur:	Stature = 2.59 Fem + 49.742 ± 3.816	
Tibia:	Stature = 2.72 Tib + 63.781 ± 3.513	

Genovés (1967) developed formulas for estimating stature among prehistoric Mesoamericans from his studies of Mexican cadavers classified as "indigenous" using morphological and serological criteria (Table 10). All measurements are maximum, except for the tibia, which does not include the tuberosity.

Numerous other formulas have been developed using different samples and approaches by Albrook (1961), Dupertuis and Hadden (1951), Jit and Singh (1956), Keen (1953), Lundy (1983), Musgrave and Harneja (1978), Nat (1931), Oliver (1963), Telkkä (1950), and Tibbetts (1981). In selecting among them, emphasis should be placed on similarity between the composition of the population represented by the archeological sample or forensic case and the population used to create the formulas. Genovés' formulas are most appropriate for Mesoamerican remains, whereas Trotter and Gleser's formulas may give more reliable results for skeletons from northern North America, especially males. Stature estimation procedures are also available by computer using the Forensic Data Base (Ousley and Jantz 1996). The results are most accurate if sex and ancestry have been estimated reliably.

To overcome the problem presented by the fragmentary condition of many bones from archeological and forensic contexts, Steele (1970) and Steele and McKern (1969) have proposed formulas for estimating the original length of a bone. Holland (1992) offers a technique that allows living stature to be estimated from measurements of the proximal end of the tibia.

ESTIMATING AGE AT DEATH

Estimation of age at death involves observing morphological features in the skeletal remains, comparing the information with changes recorded for recent populations of known age, and then estimating any sources of variability likely to exist between the unknown and the recent population furnishing the documented data. This third step is seldom recognized or discussed in osteological studies, but it represents a significant element.

To estimate age, we must utilize what is known about chronological changes in the skeleton. These changes do not occur at the same times or rates in different bones and structures. During infancy, most changes involve the appearance and growth of bones and teeth. During childhood and through adolescence, bone growth, dental eruption, and calcification continue; in addition, the epiphyses on the post-cranial skeleton develop and unite. About the age of 20, most growth is complete, most epiphyses are united, and most teeth have erupted and are fully calcified. After the age of 20, landmarks are provided by the progressive union of the cranial sutures (lines of articulation between the bones of the skull), changes in the appearance of the symphyseal face of the pubis, degenerative changes (arthritis, dental attrition), and changes in the microscopic structure of the bones and teeth.

As the preceding discussion implies, the criteria employed for estimating age at death must be relevant to the maturity of the individual. Information on dental eruption will not be useful for differentiating a 30-year-old from a 50-year-old adult. It can reveal, however, whether a child is six or eight. Thus, the first step in examining a specimen of unknown age is determining whether it is an infant, child, adolescent or adult, and selecting the criteria appropriate for defining the age within one of these categories more exactly.

The following pages describe the best methods available for age determination. They are divided into two sections: (1) criteria applicable to subadults (under 20 years old) and (2) criteria useful for adults. Anyone attempting to estimate age at death from a human skeleton should consider the variety of methods and the relative accuracy of each method. Selection should be based not only on the accuracy of the method, but also on the relative preservation of the bones and teeth, the availability of time and equipment, and the precision required by the problem being investigated. For example, it makes little sense to allocate the time and funds to prepare microscope slides when the analysis only calls for a general separation of young and old adults. Likewise, it makes little sense to attempt a detailed demographic statement about a population if the methods of estimating age incorporate errors of 20 years or more.

Subadults

The development of the teeth, length of the long bones, and union of the epiphyses are the principal criteria for estimating age in subadults. Dental development provides the most accurate results, especially between birth and 10 years; however, all skeletal data should be employed.

Dental Development. Dental calcification (tooth formation) and eruption (emergence from the gum) are the most accurate indicators of chronological age in subadults (Garn, Lewis, and Polacheck 1959, Lewis and Garn 1960). Dental development is strongly controlled by genetic factors (Glasstone 1938, 1963, 1964), with minimal influence from the environment (Paynter and Grainger 1961, 1962). Although specific diseases, such as hypo-pituitarism and syphillis, can modify the rate of dental development (Bauer 1944), most diseases affect teeth little if at all, even though parts of the skeleton may be greatly altered (Niswander and Sujaku 1965). Endocrine disorders and other maturational problems have been shown to affect teeth only one-fourth as much as the rest of the skeleton (Garn, Lewis, and Blizzard 1965).

The timing of dental formation and eruption in American Indians and other "non-white" populations is summarized on Figure 71, compiled from data published by Anderson, Thompson, and Popovich (1976), Banerjee and Mukherjee (1967), Christensen and Kraus (1965), Coughlin and Christensen (1966), Dahlberg and Menegaz-Boch (1958), Demisch and Wartmann (1956), Glister, Smith, and Wallace (1964), Hurme (1948), Kraus (1959), Lunt and Law (1974), Meredith (1946), Moorrees (1965), Moorrees, Fanning, and Hunt (1963a, 1963b), Nolla (1960), Robinow, Richards, and Anderson (1942), and Steggerda and Hill (1942). These

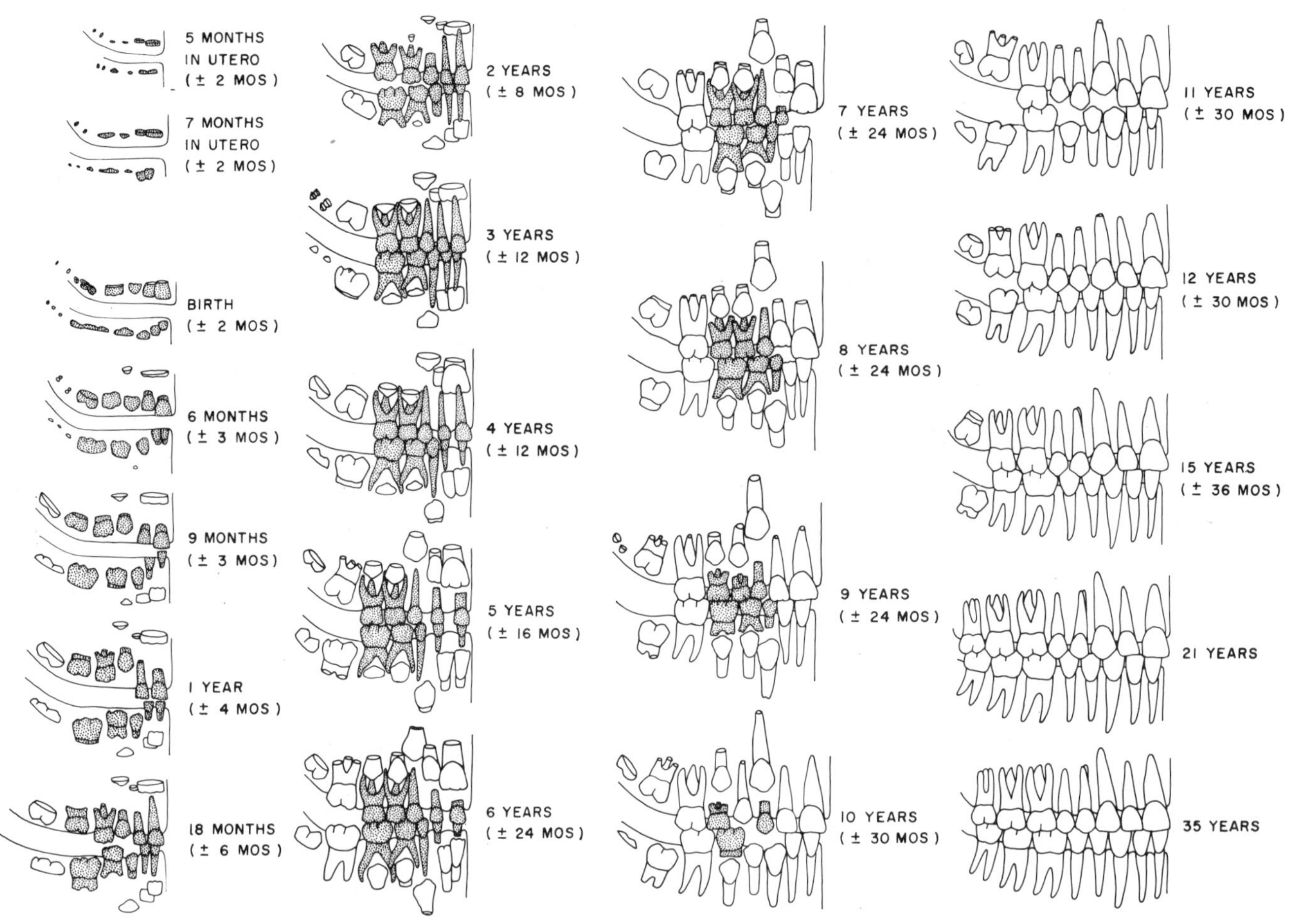

Fig. 71. The sequence of formation and eruption of teeth among American Indians. These changes are the most accurate method of establishing the age of subadult individuals at death (See p. 63 for the sources).

studies should be consulted for more exact information on individual teeth.

Unfortunately, data on dental development in American Indians are available only for eruption of the permanent teeth (Dahlberg and Menegaz-Boch 1958, Steggerda and Hill 1942). The chronologies of deciduous eruption and the calcification of deciduous and permanent teeth are derived from non-Indians, mostly United States Whites. Some studies suggest that teeth probably form and erupt earlier among Indians and I have therefore used data from the "early" end of the published variation in preparing the chart. Until better information is obtained from living Indian children, the chart is probably the best approximation available for inferring age from dental development in prehistoric and contemporary non-white subadults.

Data for males and females have been combined in Figure 71 because sex cannot be estimated reliably from skeletons of subadults. The canine tooth shows the greatest sex differences and should be avoided if possible when estimating age. Note that each stage of dental development is accompanied by a plus-and-minus factor. Although this expresses most of the variability reported in the literature, an individual estimate may be off as much as five years, especially in the older categories. Tooth formation appears to be a more reliable indicator than tooth eruption. Eruption refers to emergence through the gum, not to emergence from the bone or to reaching the occlusal plane (point of contact between the upper and lower teeth).

Standards for estimating age among modern White children have been generated by Moorrees, Fanning, and Hunt (1963a, 1963b) from studies of large samples from Ohio. They identify several stages in the formation of the crown and root in deciduous and permanent mandibular canines and molars, and root resorption in deciduous teeth (Table 11, Figs. 72–75). The range of variation is least for the crown and greatest for the root-apex closure. There are also significant differences between males and females.

The means and standard deviations for the formation of deciduous teeth are reproduced in Figures 76–77. The first standard deviation incorporates about 67 percent of the variation and the second about 95 percent. To use these norms, one must correctly identify the tooth, assess the stages of development of the crown and root, and compare the result with the chart. Averaging the scores obtained from several teeth improves the accuracy of the estimated age of the individual.

Norms for the formation of permanent mandibular canines, premolars, and molars, and the root resorption of deciduous mandibular canines are reproduced in Appendix 1, Tables 5–7.

Length of Long Bones. When the teeth are missing, age can be estimated using the lengths of the long bones. This method is not very exact, because growth rates vary widely among populations and even among individuals of the same social group (consider a class of second-graders, for example). The error of estimation is compounded by the fact that most of the data have been recorded from living children, whereas prehistoric estimates are made on dry bones. Studies of growth based on skeletal remains have been confined mainly to archeological material for which age at death has been inferred from the dentition. Thus, an estimate of age derived from long-bone length not only includes errors resulting from variability in growth, but also the errors incorporated in the original estimates of age based on dentition.

Table 11. Stages of tooth formation and their symbols (after Moorrees, Fanning, and Hunt 1963a: Table 1).

Stage	Coded Symbol
Initial cusp formation	C_i
Coalescence of cusps	C_{co}
Cusp outline complete	C_{oc}
Crown ½ complete	$Cr._{1/2}$
Crown ¾ complete	$Cr._{3/4}$
Crown complete	$Cr._c$
Initial root formation	R_i
Initial cleft formation	$Cl._i$
Root length ¼	$R_{1/4}$
Root length ½	$R_{1/2}$
Root length ¾	$R_{3/4}$
Root length complete	R_c
Apex ½ closed	$A_{1/2}$
Apical closure complete	A_c

Bone size is especially useful for predicting age at death of fetuses and very young infants. Numerous publications correlate perinatal age with crown-rump length and crown-heel length (Mall 1914, Oliver and Pineau 1958, Scammon 1937, Scammon and Calkins

Fig. 72. Stages of formation of the crown, root, and apex of deciduous mandibular canines (after Moorrees, Fanning, and Hunt 1963a: Fig. 1).

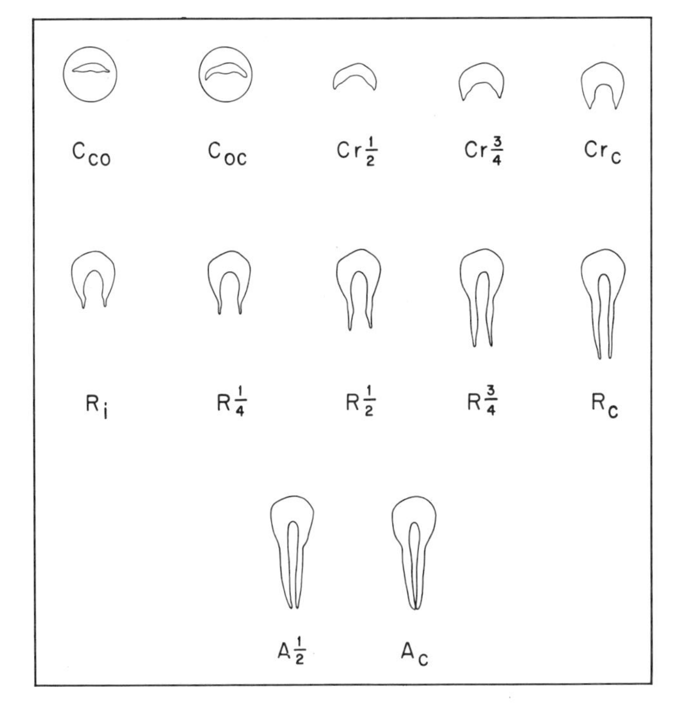

Fig. 73. Stages of formation of the crown, root, and apex of deciduous mandibular molars (after Moorrees, Fanning, and Hunt 1963a: Fig. 2).

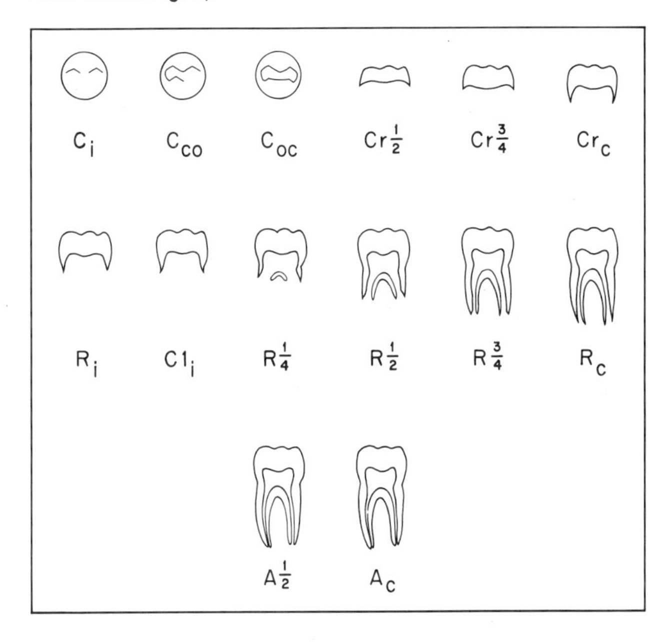

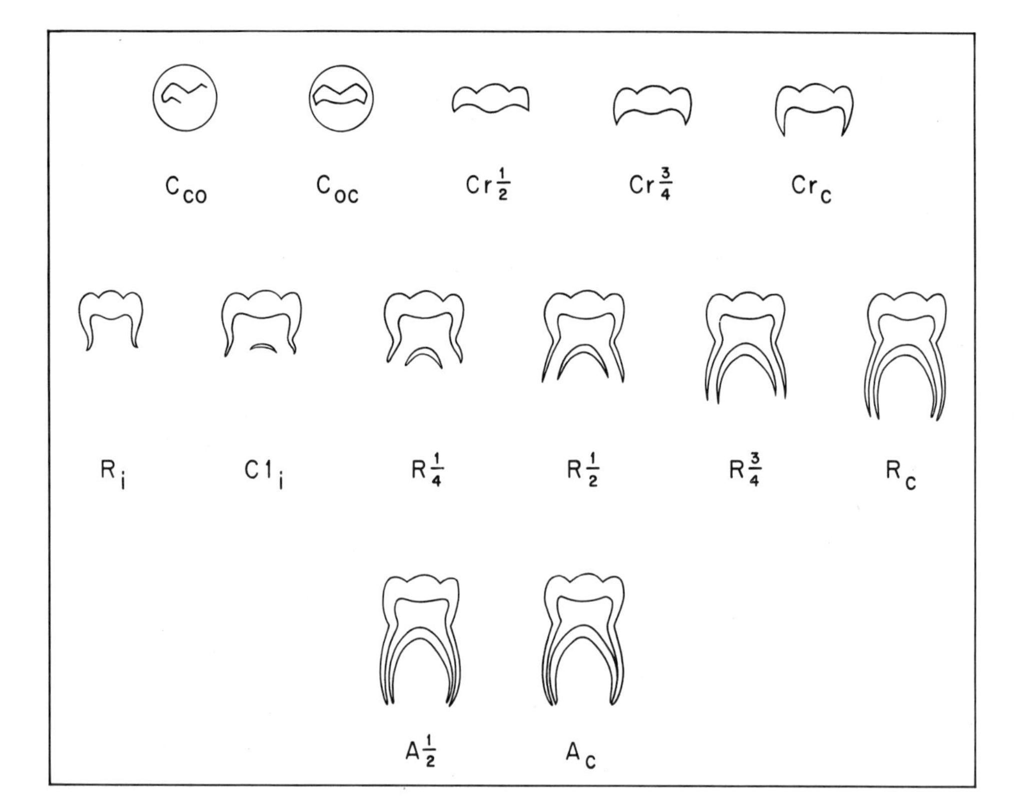

Fig. 74. Stages of formation of the crown, root, and apex of permanent mandibular molars (after Moorrees, Fanning, and Hunt 1963b: Fig. 2).

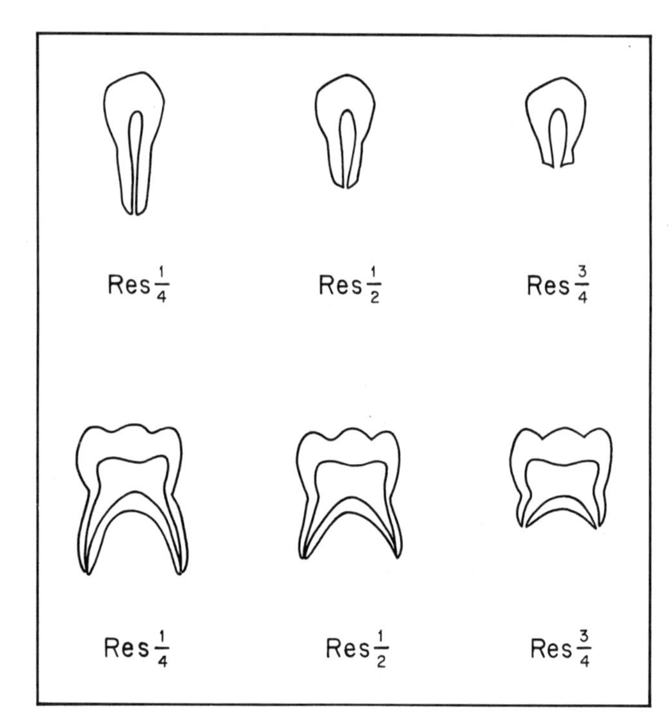

Fig. 75. Stages of root resorption for deciduous mandibular canines and molars (after Moorrees, Fanning, and Hunt, 1963a: Fig. 3).

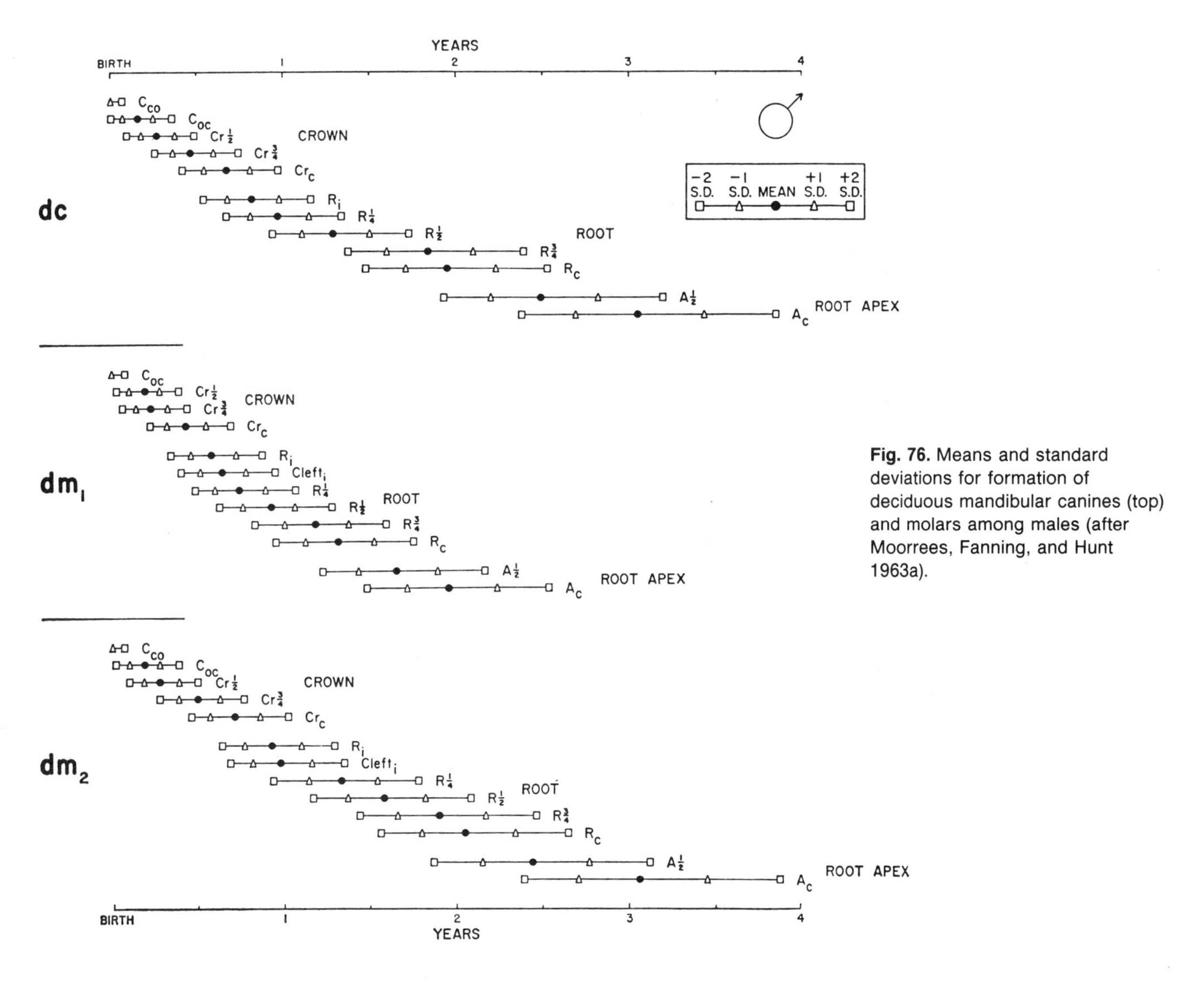

Fig. 76. Means and standard deviations for formation of deciduous mandibular canines (top) and molars among males (after Moorrees, Fanning, and Hunt 1963a).

1923a, 1923b, 1925, 1929). Standards for estimating fetal length (and subsequently age) from individual bones have been developed by Oliver and Pineau (1960) and Fazekas and Kosa (1978). Scheuer et al (1980) provide regression equations for calculating age at death directly from the length of the femur, tibia, humerus, radius, and ulna. Their application to samples from England yields standard errors as low as 1.24 weeks for the humerus when the sexes are combined. Applying these and other regressions to different bones of four ancient European fetal skeletons was less successful, however, producing within-skeleton estimates that varied by as much as 13 weeks.

The standards developed by Fazekas and Kosa (1978) are especially valuable because they are based on 138 fetal skeletons ranging in age from the third to the twelfth lunar month and employ 67 measurements on 37 bones, including not only the major components of the skeleton but also the inferior concha, vomer, and auditory ossicles. Measurements are clearly defined and regression equations produce estimates of body length in centimeters that can be converted to fetal age (Tables 12–13).

Although Fazekas and Kosa claim that error never exceeds half a lunar month, their equations appear to be considerably less accurate when applied to samples from different populations and from archeological and forensic contexts (Ubelaker 1987b). Estimates obtained from several bones of a single skeleton yield ages varying as much as 5.5 fetal months. Long-bone measurements seem to produce the most accurate results. If possible, estimates from several long bones should be averaged to provide an assessment of age at death.

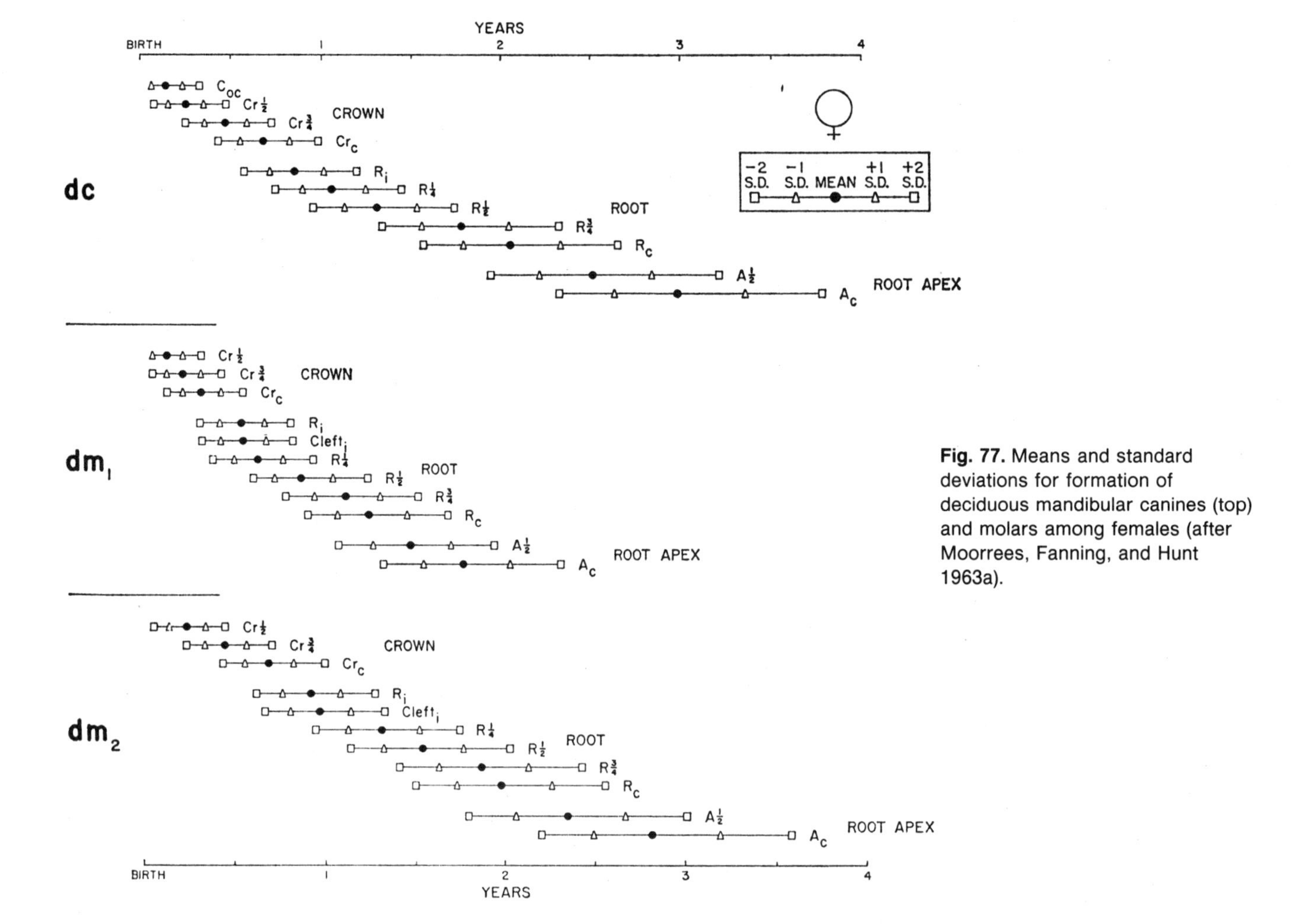

Fig. 77. Means and standard deviations for formation of deciduous mandibular canines (top) and molars among females (after Moorrees, Fanning, and Hunt 1963a).

Table 12. Regression equations for estimating fetal body length from long-bone length (After Fazakas and Kosa 1978).

Humerus	length (cm)	×	7.52 + 2.47
Humerus	width (cm)	×	28.30 + 3.95
Radius	length (cm)	×	10.61 − 2.11
Ulna	length (cm)	×	8.20 + 2.38
Femur	length (cm)	×	6.44 + 4.51
Femur	width (cm)	×	22.63 + 7.57
Tibia	length (cm)	×	7.24 + 4.90
Fibula	length (cm)	×	7.59 + 4.68

Table 13. Correlations between average fetal body length and age in lunar months (after Fazekas and Kosa 1978).

Average Body Length (cm)	Age in Lunar Months
9.5	3
12.3	3½
17.3	4
22.0	4½
25.6	5
27.3	5½
30.6	6
32.6	6½
35.4	7
37.5	7½
40.0	8
42.4	8½
45.6	9
48.0	9½
51.5	10

Comparative data on bone growth are available for five North American groups, representing Indian, Eskimo, and White populations. Specifically, these are Archaic Period skeletons from Indian Knoll, Kentucky (Johnston 1962, Sundick 1972); Late Woodland skeletons from Illinois (Walker 1969); protohistoric Arikara from South Dakota (Merchant and Ubelaker 1977); relatively recent Eskimos (Stewart 1968) and recent Whites (Anderson and Green 1948, Ghantus 1951, Hoffman 1977, and Maresh 1943, 1955). Only the last investigators had access to documented ages and long-bone lengths measured from radiograms of the living. In the other studies, ages were estimated mainly from dental development.

The Arikara data are probably the most accurate because they are based on the most exact method of age determination from dental-calcification standards (developed by Moorrees, Fanning, and Hunt 1963a, 1963b). The correlations between chronological age and the maximum diaphyseal length (without epiphyses) of each long bone and maximum width of the ilium are presented on Table 14. To use this table, simply record the appropriate measurement from a bone, turn to the appropriate place on the table, and observe the corresponding age. Note that the long-bone standards use diaphyseal length (excluding the epiphyses). Technically, these standards should be applied only to the bones of protohistoric Arikara or to related Plains Indians. However, they can be used to obtain a general estimate of age at death for subadults of any population as long as the potential variability is kept in mind.

Generalized growth curves have been reconstructed for subadults from Arikara, Indian Knoll, and other populations using different bones and different methods of estimating age at death (Figs. 78–84). The curves show little variability in rates of growth among the Indian populations, when allowance is made for variation attributable to the use of different methods of aging.

In general, the growth rates of Indians are slower than those of Whites and faster than those of Eskimos (Fig. 81), as one might predict from the statures of adults among these groups. Comparative data for United States Whites have been provided by Anderson and Green (1948), Ghantus (1951), Hoffman (1979), and Maresh (1943, 1955). All except Ghantus are longitudinal studies.

The importance of considering population variation in selecting appropriate standards of long-bone growth cannot be overemphasized. When measurements of six femurs ranging from 19 to 38 centimeters were converted to estimates of age at death using eleven different standards (Ubelaker 1987b:1260), the results differed by as much as 8.5 years (Table 15).

Appearance and Union of Epiphyses. Until the teenage years, the shafts (diaphyses) of the long bones are separated from their bony caps (epiphyses) on both ends (and sometimes on certain structures on the shafts). At about puberty, the epiphyses unite with the diaphyses, terminating longitudinal growth of the bone and increase in stature. Because these unions occur at different times on different bones, they are useful for estimating age, especially between 10 and 20 years when the data on dentition and long-bone length are of limited value.

Union of epiphyses is easy to observe because the ununited diaphyseal surface is characteristically rough and irregular in appearance. Three stages in the process are shown on Figure 85. At one extreme (left) is the proximal end of a femur with the head epiphysis still separate. At the middle, the same area is depicted with the epiphysis united, but with the junction still visible as a line. In the final state (right), union is complete and the line is obliterated. Although considerable knowledge of osteology is required to recognize incomplete union from fragments, this observation is easy to make if the bones are complete.

Data on the appearance and union of epiphyses are available from many sources in the literature. Standards for the clavicle have been provided by Todd and D'Errico (1928), the hand and wrist by Greulich and Pyle (1950), and the knee by Pyle and Hoerr (1955). McKern and Stewart (1957) contribute data on the union of a variety of epiphyses in their study of young American males who died in the Korean conflict. General summaries of these and other works can be found in Krogman (1962) and Stewart (1979).

All these works have documented a marked sex difference in the timing of epiphyseal union. Lewis and Garn (1960) noted that girls were advanced over boys by about 25 percent in the appearance of 36 ossification centers. The difference was about 19 percent in the timing of knee ossification. Krogman (1962) and Stewart (1979) have shown that union of most epiphyses occurs about one to two years earlier in females than in males. Thus, if possible, sex should be determined before estimating age from epiphyseal union. If the sex is unknown, the estimate should employ both male and female standards and should include an appropriate margin of error.

Several problems should be noted when applying data on epiphyseal union to forensic cases. Standards such as those of Greulich and Pyle (1950) and Pyle and Hoerr (1955) provide clear definitions of average development, but only McKern and Stewart (1957) document the range of variation, and then only for males. An indication of the variation in timing of epiphyseal union among different populations has been provided by Stewart (1979). Comparing the data of various investigators revealed differences of two or more years for most of the major epiphyses. Stewart also noted that gross inspection of union generally yields slightly higher estimates than radiographic assessment.

An important contribution of the McKern and Stewart (1957) study is their demonstration that several years normally elapse between initiation and final closure. They emphasize the importance of defining the exact stage of union for each epiphysis, rather than using a simple score of "united" or "ununited." Their study also shows that epiphyses are not of equal value for estimating age. The best indicators are the proximal humerus, medial epicondyle, distal radius, femoral head, distal femur, iliac crest, medial clavicle, sacrum 3/4 joint, and lateral sacral joints. They recommend assessing the total pattern of maturation of the skeleton and provide a useful method for this purpose.

In summary, four factors must be considered when estimating age from epiphyseal union: (1) the exact stage of union of each epiphysis, (2) the sex of the individual, (3) the range of variation in timing of the union, and (4) the possible differences between gross examination and radiographic methods. The latter point is especially relevant since many anthropologists lack radiographic experience and may be misled by the presence of lines that mimic incomplete union.

The approximate ages at which initial union of the epiphyses occurs for most of the major bones of the body are given on Table 16. These ages vary among different populations and between the sexes. Note that (1) union begins earlier in females than in males, and (2) in both sexes there is a variation of two to six years between individuals. Union occurs earliest in the ankle and hip, proceeds to the knee and elbow, and finally to the shoulder and wrist.

Table 14. Correlations between chronological age estimates and the maximum diaphyseal length of long bones and the maximum width of the ilium. Bones are listed in anatomical order.

Estimated Age (years)	Size of Sample	Mean Length (mm)	Standard Deviation	Range of Variation (mm)
		HUMERUS		
NB– 0.5	49	70.5	5.2	63.5– 89.0
0.5– 1.5	37	102.3	8.9	84.0–119.0
1.5– 2.5	11	129.5	5.9	121.0–138.0
2.5– 3.5	10	139.5	12.8	118.0–157.0
3.5– 4.5	2	156.5	3.5	154.0–159.0
4.5– 5.5	4	167.6	8.8	161.0–179.5
5.5– 6.5	7	180.1	6.5	172.5–192.0
6.5– 7.5	4	192.1	7.9	187.5–204.0
7.5– 8.5	2	211.8	7.4	206.5–217.0
8.5– 9.5	0			
9.5–10.5	5	228.6	4.2	225.0–235.0
10.5–11.5	1	245.0		
11.5–12.5	2	254.5	5.0	251.0–258.0
12.5–13.5	0			
13.5–14.5	0			
14.5–15.5	1	255.5		
15.5–16.5	0			
16.5–17.5	0			
17.5–18.5	0			
		RADIUS		
NB– 0.5	47	57.4	4.9	49.0– 73.5
0.5– 1.5	31	81.0	6.1	67.0– 92.0
1.5– 2.5	14	97.1	5.5	84.0–104.0
2.5– 3.5	9	106.3	9.8	93.5–119.0
3.5– 4.5	2	118.3	3.2	116.0–120.5
4.5– 5.5	4	128.1	3.4	125.0–132.5
5.5– 6.5	5	140.6	5.4	134.5–149.0
6.5– 7.5	3	149.5	3.5	146.0–153.0
7.5– 8.5	1	168.0		
8.5– 9.5	0			
9.5–10.5	3	185.7	9.3	178.0–196.0
10.5–11.5	1	189.0		
11.5–12.5	4	190.9	14.3	169.5–200.0
12.5–13.5	0			
13.5–14.5	0			
14.5–15.5	0			
15.5–16.5	0			
16.5–17.5	0			
17.5–18.5	0			
		ULNA		
NB– 0.5	47	66.1	5.0	60.0– 82.5
0.5– 1.5	22	92.1	7.7	74.5–103.0
1.5– 2.5	13	108.5	6.8	94.0–116.0
2.5– 3.5	9	117.9	10.9	100.0–129.5
3.5– 4.5	2	129.8	4.6	126.5–133.0
4.5– 5.5	4	142.8	2.9	140.0–145.5
5.5– 6.5	6	153.8	7.9	145.0–166.0
6.5– 7.5	4	167.1	6.1	161.0–175.0
7.5– 8.5	2	180.0	5.7	176.0–184.0
8.5– 9.5	0			

Estimated Age (years)	Size of Sample	Mean Length (mm)	Standard Deviation	Range of Variation (mm)
9.5–10.5	3	201.5	10.0	194.5–213.0
10.5–11.5	0			
11.5–12.5	2	217.5	2.1	216.0–219.0
12.5–13.5	0			
13.5–14.5	0			
14.5–15.5	0			
15.5–16.5	0			
16.5–17.5	0			
17.5–18.5	0			
		FEMUR		
NB– 0.5	51	82.2	8.7	62.5–106.0
0.5– 1.5	37	126.9	14.6	92.5–161.0
1.5– 2.5	14	167.1	12.2	141.0–186.0
2.5– 3.5	9	185.1	20.7	155.0–215.0
3.5– 4.5	2	213.0	7.1	208.0–218.0
4.5– 5.5	3	234.3	9.0	225.0–243.0
5.5– 6.5	8	248.6	14.5	236.0–277.0
6.5– 7.5	4	262.0	9.2	252.0–274.0
7.5– 8.5	2	292.8	11.0	285.0–300.5
8.5– 9.5	0			
9.5–10.5	2	321.0	1.4	320.0–322.0
10.5–11.5	1	342.0		
11.5–12.5	4	344.5	5.8	339.0–350.0
12.5–13.5	0			
13.5–14.5	0			
14.5–15.5	2	356.5	16.3	345.0–368.0
15.5–16.5	0			
16.5–17.5	0			
17.5–18.5	1	406.5		
		TIBIA		
NB– 0.5	47	71.6	7.2	59.5– 94.0
0.5– 1.5	30	104.8	11.3	81.0–131.5
1.5– 2.5	11	138.6	7.8	125.0–151.0
2.5– 3.5	9	153.8	18.8	127.0–184.0
3.5– 4.5	2	170.5	7.8	165.0–176.0
4.5– 5.5	3	190.8	10.3	181.0–201.5
5.5– 6.5	8	201.6	10.1	191.0–222.0
6.5– 7.5	4	221.4	7.2	212.0–229.5
7.5– 8.5	2	242.5	21.9	227.0–258.0
8.5– 9.5	0			
9.5–10.5	3	272.3	11.6	261.5–284.5
10.5–11.5	1	285.0		
11.5–12.5	4	287.5	8.3	279.0–296.0
12.5–13.5	1	299.0		
13.5–14.5	0			
14.5–15.5	2	306.5	17.7	294.0–319.0
15.5–16.5	0			
16.5–17.5	0			
17.5–18.5	1	334.5		

Estimated Age (years)	Size of Sample	Mean Length (mm)	Standard Deviation	Range of Variation (mm)
		FIBULA		
NB– 0.5	37	68.9	6.6	60.0– 88.0
0.5– 1.5	27	103.0	11.7	75.0–122.0
1.5– 2.5	13	133.2	9.1	111.5–142.5
2.5– 3.5	7	152.3	19.9	124.0–182.0
3.5– 4.5	2	168.5	7.8	163.0–174.0
4.5– 5.5	3	185.8	7.8	178.0–193.5
5.5– 6.5	6	194.4	5.3	188.0–201.0
6.5– 7.5	4	216.9	7.9	209.0–227.0
7.5– 8.5	1	246.0		
8.5– 9.5	0			
9.5–10.5	3	264.0	10.5	255.0–275.5
10.5–11.5	1	280.0		
11.5–12.5	3	285.0	10.4	273.0–292.0
12.5–13.5	1	291.5		
13.5–14.5	0			
14.5–15.5	3	299.0	11.5	287.0–310.0
15.5–16.5	1	332.5		
16.5–17.5	0			
17.5–18.5	1	330.0		
		ILIUM		
NB– 0.5	38	37.0	3.0	32.5– 44.5
0.5– 1.5	34	55.8	4.4	46.0– 65.0
1.5– 2.5	13	69.3	4.5	60.0– 74.5
2.5– 3.5	7	73.4	6.1	64.0– 82.0
3.5– 4.5	2	80.3	1.8	79.0– 81.5
4.5– 5.5	5	83.5	8.3	69.0– 89.0
5.5– 6.5	5	92.8	2.2	90.5– 96.0
6.5– 7.5	4	97.4	1.6	95.0– 98.5
7.5– 8.5	2	108.5	5.0	105.0–112.0
8.5– 9.5	0			
9.5–10.5	3	119.2	2.6	117.0–122.0
10.5–11.5	1	123.0		
11.5–12.5	4	119.1	5.0	114.0–126.0
12.5–13.5	4	137.8	9.7	129.5–148.0
13.5–14.5	0			
14.5–15.5	3	126.0	4.4	123.0–131.0
15.5–16.5	1	144.0		
16.5–17.5	0			
17.5–18.5	1	141.0		

Fig. 78. Growth curves of prehistoric Arikara and Indian Knoll subadult humerii calculated using measurements of maximum diaphyseal length and different methods of estimating age at death. The results are similar up to about the age of nine; then the growth rate of the Indian Knoll children appears to lag behind that of the Arikara children.

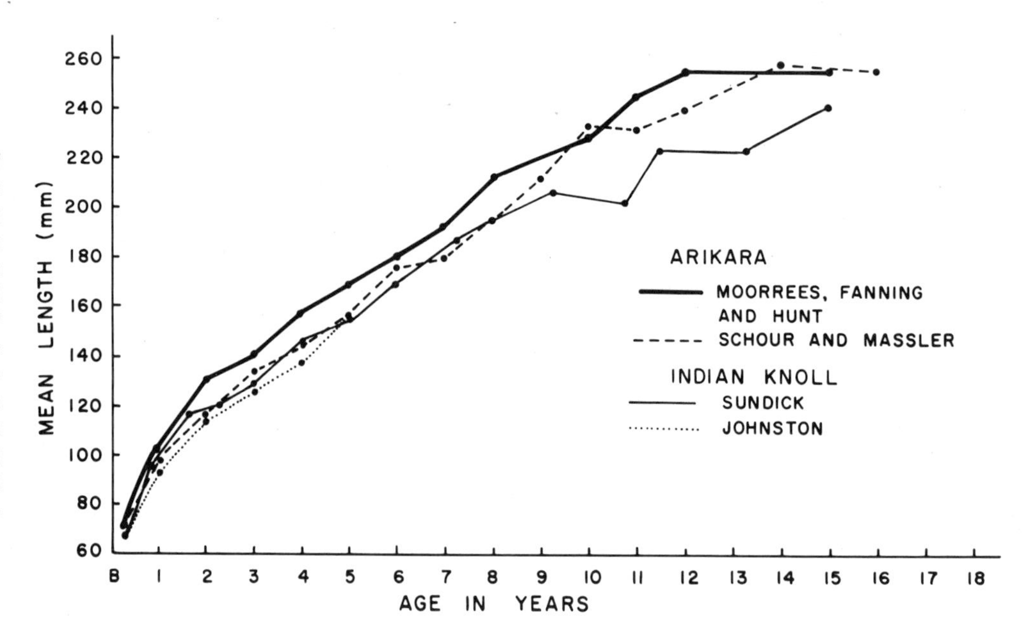

Fig. 79. Growth curves of prehistoric subadult Arikara and Indian Knoll radii calculated using measurements of maximum diaphyseal length and different methods of estimating age at death. The results are most similar before the age of seven.

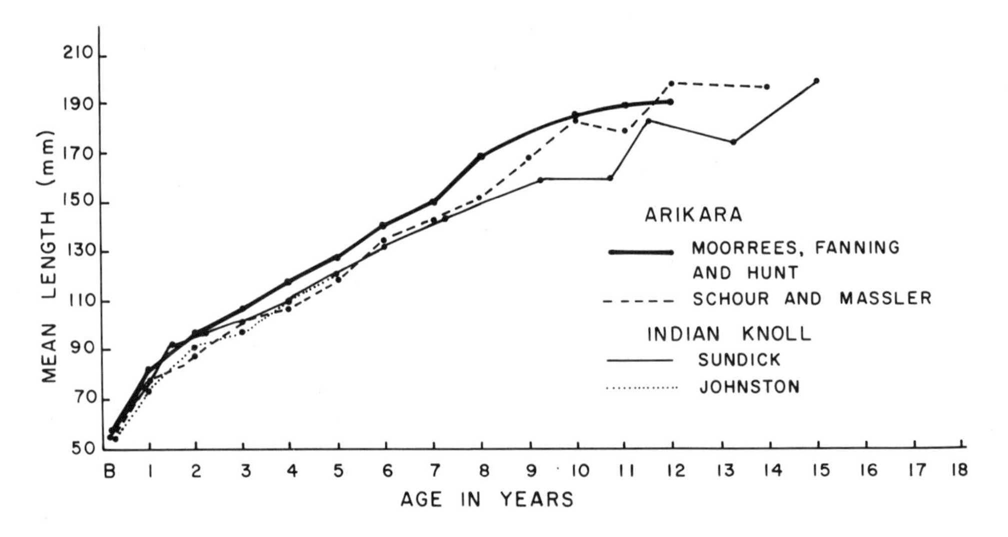

Fig. 80. Growth curves of prehistoric subadult Arikara and Indian Knoll ulnae calculated using measurements of maximum diaphyseal length and different methods of estimating age at death. The results are very similar to those obtained by applying the same formulas to the radii (Fig. 79).

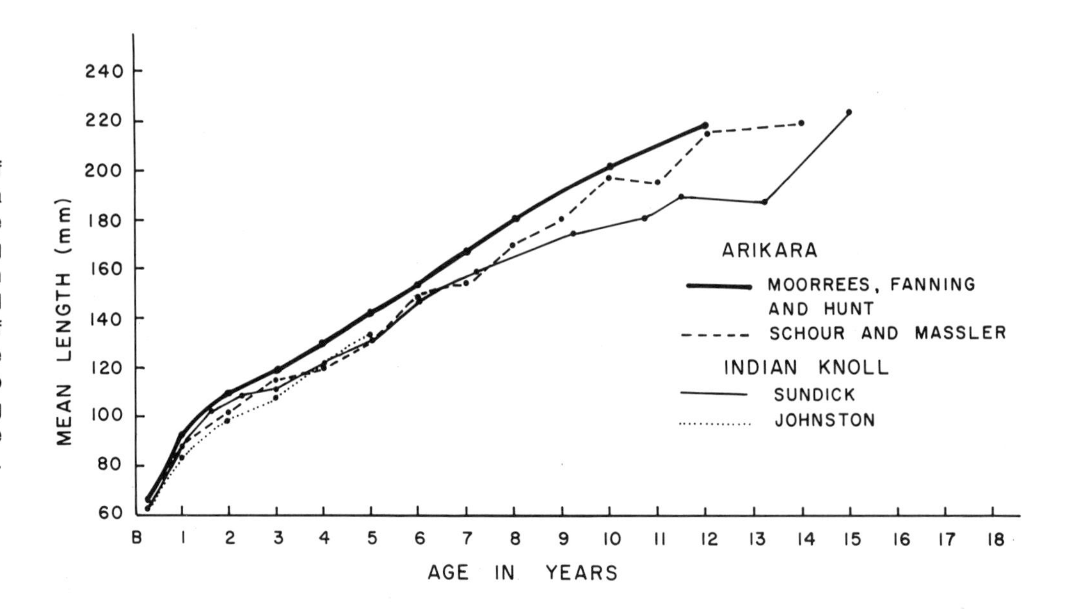

Fig. 81. Growth curves of prehistoric subadult Arikara and Indian Knoll femora calculated using measurements of maximum diaphyseal length and different methods of estimating age at death. Samples from White, Eskimo, and Late Woodland Indian populations are included for comparison.

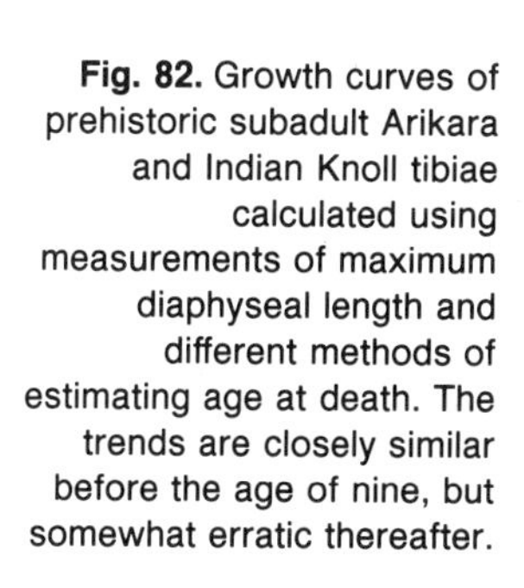

Fig. 82. Growth curves of prehistoric subadult Arikara and Indian Knoll tibiae calculated using measurements of maximum diaphyseal length and different methods of estimating age at death. The trends are closely similar before the age of nine, but somewhat erratic thereafter.

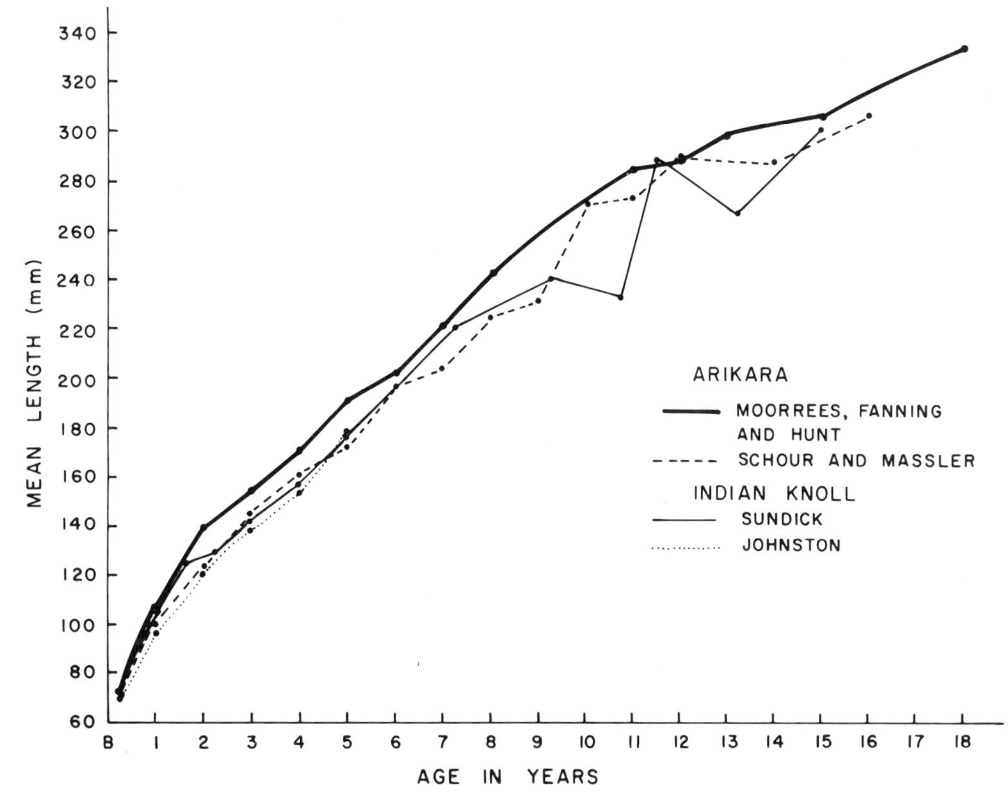

Fig. 83. Growth curves of prehistoric subadult Arikara and Indian Knoll fibulae calculated using measurements of maximum diaphyseal length and different methods of estimating age at death. The trends are generally similar to those obtained from the tibiae.

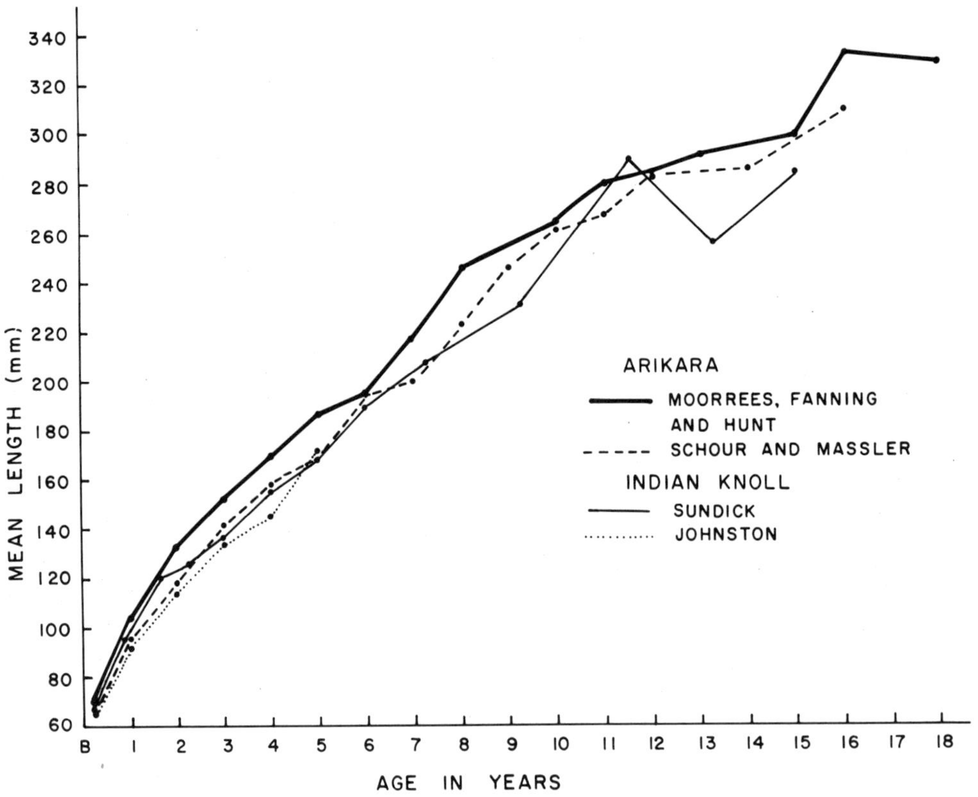

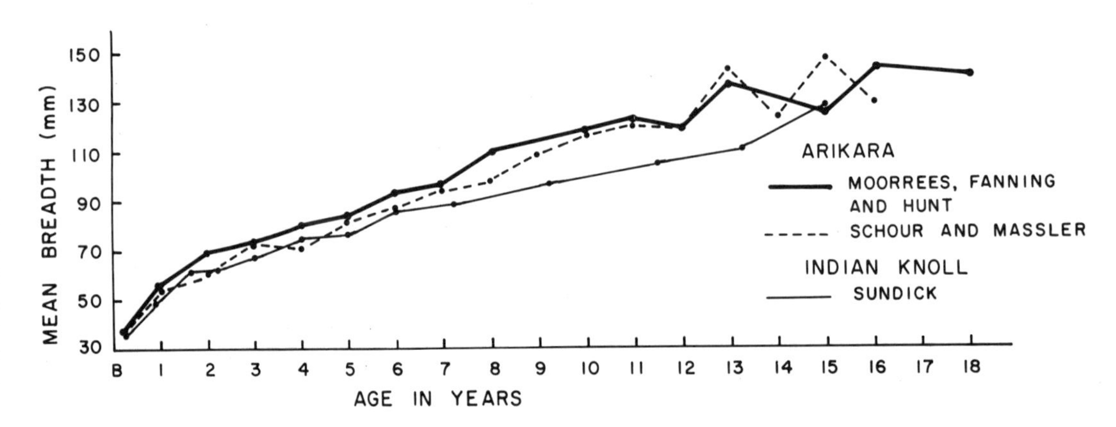

Fig. 84. Growth curves of prehistoric subadult Arikara and Indian Knoll ilia calculated using measurements of maximum breadth and different methods of estimating age at death. The Indian Knoll children appear from these data to be shorter than the Arikara children between the ages of about 8 and 15, when their statures tend to equalize.

Adults: Macroscopic Methods

By about the age of 20, most teeth are completely formed and erupted, most epiphyses are united, and longitudinal bone growth is complete. Consequently, other criteria must be employed to estimate age in adults. Two types of methods are available: macroscopic and microscopic. Macroscopic methods are faster and do not involve destruction of the specimen. Microscopic methods require more time, equipment, and knowledge, and necessitate some destruction, but give much more accurate results.

The principal progressive macroscopic changes are metamorphosis of the pubic symphysis; closure and obliteration of the sutures in the skull; degenerative alterations in the spine, joints, and skull; resorption of cancellous bone, and loss of teeth. The landmarks

Table 15. Range of ages at death estimated for six femoral lengths using eleven different growth standards (after Ubelaker 1987b: Table 2).

Femur Length, cm	Mean Age Estimate, years	Range of Age Estimates, years	Years Within Range
19	3.8	2.0–5.5	3.5
24	5.8	3.5–8.0	4.5
28	8.3	5.5–11.0	5.5
32	9.3	6.0–12.5	6.5
35	11.2	7.8–14.5	6.7
38	13.8	9.5–18+	8.5

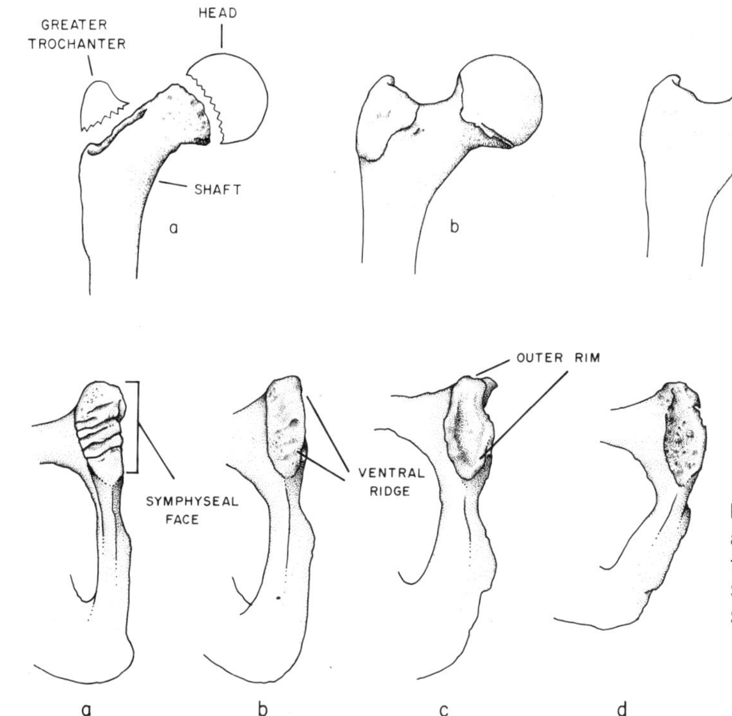

Fig. 85. Stages in the ossification of the proximal end of the femur. **a,** Shaft and epiphyses completely separate. **b,** Shaft and epiphyses united, but their junction clearly defined by a line. **c,** Completed union, with obliteration of lines of junction.

Fig. 86. Age changes in the symphyseal face of the pubis in adults. **a.** Ridges and deep furrows (age 21). **b,** Ridges and furrows semi-obliterated and ventral ridge initiated (age 29). **c,** Surface smooth and an outer rim complete (age 56). **d,** Surface deteriorated (age 90).

Table 16. Age of initial union for epiphyses of several bones.

Epiphysis	Age of Initial Union	
	Males	Females
Clavicle: medial end	18-22	17-21
Scapula: acromial process	14-22	13-20
Humerus: Head	14-21	14-20
Greater tubercle	2-4	2-4
Trochlea	11-15	9-13
Lateral epicondyle	11-17	10-14
Medial epicondyle	15-18	13-15
Radius: Head	14-19	13-16
Distal end	16-20	16-19
Ulna: Distal end	18-20	16-19
Ilium: Iliac crest	17-20	17-19
Ischium-Pubis	7-9	7-9
Ischial tuberosity	17-22	16-20
Femur: Head	15-18	13-17
Greater trochanter	16-18	13-17
Lesser trochanter	15-17	13-17
Distal end	14-19	14-17
Tibia: Proximal end	15-19	14-17
Distal end	14-18	14-16
Fibula: Proximal end	14-20	14-18
Distal end	14-18	13-16

are of different reliability for estimating age, as the following discussions will show.

Pubic Symphysis: Males. Adult age estimates may be generated from observations of the symphyseal face of the pubis (that is, the surface where one pubis joins the other). In early adulthood, this area appears very rough, with ridges and deep furrows (Fig. 86a). As the furrows gradually fill to create a smooth surface, a ridge forms on the outer (ventral) surface (Fig. 86b). After this ridge is complete and the surface is smooth (Fig. 86c), a rim of bone forms along the outer circumference of the face. Finally, the symphyseal face begins to deteriorate (Fig. 86d).

T. W. Todd studied a collection of white male skeletons of known age and identified 10 stages between the ages of 18 and 50 years (Fig. 87). He characterized these as follows (for his complete description, see Todd 1920:301–314).

I. First post-adolescent phase: Age 18–19. Symphysial surface rugged, traversed by horizontal ridges separated by well marked grooves; no ossific (epiphysial) nodules fusing with the surface; no definite delimiting margin; no definition of extremities.

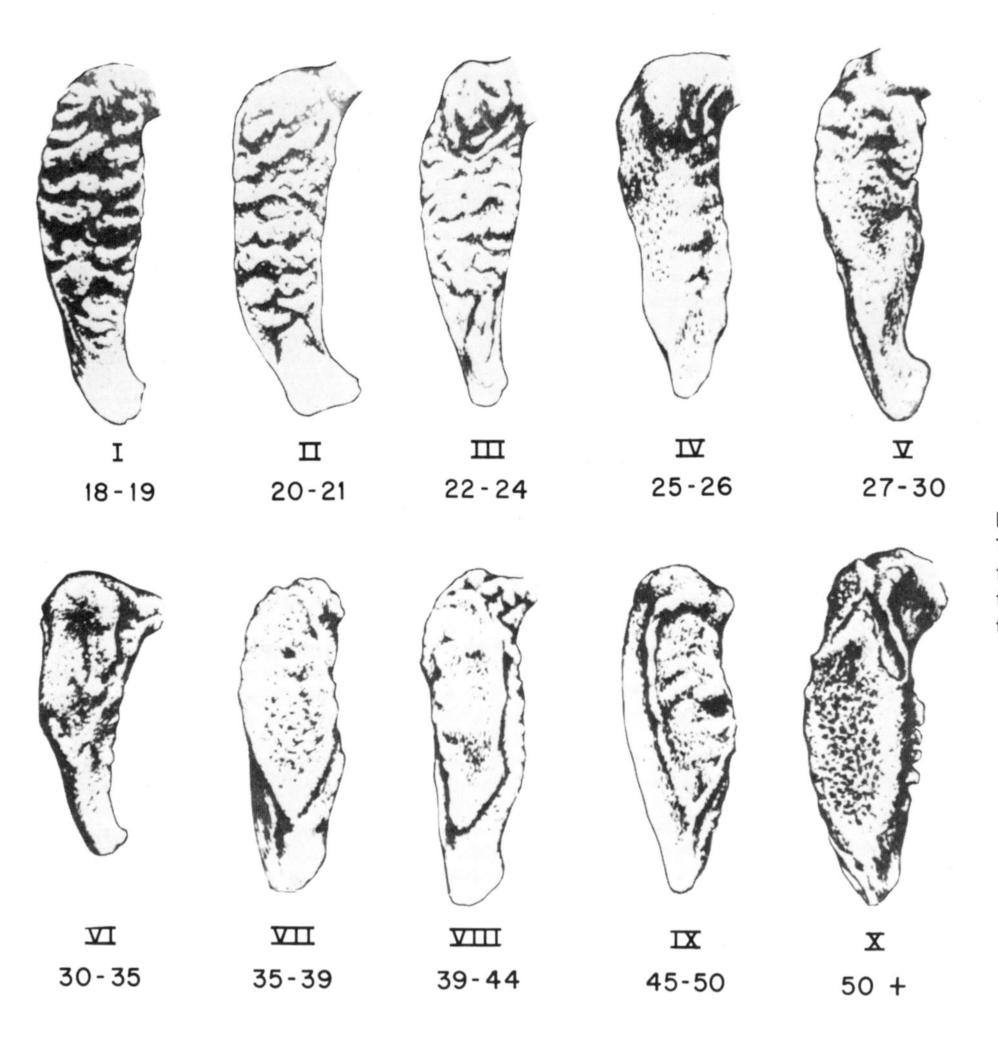

Fig. 87. Ten stages defined by Todd for estimating age at death from changes in the symphyseal face of the pubis in adults between the ages of 18 and 50 years.

II. Second post-adolescent phase: Age 20–21. Symphysial surface still rugged, traversed by horizontal ridges, the grooves between which are, however, becoming filled near the dorsal limit with a new formation of finely textured bone. This formation begins to obscure the hinder extremities of the horizontal ridges. Ossific (epiphysial) nodules fusing with the upper symphysial face may occur; dorsal limiting margin begins to develop; no delimitation of extremities; foreshadowing of ventral bevel.

III. Third post-adolescent phase: Age 22–24. Symphysial face shows progressive obliteration of ridge and furrow system: commencing formation of the dorsal plateau; presence of fusing ossific (epiphysial) nodules; dorsal margin gradually becoming more defined; beveling as a result of ventral rarefaction becoming rapidly more pronounced; no delimitation of extremities.

IV. Fourth phase: Age 25–26. Great increase of ventral beveled area; corresponding diminution of ridge and furrow formation; complete definition of dorsal margin through the formation of the dorsal plateau; commencing delimitation of lower extremity.

V. Fifth phase: Age 27–30. Little or no change in symphysial face and dorsal plateau except that sporadic and premature attempts at the formation of a ventral rampart occur; lower extremity, like the dorsal margin, is increasing in clearness of definition; commencing formation of upper extremity with or without the intervention of a bony (epiphysial) nodule.

VI. Sixth phase: Age 30–35. Increasing definition of extremities; development and practical completion of ventral rampart; retention of granular appearance of symphysial face and ventral aspect of pubis; absence of lipping of symphysial margin.

VII. Seventh phase: Age 35–39. Changes in symphysial face and ventral aspect of pubis consequent upon diminishing activity; commencing bony outgrowth into attachments of tendons and ligaments, especially the gracilis tendon and sacro-tuberous ligament.

VIII. Eighth phase: Age 39–44. Symphysial face generally smooth and inactive; ventral surface of pubis also inactive; oval outline complete or approximately complete; extremities clearly defined: no distinct "rim" to symphysial face; no marked lipping of either dorsal or ventral margin.

IX. Ninth phase: Age 45–50. Symphysial face presents a more or less marked rim; dorsal margin uniformly lipped; ventral margin irregularly lipped.

X. Tenth phase: Age 50 and upward. Symphysial face eroded and showing erratic ossification; ventral border more or less broken down; disfigurement increases with age.

Later, Todd (1921) reported that the same general changes characterize Negro males and Negro and White females, but they occur two to three years earlier in these populations than in White males. This variation should be kept in mind when the criteria are employed for estimating age.

An alternative method for estimating the ages of adult males at death from the symphyseal face of the pubis has been presented by McKern and Stewart (1957). Their system focuses on three aspects of the symphyseal face: the dorsal demi-face, the ventral rampart, and the symphyseal rim. They showed that these components change independently at different rates, and that Todd's method over-simplifies the changes at the expense of accuracy. The McKern and Stewart system involves ranking each component on a scale of 0 to 5 and adding the three numerical values to provide a total score that can be converted to an age estimate. McKern and Stewart (1957:74–79) describe the pubic symphyseal metamorphosis as follows:

Component I. Dorsal Plateau (Fig. 88). Between the ages of 17–18, the grooves near the dorsal margin begin to fill in with finely textured bone and the ridges show the first evidence of resorption. Coincident with this process, a delimiting dorsal margin appears which eventually outlines the entire demi-face.

Starting in the same general area, the interacting processes of resorption and fill-in spread over the dorsal demi-face until the ridge and groove pattern has been obliterated. Ultimately, this gives to the demi-face a flat, platform-like aspect and for this reason, the component has been given the name, dorsal plateau.

Attention is called to the dorsal nodules (not described by Todd), sometimes associated with the early metamorphosis of the dorsal demi-face. When present, they are found in the lower third of the demi-face. They are not simply enlarged ridges but round lumps of bone incorporated in the ridges. Since they do not aid in delimiting the lower symphyseal extremity and appear in only a small number of cases, we have not regarded them as a distinctive part of Component I.

The 6 (0–5) stages of Component I follow:

0. Dorsal margin absent.
1. A slight margin formation first appears in the middle third of the dorsal border.
2. The dorsal margin extends along entire dorsal border.
3. Filling in of grooves and resorption of ridges to form a beginning plateau in the middle third of the dorsal demi-face.
4. The plateau, still exhibiting vestiges of billowing, extends over most of the dorsal demi-face.
5. Billowing disappears completely and the surface of the entire demi-face becomes flat and slightly granulated in texture.

Component II. Ventral rampart (Fig. 89). Early in the development of Component I, differentiation of dorsal and ventral demi-faces becomes pronounced due to the breakdown, by rarefaction, of the ventral half. Over this porous,

Fig. 88. Characteristic age changes in the dorsal plateau of the male pubic symphysis. Six stages (the earliest not shown) are defined by progressive reduction in the prominence of the ridges along the dorsal (left) margin. This is one of the three aspects or components of the symphyseal face used by McKern and Stewart to obtain a general score that can be converted into an estimate of age at death.

beveled surface, an elongated and more or less complete epiphysis or rampart forms. This rampart is produced by the extension of ossification from upper and lower extremities aided, at times, by independent ossicles along the line of the future ventral margin. Obviously, however, the pattern is variable and the rampart may remain incomplete even in later age groups (the hiatus is usually in the middle two-thirds of the ventral border) or may bridge only certain portions of the beveled surface.

The 6 (0–5) stages of Component II are as follows:

0. Ventral beveling is absent.
1. Ventral beveling is present only at superior extremity of ventral border.
2. Bevel extends inferiorly along ventral border.
3. The ventral rampart begins by means of bony extensions from either or both of the extremities.
4. The rampart is extensive but gaps are still evident along the earlier ventral border, most evident in the upper two-thirds.
5. The rampart is complete.

Component III. Symphyseal rim (Fig. 90). The final stages of symphyseal maturation are characterized by the formation of a distinct and elevated rim surrounding the now level face. At the same time, the bony texture of the face begins to change from a somewhat granular to a more finely grained or dense bone and, although vestiges of the ridge and groove pattern still may be recognized in the lower third of the dorsal demi-face, it is sometimes difficult to tell whether they are merely regular undulations of the smooth bony surface or true remnants of the earlier ridge and groove pattern.

Following the completion of the symphyseal rim, there is a period during which changes are minute and infrequent. Ultimately the rim is worn down or resorbed and a smooth surface extends to the margins. As the face levels off it undergoes erosion and erratic ossification, the bone becomes more porous and the margins may be lipped.

Metamorphosis of the symphysis in the last decades of life is characterized by further breakdown of the bony tissue. However, because of the small number of older individuals present, the series under discussion does not enable us to define the last stages clearly.

The 6 (0–5) stages of Component III are as follows:

0. The symphyseal rim is absent.
1. A partial dorsal rim is present, usually at the superior end of the dorsal margin, it is round and smooth in texture and elevated above the symphyseal surface.
2. The dorsal rim is complete and the ventral rim is beginning to form. There is no particular beginning site.
3. The symphyseal rim is complete. The enclosed symphyseal surface is finely grained in texture and irregular or undulating in appearance.
4. The rim begins to break down. The face becomes smooth and flat and the rim is no longer round but sharply defined. There is some evidence of lipping on the ventral edge.
5. Further breakdown of the rim (especially along superior ventral edge) and rarefaction of the symphyseal face. There is also disintegration and erratic ossification along the ventral rim.

The correlations of the total scores for the three components with age are presented on Table 17. Note the range of variability around the mean, expressed both in the standard deviation and the ranges of the scores. This variability existed among the 349, mostly White, North American males that composed the sample. The probability of erroneous results would increase if the method were used to estimate age from pubic bones of different geographical origin, racial affiliation or sex. The accuracy of the method increases if the analyst is very familiar with the structural morphology of the pubic symphyseal face. Use of plastic casts of the various stages of the three components for comparison also enhances the accuracy of the scoring.

Applying the Todd and the McKern-Stewart standards to 739 males of known age from the Los Angeles, California area revealed so much variability that Angel et al (1986) recommend using for males the modifications of the Todd system described in Table 18. Studies by Suchey et al (1986) also indicate that the male symphyseal face is not a reliable indicator for ages above 40.

Table 17. Mean age, standard deviation, and age ranges of males obtained from the total scores calculated using the McKern and Stewart symphyseal formulas (after McKern and Stewart 1957:85).

Total Score	Age Range	Mean Age	Standard Deviation
0	−17	17.29	.49
1−2	17−20	19.04	.79
3	18−21	19.79	.85
4−5	18−23	20.84	1.13
6−7	20−24	22.42	.99
8−9	22−28	24.14	1.93
10	23−28	26.05	1.87
11−13	23−39	29.18	3.33
14	29+	35.84	3.89
15	36+	41.00	6.22

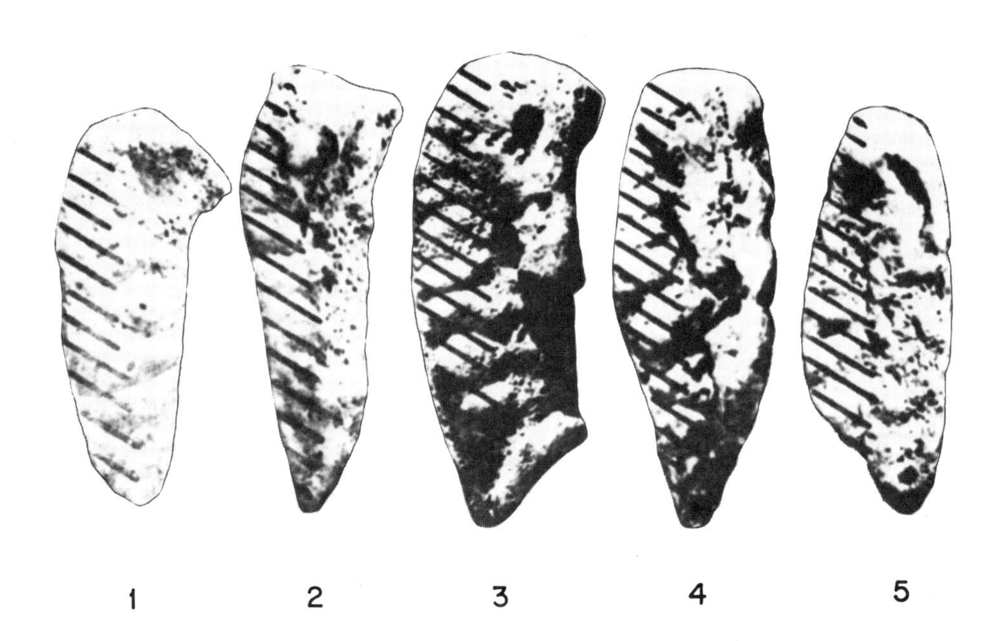

Fig. 89. Characteristic age changes in the ventral rampart of the male pubic symphysis. Six stages (the earliest not shown) are defined by the appearance of bevelling on the upper ventral (right) border and its gradual development into a rampart. This is the second of the three components used by McKern and Stewart to obtain a general score that can be converted into an estimate of age at death.

Fig. 90. Characteristic age changes in the development and breakdown of the symphyseal rim around the margin of the male pubic symphysis. Six stages (the earliest not shown) have been recognized by McKern and Stewart for this component. The score is combined with those obtained for Components I and II (Figs. 88–89) to estimate age at death.

Table 18. Modifications of the Todd system of estimating age using the pubic symphysis (After Suchey, Wiseley, and Katz 1986).

New phases	Todd's phases	Ageing rule	Assessment of rule
Todd A	I, II, III	24 years or under	Covers all of sample variability
Todd B	IV, V	19–30 years	Leaves out six advanced cases from 30–45; 93% sample variability is included
Todd C	VI, VII	22–50 years	Leaves out 13 advanced cases; 94% of sample variability is included
Todd D	IX	30 years or older	Leaves out 13 young cases; 95% of sample variability is included
Todd E	X	45 years or older	Leaves out five young cases; 95% of sample variability is included

Pubic Symphysis: Females. Gilbert and McKern (1973) have suggested that sex differences in pubic symphyseal metamorphosis are more marked than indicated by Todd. Their study of the pubic bones from 103 individuals of documented age at death showed that when standards derived from males were applied to females the estimates were too old. The timing of metamorphosis in different aspects of the pubis was also inconsistent. For example, "compared to the male os pubis, that of a female of the same age may appear to be ten years younger based upon the ventral rampart, and ten years older based upon the dorsal plateau" (Gilbert and McKern 1973: 31). To overcome these discrepancies, they adopted the triple-component approach devised by McKern and Stewart (1957) and established different correlations and definitions of stages for females. Their six categories of development of the three components (Fig. 91) are defined as follows (Gilbert and McKern 1973:33–34):

Component I [Dorsal demi-face]

0. Ridges and furrows very distinct, ridges are billowed, dorsal margin undefined.
1. Ridges begin to flatten, furrows to fill in, and a flat dorsal margin begins in mid-third of demi-face.
2. Dorsal demi-face spreads ventrally, becomes wider as flattening continues, dorsal margin extends superiorly and inferiorly.
3. Dorsal demi-face is quite smooth, margin may be narrow or indistinct from face.
4. Demi-face becomes complete and unbroken, is broad and very fine grained, may exhibit vestigial billowing.
5. Demi-face becomes pitted and irregular through rarefaction.

Component II [Ventral rampart]

0. Ridges and furrows very distinct. The entire demi-face is beveled up toward the dorsal demi-face.
1. Beginning inferiorly, the furrows of the ventral demi-face begin to fill in, forming an expanding beveled rampart, the lateral edge of which is a distinct, curved line extending the length of the symphysis.
2. Fill in of furrows and expansion of demi-face continue from both superior and inferior ends, rampart spreads laterally along its ventral edge.
3. All but about one-third of the ventral demi-face is filled in with fine grained bone.
4. The ventral rampart presents a broad, complete, fine grained surface from the pubic crest to the inferior ramus.
5. Ventral rampart may begin to break down, assuming a very pitted and perhaps cancellous appearance through rarefaction.

Component III [Symphyseal rim]

0. The rim is absent.
1. The rim begins in the mid-third of the dorsal surface.
2. The dorsal part of the symphyseal rim is complete.
3. The rim extends from the superior and inferior ends of the symphysis until all but about one-third of the ventral aspect is complete.
4. The symphyseal rim is complete.
5. Ventral margin of dorsal demi-face may break down so that gaps appear in the rim, or it may round off so that there is no longer a clear dividing line between the dorsal demi-face and the ventral rampart.

The Gilbert and McKern method should be used in the same manner as the McKern and Stewart method. Each component of the symphyseal face should be classified on a scale of 0 to 5. The total score of the three components should be compared with Table 19 to obtain the estimated age (mean age on the table). For example, a total score of 9 would indicate an age of 33 years. Note, however, that the ages of individuals with that total score range from 22 to 40, with a standard deviation value of 7.75. This value means that only in two-thirds of the cases where a score of 9 was obtained would the actual age of the woman be within 7.75 years of the mean age of 33. This example illustrates the variability of the age changes and the limited accuracy of the method.

The need for caution in applying existing female standards is emphasized by the high rate of error ob-

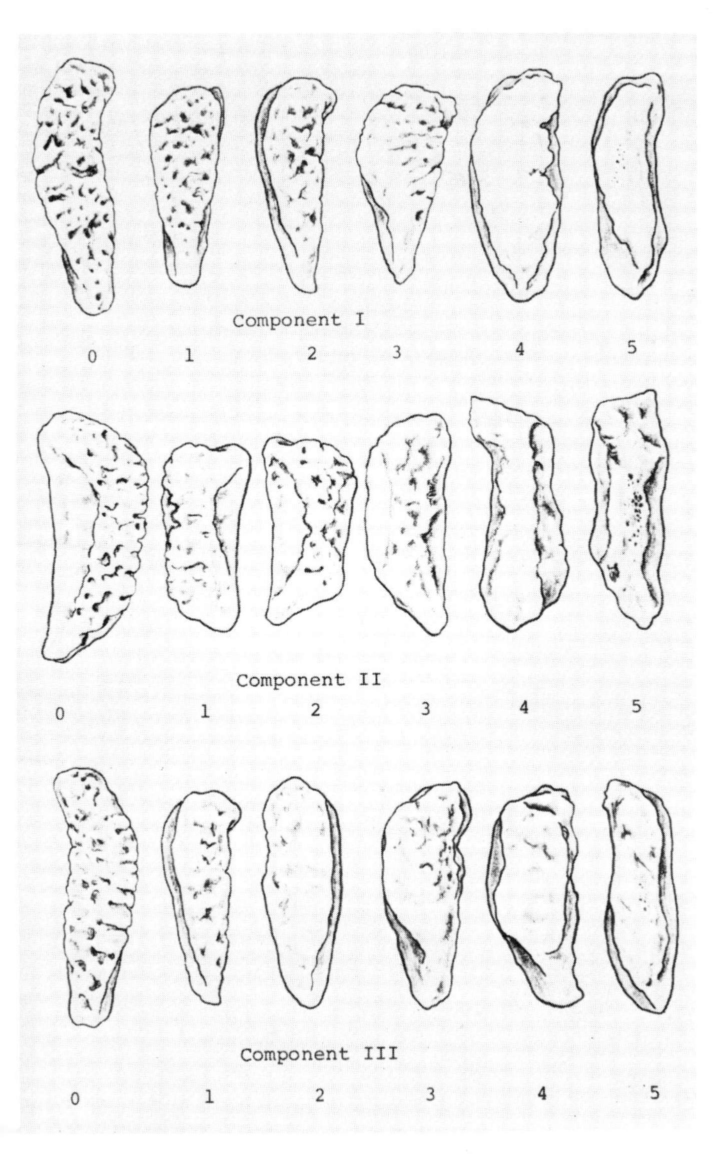

Fig. 91. Age changes in the female pubic symphysis recognized by Gilbert and McKern. The three components are the same ones used by McKern and Stewart for estimating age in males: I, The dorsal demi-face; II, The ventral rampart, and III, The symphyseal rim. The initial stage, not illustrated in Figures 88–90, is shown.

tained for a large California series of pubic bones of known age (Suchey 1979, Angel et al 1986). Angel et al (1986:213–214) and Suchey et al (1986) found that

certain traits are useful in the ageing of specific female pubic bones and these general rules are helpful in those cases:

1. Deep, distinct ridges on both the dorsal and ventral demi-face. Less than 24 years old.

Table 19. Mean age, standard deviation, and age ranges of females obtained from the total scores calculated using the Gilbert and McKern symphyseal formulas (Gilbert, personal communication)

Total Score	Age Range	Mean Age	Standard Deviation
0	14–18	16.00	2.82
1	13–24	19.80	2.62
2	16–25	20.15	2.19
3	18–25	21.50	3.10
4–5	22–29	26.00	2.61
6	25–36	29.62	4.43
7–8	23–39	32.00	4.55
9	22–40	33.00	7.75
10–11	30–47	36.90	4.94
12	32–52	39.00	6.09
13	44–54	47.75	3.59
14–15	52–59	55.71	3.24

2. The ventral rampart has begun its formation, but is not complete. The symphysial rim has begun its development but is not complete. Range of 20 to 49 years.
3. Ossific nodules are present. Less than 28 years old.

Additional discussion of revisions in age estimates is available in Brooks and Suchey (1990).

Auricular Surface of the Ilium. The auricular area is the iliac portion of the sacro-iliac joint. Several components that appear to undergo regular changes with age were identified by Lovejoy et al (1985a) from a study of a large sample in the Todd Collection in Cleveland OH. The components are defined as follows (Fig. 92):

Apex. Portion of the auricular surface that articulates with the posterior aspect of the arcuate line.
Superior demiface. Portion of the auricular area above the apex.
Inferior demiface. Portion of the auricular area below the apex.
Retroarticular area. Region between the auricular surface and the posterior inferior iliac spine.

The descriptions of age changes employ the following qualitative terms:

Billowing. Transverse ridging.
Granularity. Appearance of the compact fine structure on the surface; a heavily "grained" surface resembles fine sandpaper.
Density. Compactness; subchondral bone appears smooth and shows marked absence of grain.
Porosity. Perforations ranging from barely visible to 10 millimeters diameter.

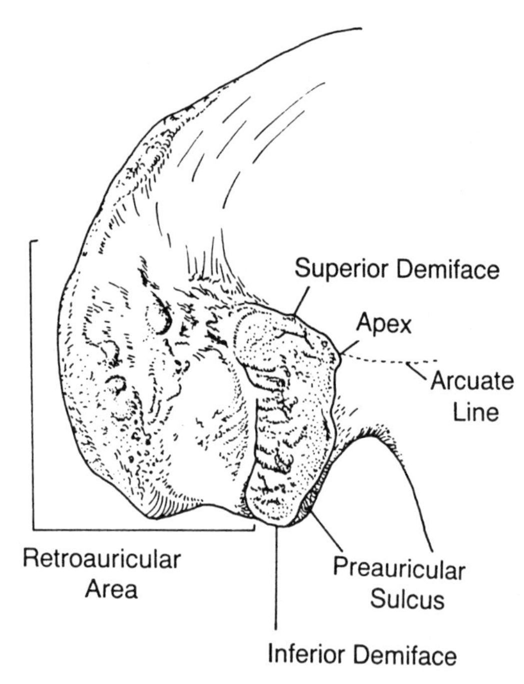

Fig. 92. Components of the auricular surface of the ilium used for estimating age at death (After Lovejoy et al 1985: Fig. 1).

20–24

The surface displays fine granular texture and marked transverse organization. There is no retro-auricular activity, apical activity, or porosity. The surface appears youthful because of broad and well-defined billows, which impart the definitive transverse organization. Billows are well-defined and cover most of the surface. Any subchondral defects are smooth-edged and rounded. Note distinct transverse billows and very fine granularity.

25–29

Changes from the previous phase are not marked and are mostly reflected in slight to moderate loss of billowing, with replacement by striae. There is no apical activity, porosity, or retroauricular activity. The surface still appears youthful owing to marked transverse organization. Granulation is slightly more coarse. . . .

30–34

Both faces are largely quiescent with some loss of transverse organization. Billowing is much reduced and replaced by (definite) striae. The surface is more coarsely and recognizably granular than in previous phase, with no significant changes at apex. Small areas of microporosity may appear. Slight retroauricular activity may occasionally be present. In general, coarse granulation supercedes and replaces billowing. Note smoothing of surface by replacement of billows by fine striae, but distinct retention of slight billowing. Loss of transverse organization and coarsening of granularity is evident.

35–39

Both faces are coarsely and uniformly granulated, with marked reduction of both billowing and striae, but striae may still be present under close examination. Transverse organization is present but poorly defined. There is some activity in the retroauricular area but this is usually slight. Minimal changes are seen at the apex, microporosity is slight, and there is no macroporosity. This is the primary period of uniform granularity. . . .

40–44

No billowing is seen. Striae may be present but very vague. The face is still partially (coarsely) granular and there is a marked loss of transverse organization. Partial densification (which may occur in islands) of the surface with commensurate loss of grain is present along with slight to moderate activity in the retroauricular area. Occasional macroporosity is seen, but this is not typical. Slight changes are usually present at apex. Some increase in microporosity is seen, depending upon the degree of densification. The primary feature is the transition from a granular to a dense surface. . . .

45–49

Significant loss of granulation is seen in most specimens, with replacement by dense bone. No billows or striae are present. Changes at apex are slight to moderate but are almost always present. There is a distinct tendency for the surface to become dense. No transverse organization is evident. Most or all of any microporosity is lost to densification process. There is increased irregularity of margins with moderate retroauricular activity and little or no macroporosity. Note distinct densification of these three specimens and the almost complete lack of transverse organization. . . .

50–60

This is a further elaboration of previous stage, in which marked surface irregularity becomes paramount feature. Topography, however, shows no transverse or other form of organization. Moderate granulation is occasionally retained, but is usually lost during previous phase and is generally absent. No striae or billows are present. The inferior face generally is lipped at inferior terminus, so as to extend beyond the body of the innominate bone. Apical changes are almost invariable and may be marked. Increasing irregularity of margins is seen. Macroporosity is present in some cases but it is not requisite. Retroauricular activity is moderate to marked in most cases. . . .

60+

The paramount feature is a nongranular, irregular surface, with distinct signs of subchondral destruction. No transverse organization is seen and there is a definitive absence of any youthful criteria. Macroporosity is present in about one-third of all cases. Apical activity is usually marked but is not requisite for this age category. Margins become dramatically irregular and lipped, with typical degenerative joint change. The retroauricular area becomes well defined with profuse osteophytes of low to moderate relief. There is clear destruction of subchondral bone, absence of transverse organization, and increased irregularity (Lovejoy et al 1985a:21–26).

The diagnostic features of each age category are summarized by Lovejoy et al (1985a:27) as follows:

1. 20–24: billowing and very fine granularity.
2. 25–29: reduction of billowing, but retention of youthful appearance.
3. 30–34: general loss of billowing, replacement by striae, and distinct coarsening of granularity.
4. 35–39: uniform coarse granularity.
5. 40–44: transition from coarse granularity to dense surface; this may take part over islands of the surface of one or both faces.
6. 45–49: completion of densification with complete loss of granularity.
7. 50–59: dense irregular surface of rugged topography and moderate to marked activity in periauricular areas.
8. 60+: breakdown with marginal lipping, macroporosity, increased irregularity, and marked activity in periauricular areas.

Cranial Suture Closure. Sutures are the lines or joints between the 22 bones forming the skull (Figs. 162–163). In young adults and subadults, they are clearly visible. During adult life, they gradually disappear as adjacent bones unite. In old individuals, many become obliterated completely.

Todd and Lyon (1924, 1925a, 1925b, 1925c) attempted to quantify the changes in suture closure by a detailed examination of each suture on 514 skulls of White and Black males and females of known ages. They observed the same general changes in most of the sutures regardless of sex or race. Closure usually begins endocranially (inside the vault) and proceeds ectocranially (toward the exterior). Although they were able to correlate patterns with age, they cautioned that "the individual variability in progress of suture union makes it unwise to depend too much upon the stage as an age maker, valuable as the indications may be when linked up with other features" (Todd and Lyon 1924:383).

More detailed data on the age progression of suture closure were obtained by McKern and Stewart (1957) from their large series of North American males. They also concluded that "progress of closure has only a very general relationship with age."

Refinements developed by Meindl and Lovejoy (1985) from studying 236 crania in the Hamann-Todd Collection in Cleveland OH improve the accuracy of cranial suture closure as an age indicator. In estimating age from that sample, they found the reliability of lateral-anterior sutures superior to sutures of the vault, and ectocranial closure superior to endocranial closure. They also found age changes to be the same for both sexes and different racial groups. The method they propose employs 10 sites (Fig. 93) and 4 stages of closure. The sites are grouped into two systems:

Vault System

1. Midlambdoid. Midpoint of each half of the lambdoid suture.
2. Lambda.
3. Obelion.
4. Anterior sagittal. Juncture of the anterior third and posterior two-thirds of the length of the sagittal suture.
5. Bregma.
6. Midcoronal. Midpoint of each half of the coronal suture.
7. Pterion. Usually the point at which the parietosphenoid suture meets the frontal bone.

Lateral-Anterior System

6. Midcoronal.
7. Pterion.
8. Sphenofrontal. Midpoint of the sphenofrontal suture.
9. Inferior sphenotemporal. Intersection of the sphenotemporal suture with a line connecting both articular tubercles of the temporomandibular joint.
10. Superior sphenotemporal. Point two centimeters below the junction with the parietal.

Each site is defined as a one-centimeter area, which is classified into one of the following stages of closure:

0. Open; no evidence of ectocranial closure.
1. Minimal, ranging from a single body bridge across the suture to about 50 percent synostosis.
2. Significant; a marked degree of closure but some portion remains incompletely fused.
3. Complete obliteration.

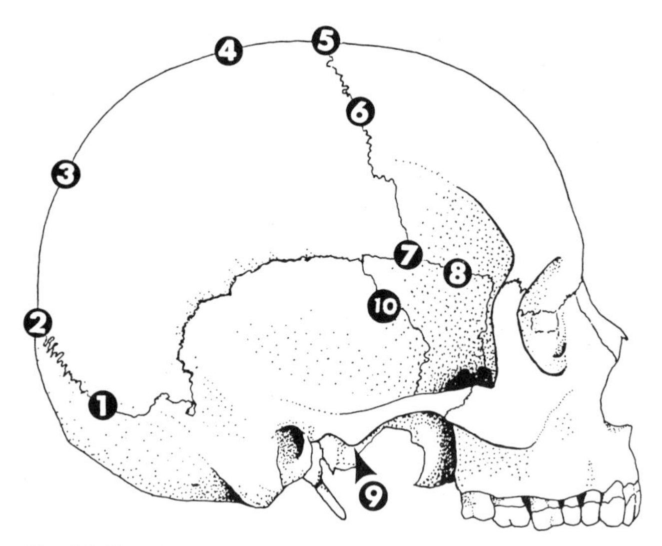

Fig. 93. Ten sites at which suture closure is scored to estimate age at death (After Meindl and Lovejoy 1985: Fig. 1).

To estimate age at death, each side in either or both systems must be assigned a score of 0 to 3. The sum of the scores for either system should be compared with Tables 20-21 to estimate age. For example, a composite score of 10 in the lateral-anterior system suggests a mean age of 51.9 years, with a standard deviation of 12.5 and a total range from 33 to 76 years. (This and other studies based on the Hamann-Todd Collection have been criticized because of possible errors in the ages at death; Hoffman 1987:730). Additional approaches have been presented by Acsadi and Nemeskeri (1970), Masset (1982), and Baker (1984). An independent evaluation by Galera, Ubelaker, and Hayek (1998) suggests endocranial closure provides more reliable data and the methods of Acsadi and Nemeskeri (1970) and Masset (1982) produce slightly more accurate results.

Degenerative Changes. Degenerative changes in the skeleton serve only as very general indicators of age at death, but three features are worth noting.

Stewart (1958) has called attention to the utility of the development of vertebral osteoarthritis (lipping) as an indicator of general age. As age increases, bony outgrowths called osteophytes extend from the margins of the vertebral centra (bodies or rounded por-

Table 20. Estimated age at death using ectocranial lateral-anterior suture closure (after Meindl and Lovejoy 1985: Table 6).

Composite score	No.	Mean age	Standard dev.	Mean dev.	Inter-decile range	Range
0 (Open)	42				-43	-50
1	18	32.0	8.3	6.7	21–42	19–48
2	18	36.2	6.2	4.8	29–44	25–49
3, 4, 5	56	41.1	10.0	8.3	28–52	23–68
6	17	43.4	10.7	8.5	30–54	23–63
7, 8	31	45.5	8.9	7.4	35–57	32–65
9, 10	29	51.9	12.5	10.2	39–69	33–76
11, 12, 13, 14	24	56.2	8.5	6.3	49–65	34–68
15 (Closed)	1					
	236					

Table 21. Estimated age at death using ectocranial vault suture closure (after Meindl and Lovejoy 1985: Table 7).

Composite score	No.	Mean age	Standard dev.	Mean dev.	Inter-decile range	Range
0 (Open)	24				-35	-49
1, 2	12	30.5	9.6	7.4	19–44	18–45
3, 4, 5, 6	30	34.7	7.8	6.4	23–45	22–48
7, 8, 9, 10, 11	50	39.4	9.1	7.2	28–44	24–60
12, 13, 14, 15	50	45.2	12.6	10.3	31–65	24–75
16, 17, 18	31	48.8	10.5	8.3	35–60	30–71
19, 20	26	51.5	12.6	9.8	34–63	23–76
21 (Closed)	13				43–	40–
	236					

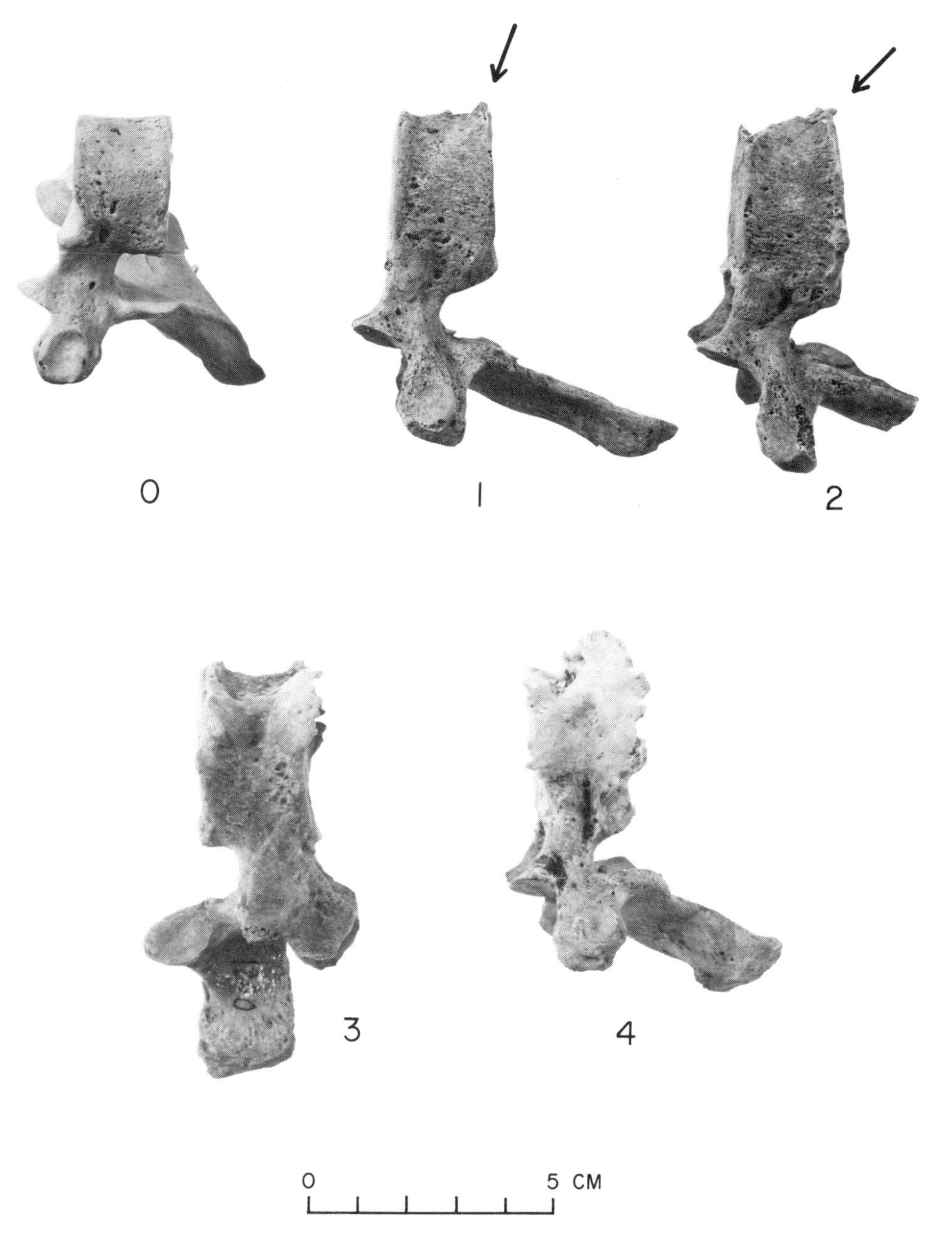

Fig. 94. Progressive development of bony outgrowths (osteophytes) on the margins of the rounded center of a vertebra. Stewart has recognized five stages, which can be used for estimating age at death.

tions of the vertebrae), especially where intervertebral-joint motion is greatest. Stewart classified these structures on a scale of 0 to 4, 0 indicating no lipping and 4 indicating maximum lipping (Fig. 94). A large series of documented White United States males from the Smithsonian's Terry Collection and from Korean War dead (McKern and Stewart 1957) were examined and the age changes in the cervical, thoracic, and lumbar vertebrae were scored separately. In compiling the data, Stewart first averaged the scores for the vertebrae within each region for each individual. He then computed the percentage of individuals in each of the five categories of osteophytosis for each age group. The graphs (Figs. 95–97) show a general correlation between degree of osteophytosis and age, but also a high degree of variability that limits the usefulness of this feature for aging single specimens.

Similar types of ossification and bony extensions appear elsewhere on the skeleton with advancing age. Most of the joints (especially the elbow and knee) develop small deposits of bone (Fig. 98b) or pits (Fig. 98c) on the articular surfaces. Occasionally, these deposits grow large enough to destroy the cartilage. When this happens, the bones come into contact producing abrasion or polishing on the surfaces. This polishing effect is called eburnation (Fig. 98d).

Bone extensions can develop on other surfaces, usually through the ossification of cartilage. They are

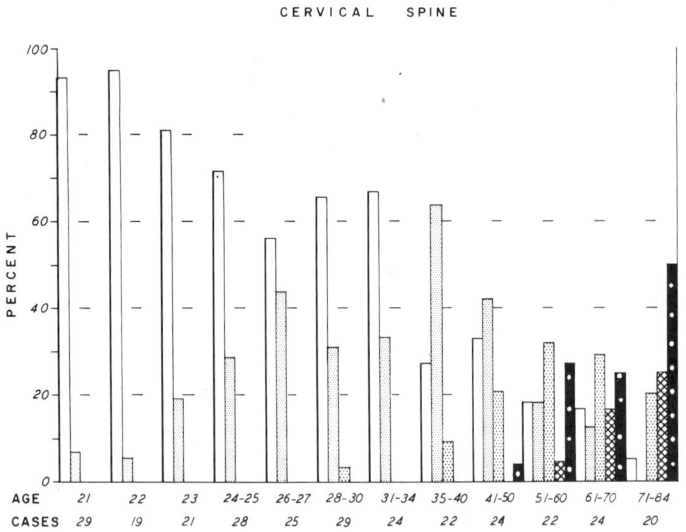

Fig. 95. The percentage frequency of each stage of osteophytosis shown on Figure 94 in the cervical vertebrae of a sample of United States males whose age at death was known. The vertical bars show the number of individuals classified in each stage at each age interval. At age 21, most fell in stage 0 (no lipping). By age 28–30, the majority were still in stage 0, but about 30 percent were scored in stage 1 and 4 percent in stage 2. By the ages of 71–84, 50 percent were classified in stage 4 (see Fig. 97 for key).

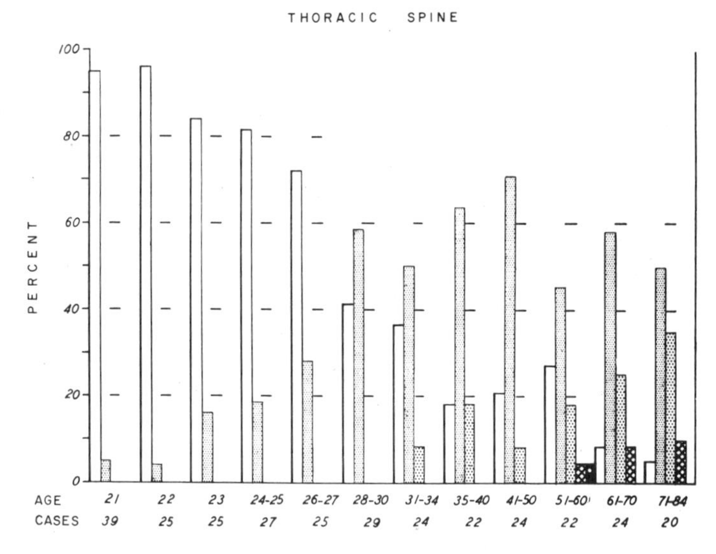

Fig. 96. The percentage frequency of each stage of osteophytosis in the thoracic vertebrae of a sample of United States males whose age at death was known. Changes are more rapid than in the cervical spine, so that by age 28–30 nearly twice as many individuals were classified in stage 1.

frequently found on the ischium and calcaneous, and especially on the sternal ends of the ribs. The sternal (anterior) ends of the ribs are connected to the sternum by cartilage. In early life, the ends of the ribs are relatively blunt with rounded margins (Fig. 99a). By middle age, they have become progressively sharper as the edges of the cartilage become ossified (Fig. 99b). Eventually, these extensions enlarge until the sternal end of the rib assumes a ragged appearance (Fig. 99c). A similar progression occurs on the sternum (Fig. 100).

The occurrence of marked, scooped-out depressions on the parietals is a good indication of advanced age (Fig. 101). This condition results from thinning of the bone and seldom occurs before the age of 60 (Kerley 1970). Since it is uncommon, its absence does not indicate that an individual is younger than 60.

All of the arthritic changes described are general indicators of advancing age. Care must be taken, however, to avoid confusing them with the effects of local trauma. A fracture, dislocation or even infection can lead to the ossification of cartilage or to bony extensions resembling those produced during normal aging.

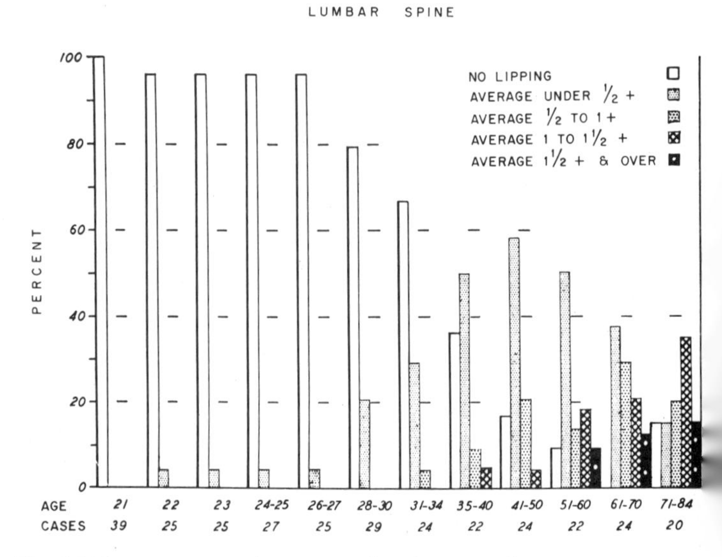

Fig. 97. The percentage frequency of each stage of osteophytosis in the lumbar vertebrae of a sample of United States males whose age at death was known. Little change occurs before the age of 28–30, in contrast to the patterns obtained from the cervical and thoracic portions of the spine.

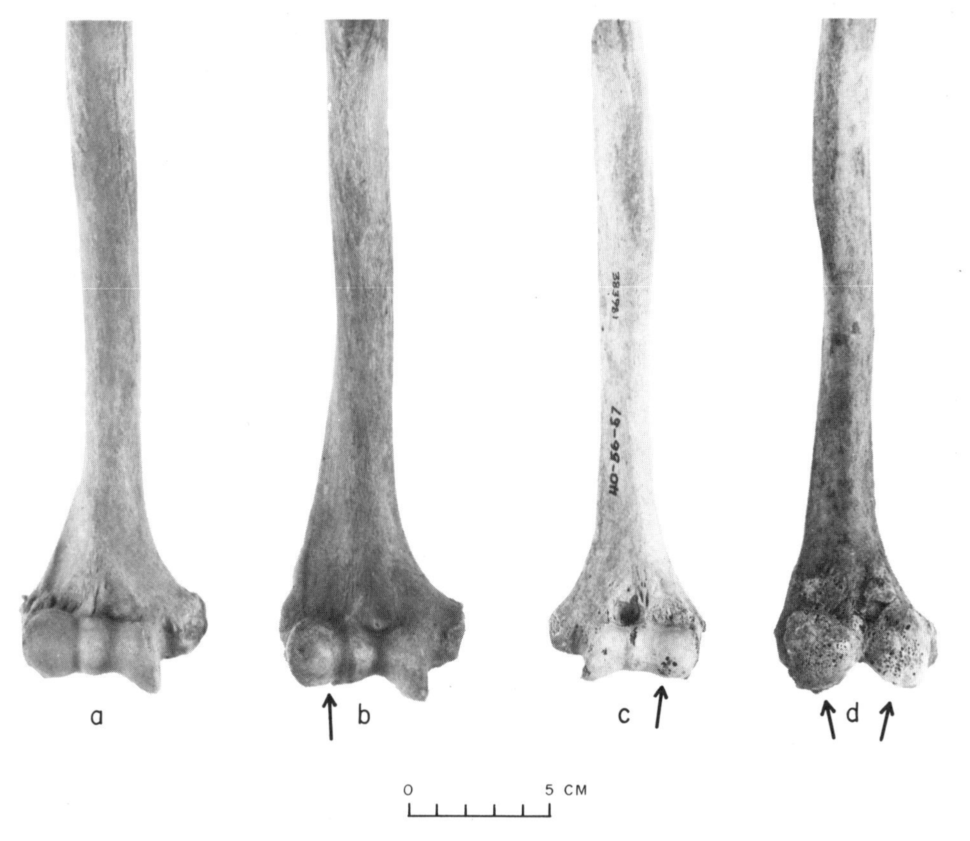

Fig. 98. Degenerative changes on the distal end of the humerus indicative of increasing age. **a,** Normal articular surface. **b,** Appearance of small deposits of bone. **c,** Small pits. **d,** Polishing (eburnation) resulting from friction between the articular surfaces as a consequence of destruction of the intervening cartilage.

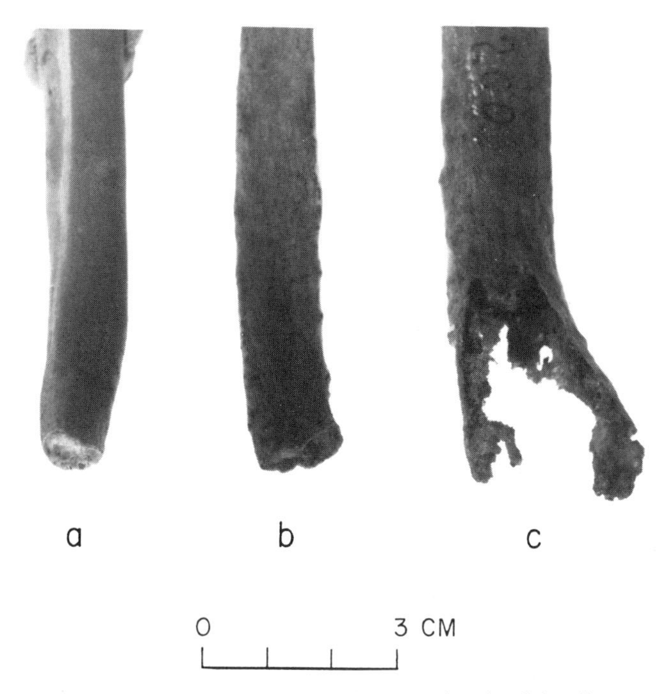

Fig. 99. Degenerative changes in the sternal ends of the ribs. **a,** Blunt end and rounded margins characteristic of youth. **b,** Sharper edges produced by ossification of the cartilage during middle age. **c,** Ragged appearance culminating the process of degeneration.

If enough of the skeleton is preserved, the fact that traumatic effects are usually confined to one area whereas age changes are distributed throughout the body provides a basis for distinction.

Resorption of Cancellous Bone. Hansen (1953–1954) and Schranz (1959) have shown that the medullary (central) cavity expands at the expense of cancellous bone in the proximal ends of the humerus and the femur. Between the ages of 41 and 50, the upper end of the medullary cavity in the femur expands to the level of the surgical neck. Between ages 61 and 74, the cavity reaches the epiphyseal line. The degree of expansion thus provides a general indication of age.

The medullary cavity also increases in diameter with age, eventually producing the extremely thin-walled, osteoporotic condition that characterizes many long bones of very old people.

Studies of radiographs of the proximal femur, proximal humerus, clavicle, and calcaneus of 130 individuals of known age at death in the Hamann-Todd Collection showed the highest correlation with age in the clavicle, followed closely by the proximal femur. The descriptions should be consulted for details (Walker and Lovejoy 1985:72).

Sternal Rib Ends: Males. A new technique for assessing age using changes in the sternal ends of ribs was introduced in 1984 by Iscan, Loth, and Wright based on examination of 230 right fourth ribs removed at autopsy from Whites of known sex and age. Both male and female standards have been blind tested and found to be reliable (Table 22), relatively easy to apply, and negligibly affected by the training and experience of the observer. Additional research suggests that age changes are population specific (Iscan, Loth, and Wright

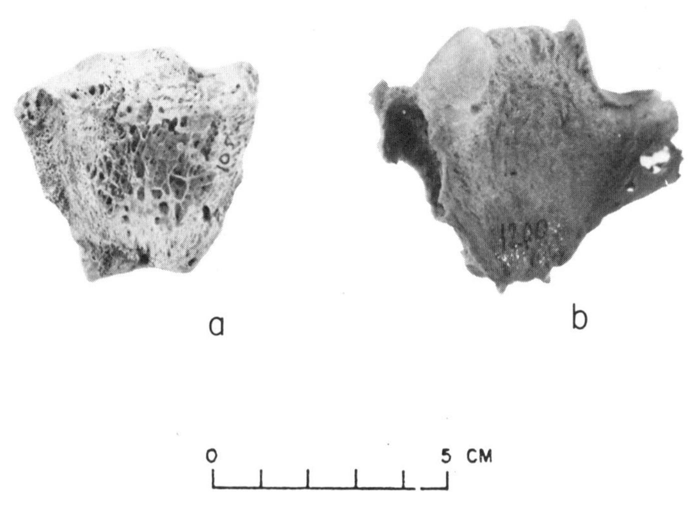

Fig. 100. Degenerative changes in the sternum caused by progressive ossification of the cartilage. **a,** Original appearance. **b,** Condition indicative of advanced age.

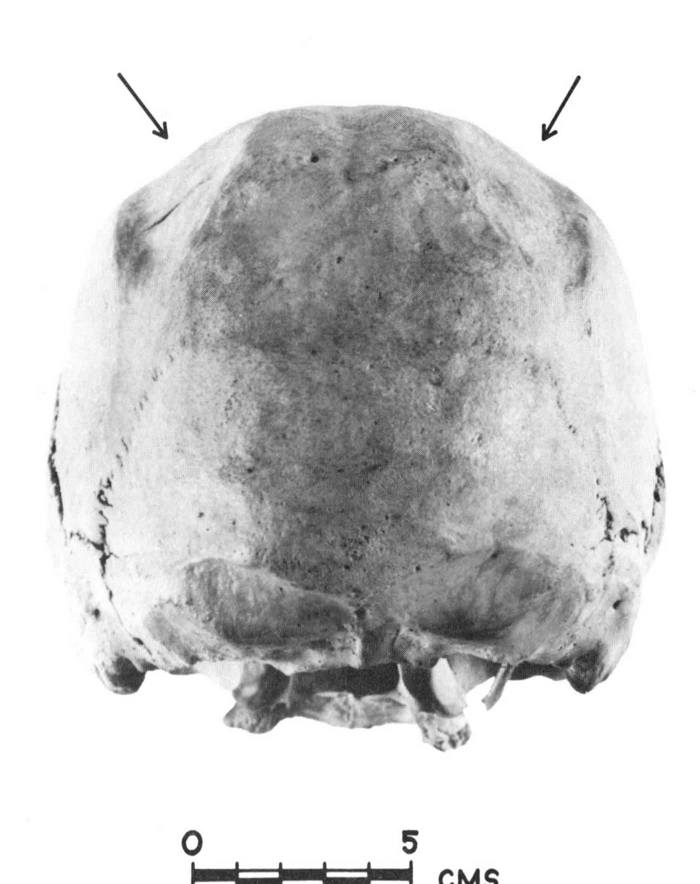

Fig. 101. Depressions caused by thinning of the parietal bones of the skull with increasing age. This skull belonged to a male who died at the age of 72.

Table 22. Descriptive statistics for the metamorphic phases correlating changes in sternal rib ends with age (After Iscan et al 1987).

	Males			Females		
Phase	Mean Age	S.D.	95% Conf. Interval	Mean Age	S.D.	95% Conf. Interval
1	17.3	0.50	16.5–18.0	14.0		
2	21.9	2.13	20.8–23.1	17.4	1.52	15.5–19.3
3	25.9	3.50	24.1–27.7	22.6	1.67	20.5–24.7
4	28.2	3.83	25.7–30.6	27.7	4.62	24.4–31.0
5	38.8	7.00	34.4–42.3	40.0	12.22	33.7–46.3
6	50.0	11.17	44.3–55.7	50.7	14.93	43.3–58.1
7	59.2	9.52	54.3–64.1	65.2	11.24	59.2–71.2
8	71.5	10.27	65.0–78.0	76.4	8.83	70.4–82.3

1987). Applying the standards to other ribs revealed differences between left and right sides as well as between the right fourth and some other ribs suggesting caution should be used if ribs other than the right fourth are examined (Yoder 1999).

The following progression of changes has been provided by Loth, modified from Iscan, Loth, and Wright (1984: 1096, 1099) and Iscan and Loth (1986: 71-72).

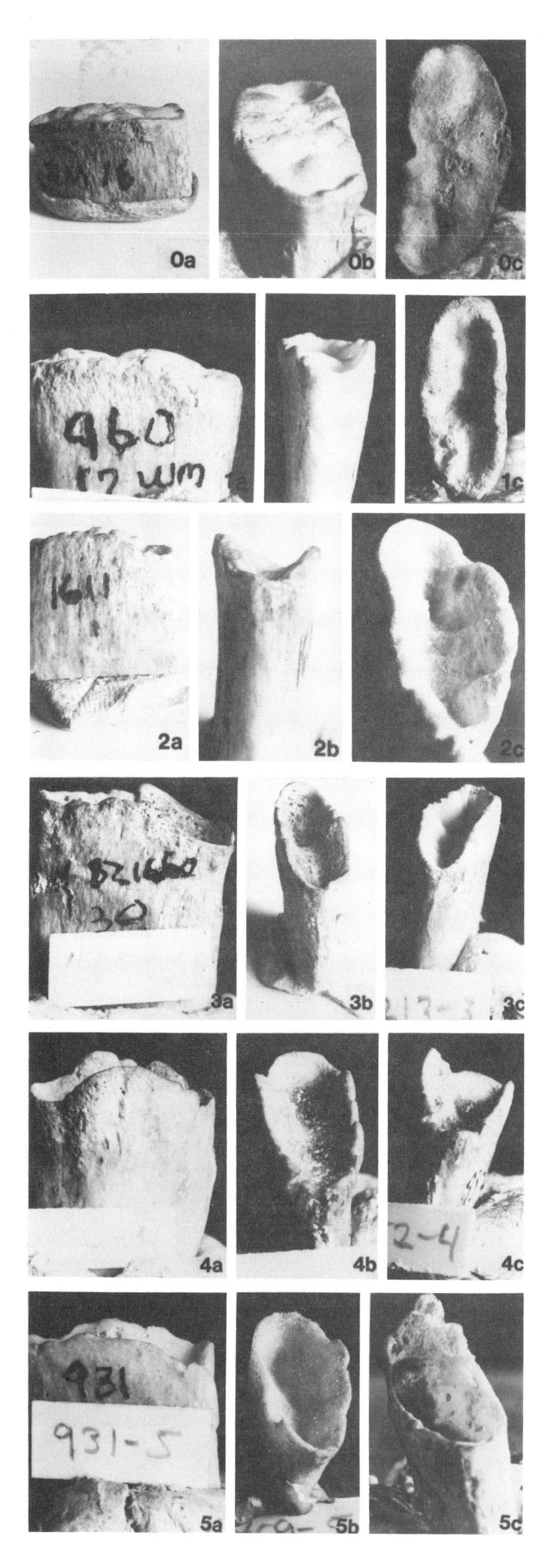

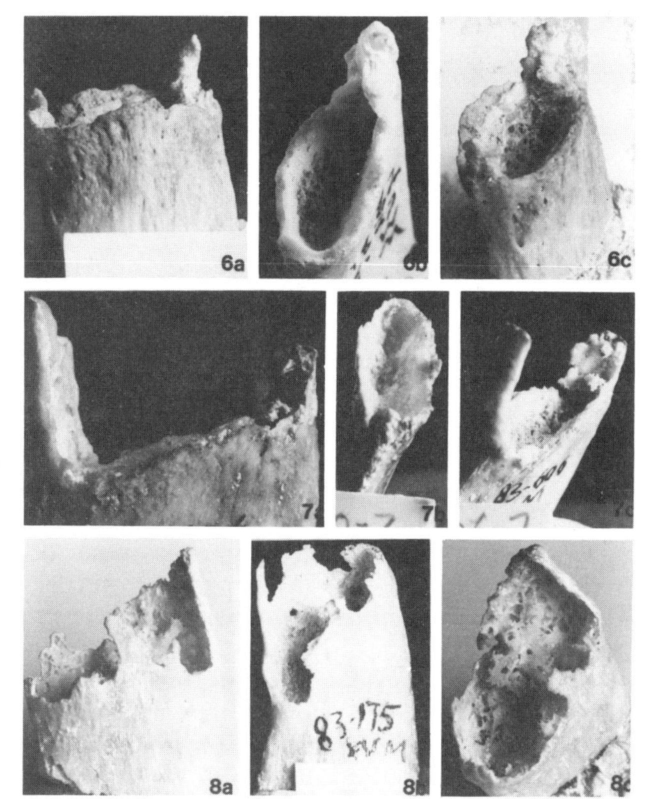

Fig. 102. Phases in progressive changes in the sternal ends of ribs of males with increasing age (After Iscan, Loth, and Wright 1984).

Phase 0 (16 and younger): The articular surface is flat or billowy with a regular rim and rounded edges. The bone itself is smooth, firm and very solid (Fig. 102, 0a-0c).

Phase 1 (17–19): There is a beginning amorphous indentation in the articular surface, but billowing may also still be present. The rim is rounded and regular. In some cases scallops may start to appear at the edges. The bone is still firm, smooth and solid (Fig. 102, 1a-1c).

Phase 2 (20–23): The pit is now deeper and has assumed a V-shaped appearance formed by the anterior and posterior walls. The walls are thick and smooth with a scalloped or slightly wavy rim with rounded edges. The bone is firm and solid (Fig. 102, 2a-2c).

Phase 3 (24–28): The deepening pit has taken on a narrow to moderately U-shape. Walls are still fairly thick with rounded edges. Some scalloping may still be present but the rim is becoming more irregular. The bone is still quite firm and solid (Fig. 102, 3a-3c).

Phase 4 (26–32): Pit depth is increasing, but the shape is still a narrow to moderately wide U. The walls are thinner, but the edges remain rounded. The rim is more irregular with no uniform scalloping pattern remaining. There is some decrease in the weight and firmness of the bone, however, the overall quality of the bone is still good (Fig. 102, 4a-4c).

Phase 5 (33–42): There is little change in pit depth, but the shape in this phase is predominantly a moderately wide U. Walls show further thinning and the edges are becoming sharp. Irregularity is increasing in the rim. Scalloping pattern is completely gone and has been replaced with irregular bony projections. The condition of the bone is fairly good, however, there are some signs of deterioration with evidence of porosity and loss of density (Fig. 102, 5a-5c).

Phase 6 (43–55): The pit is noticeably deep with a wide U-shape. The walls are thin with sharp edges. The rim is irregular and exhibits some rather long bony projections that are frequently more pronounced at the superior and inferior borders. The bone is noticeably lighter in weight, thinner and more porous, especially inside the pit (Fig. 102, 6a-6c).

Phase 7 (54–64): The pit is deep with a wide to very wide U-shape. The walls are thin and fragile with sharp, irregular edges and bony projections. The bone is light in weight and brittle with significant deterioration in quality and obvious porosity (Fig. 102, 7a-7c).

Phase 8 (65 and older: In this final phase the pit is very deep and widely U-shaped. In some cases the floor of the pit is absent or filled with bony projections. The walls are extremely thin, fragile and brittle with sharp, highly irregular edges and bony projections. The bone is very lightweight, thin, brittle, friable and porous. "Window" formation is sometimes seen in the walls (Fig. 102, 8a-8c).

Sternal Rib Ends: Females. The following progression has been supplied by Loth, modified from Iscan, Loth, and Wright (1985:855, 858) and Iscan and Loth (1986:73–74).

Phase 0 (13 and younger): The articular surface is nearly flat with ridges or billowing. The outer surface of the sternal extremity of the rib is bordered by what appears to be an overlay of bone. The rim is regular with rounded edges, and the bone itself is firm, smooth, and very solid (Fig. 103, 0a-0c).

Phase 1 (14–15): A beginning, amorphous indentation can be seen in the articular surface. Ridges or billowing may still be present. The rim is rounded and regular with a little waviness in some cases. The bone remains solid, firm, and smooth (Fig. 103, 1a-1c).

Phase 2 (16–19): The pit is considerably deeper and has assumed a V-shape between the thick, smooth anterior and posterior walls. Some ridges or billowing may still remain inside the pit. The rim is wavy with some scallops beginning to form at the rounded edge. The bone itself is firm and solid (Fig. 103, 2a-2c).

Phase 3 (20–24): There is only slight if any increase in pit depth, but the V-shape is wider, sometimes approaching a narrow U as the walls become a bit thinner. The still rounded edges now show a pronounced, regular scalloping pattern. At this stage, the anterior or posterior walls may first start to exhibit a central, semicircular arc of bone. The rib is firm and solid (Fig. 103, 3a-3c).

Phase 4 (24–32): There is a noticeable increase in the depth of the pit, which now has a wide V- or narrow U-shape with, at times, flared edges. The walls are thinner but the rim remains rounded. Some scalloping is still present, along with the central arc; however, the scallops are not as well defined and the edges look somewhat worn down. The quality of the bone is fairly good but there is some decrease in density and firmness (Fig. 103, 4a-4c).

Phase 5 (33–46): The depth of the pit stays about the same, but the thinning walls are flaring into a wider V- or U-shape. In most cases, a smooth, hard, plaque-like deposit lines at least part of the pit. No regular scalloping pattern remains and the edge is beginning to sharpen. The rim is becoming more irregular, but the central arc is still the most prominent projection. The bone is noticeably lighter in weight, density and firmness. The texture is somewhat brittle (Fig. 103, 5a-5c).

Phase 6 (43–58): An increase in pit depth is again noted, and its V- or U-shape has widened again because of pronounced flaring at the end. The plaque-like deposit may still appear but is rougher and more porous. The walls are quite thin with sharp edges and an irregular rim. The central arc is less obvious and, in many cases, sharp points project from the rim of the sternal extremity. The bone itself is fairly thin and brittle with some signs of deterioration (Fig. 103, 6a-6c).

Phase 7 (59–71): In this phase, the depth of the predominantly flared U-shaped pit not only shows no increase, but actually decreases slightly. Irregular bony growths are often seen extruding from the interior of the pit. The central arc is still present in most cases but is now accompanied by pointed projections, often at the superior and inferior borders, yet may be evidenced anywhere around the rim. The very thin walls have irregular rims with sharp edges. The bone is very light, thin, brittle, and fragile, with deterioration most noticeable inside the pit (Fig. 103, 7a-7c).

Phase 8 (70 and older): The floor of the U-shaped pit in this final phase is relatively shallow, badly deteriorated, or completely eroded. Sometimes it is filled

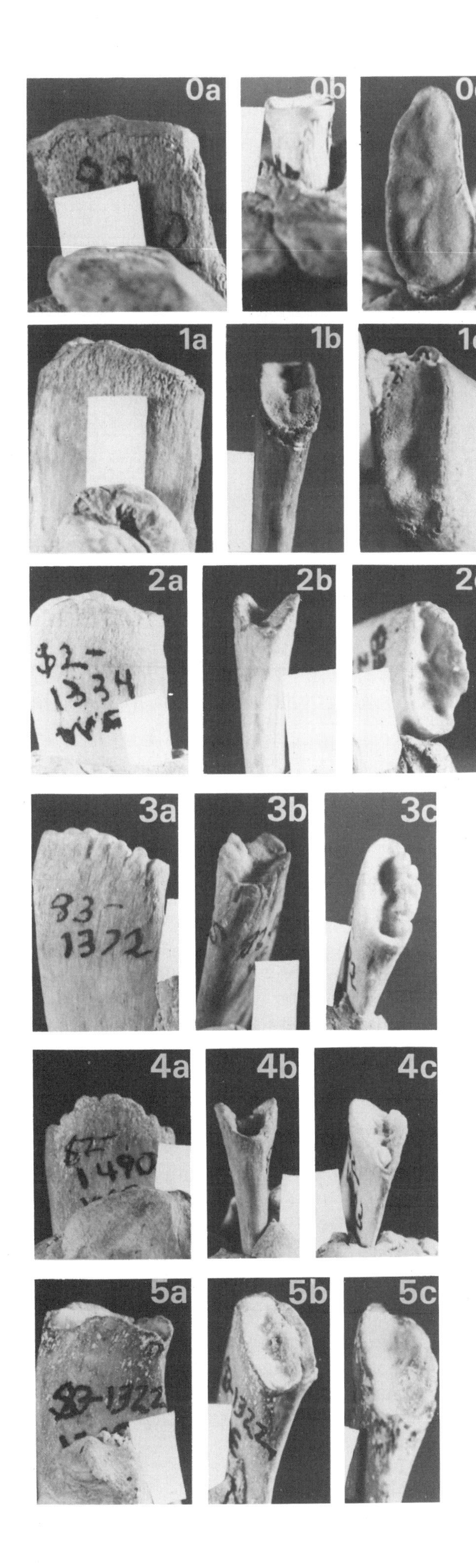

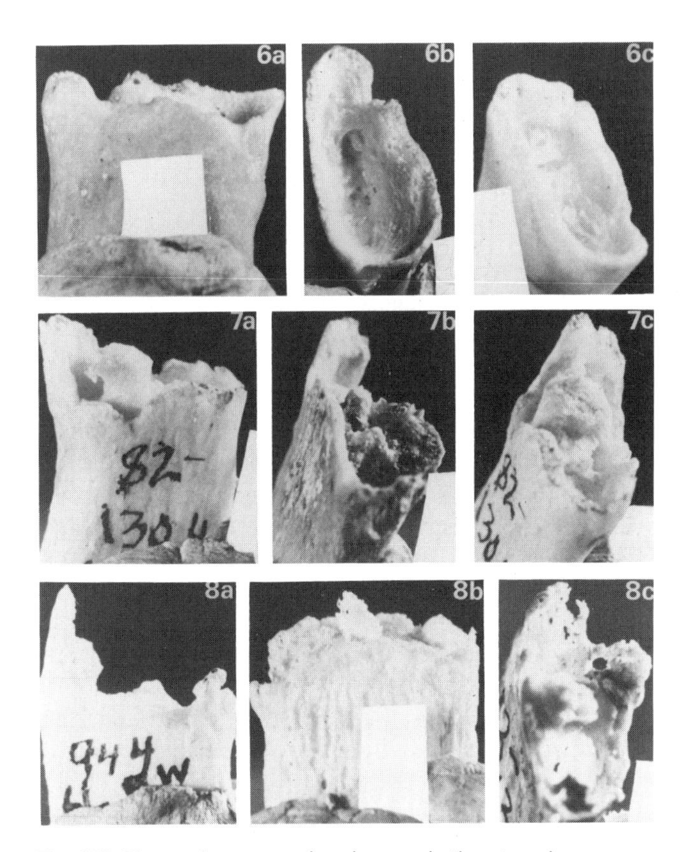

Fig. 103. Phases in progressive changes in the sternal ends of ribs of females with increasing age (After Iscan, Loth, and Wright 1985)

with bony growths. The central arc is barely recognizable. The extremely thin, fragile walls have highly irregular rims with very sharp edges, and often fairly long projections of bone at the inferior and superior borders. "Window" formation sometimes occurs in the walls. The bone itself is in poor condition—extremely thin, light in weight, brittle, and fragile (Fig. 103, 8a-8c).

Dental Attrition. Dental attrition, or wear resulting from chewing, generally proceeds continously during life and has been used to estimate age. Hrdlicka (1939:45) recognized five general stages of attrition of the permanent teeth:

1. First signs, on the tips of the cusps and edges: beginning of adult life
2. Cusps of the molars worn off: 26–33 years
3. The enamel of the masticating surfaces of the teeth worn off completely: 35–50 years
4. The crowns themselves worn off markedly: 6th-7th decade
5. Crowns worn off entirely, the roots exposed: 65 years and upward.

He added that "it should be understood that all the above applies only to the American aborigines, and that before they became more or less civilized. In every other group of humanity that is to be dealt with by the student the subject will demand a study of tooth wear in the living of that group and the establishment of separate criteria based on that study" (1939:46).

Unfortunately, Hrdlicka's caution must be applied among prehistoric American groups as well. Although prehistoric populations exhibit much more dental wear than most modern ones, there is considerable variability not only between groups, but between individuals in the same group, and even among the teeth in a single mouth. Since teeth erupt at different ages, they are exposed to attrition-producing factors for different lengths of time. For example, the first molar reaches the occlusal surface about the age of six years, the second molar erupts at 12 years, and the third molar at 18 to 21 years. As a result, the first molar is exposed to at least 12 more years of use than the third molar and thus shows considerably more wear. Attrition rates also vary greatly among populations and individuals because of differences in diet, occlusion, morphology, and even the use of teeth as tools.

Differences in extent of dental wear among the molars, as well as other teeth, can be used to derive the rate of wear for that individual. Applying the rate to the amount of attrition on the occlusal surfaces may permit an estimate of age at death. See Miles (1958, 1962, 1963a, 1963b, 1978) for explanations of this approach and Lovejoy (1985) for an application to a prehistoric skeletal sample from the Libben site in Ohio.

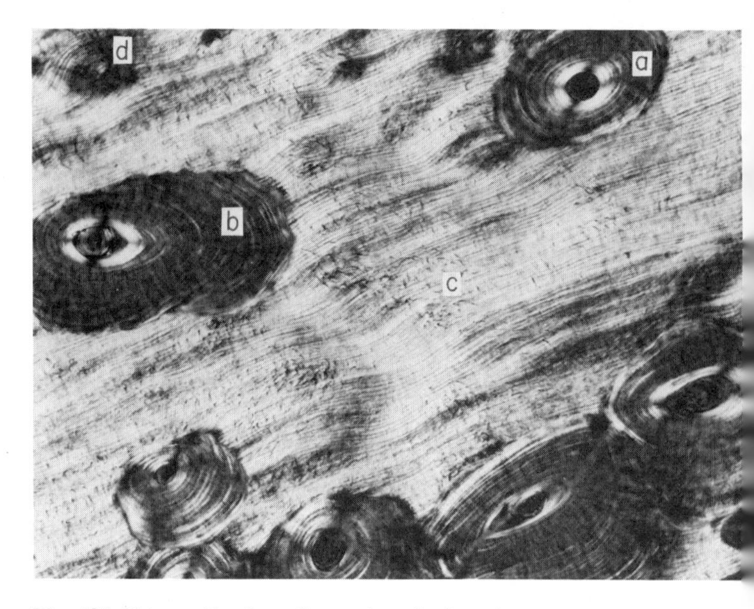

Fig. 104. Thin section from the cortex of a long bone (photographed with polarized light) showing microscopic features useful for estimating age at death. **a,** Complete osteon composed of concentric layers of bone surrounding a Haversian canal. **b,** Osteon fragment attesting to the destruction of an earlier osteon by a later one. **c,** Circumferential lamellar bone, the original layered composition of the cortex. **d,** Non-Haversian canal (Courtesy of D. Ortner).

Adults: Microscopic Methods

Macroscopic criteria provide useful, rapid methods of estimating the ages of adults at death that are accurate enough to allow sorting in the field and preliminary statements on the correlation between method of interment and age. Certain kinds of demographic analyses require more accurate methods, however, and those developed so far utilize microscopic changes in the long-bone cortex and the teeth. Although the skills and equipment are not likely to be available to most users of this manual, discussion of the procedures will indicate how much information can be potentially obtained and emphasize the importance of saving human skeletal remains, however fragmentary.

More detailed discussion of histological approaches is provided in Ubelaker (1986).

Cortical Remodeling in Long Bones. When first formed, the long-bone cortex consists mainly of thin, parallel, slightly undulating layers known as circumferential lamellar bone (Fig. 104c). Beginning in early infancy and continuing throughout life, microscopic structures called "osteoclasts" cut longitudinal tunnels or channels through the cortex, interrupting the layered structures. In cross section, these tunnels look like circular or slightly ovoid holes. During the normal aging process, these holes are gradually filled with concentric layers of bone by structures called "osteoblasts." A small opening, called the Haversian canal, remains in the center. This canal eventually carries a small blood vessel and nerve fiber, and becomes an active agent in the chemical exchange in the immediate area. The entire structure (concentric circles and Haversian canal) is known as an osteon (Fig. 104a). The outer limit of each osteon is defined by a denser layer of bone, called a "reversal line" because it marks the boundary where osteoclastic tunneling ended and osteoblastic filling began. The process of osteon formation continues throughout life, so that the number of osteons increases with age. Since spacing is uncontrolled, the more numerous the osteons, the greater the probability that new ones will cut into older ones. On a microscope slide, the result of this process is readily observable (Fig. 104b). The frequency of these fragments also increases with age.

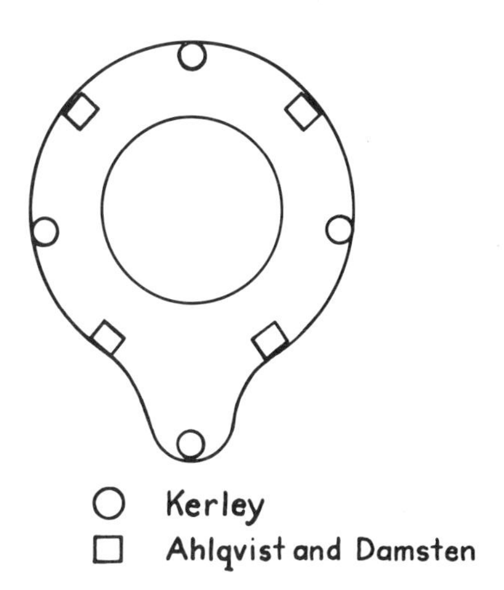

Fig. 105. Locations of the four circular fields employed by Kerley and the four square fields recommended by Ahlqvist and Damsten for observing microscopic structure to estimate age at death.

In 1965, Kerley described a method of estimating age based on the process of osteon formation, derived from studying 126 cross sections cut from femora, tibiae, and fibulae of individuals whose known ages at death ranged from birth to 95 years. The features observed are: (1) the number of whole osteons, (2) the number of fragments of old osteons, (3) the percentage of circumferential lamellar bone remaining, and (4) the number of non-Haversian canals. Osteons, osteon fragments, and circumferential lamellar bone were defined earlier. Non-Haversian canals are vascular canals formed by "the inclusion of small, peripheral blood vessels into the bone by the rapid expansion of the cortex in diameter" (Kerley 1965:152). These structures can be distinguished from normal osteons by the presence of a denser reversal line and usually by the fact that the surrounding lamellae bend around them (Fig. 104d), whereas osteons cut through the lamellae (Fig. 104a).

The Kerley method involves counting these four types of structures (osteons, osteon fragments, percentage of lamellar bone, and non-Haversian canals) under 100-power magnification in four standard locations along the extreme outer (periosteal) edge of a cross section of the midshaft of the bone (Fig. 105). The locations are anterior, posterior, medial, and lateral. According to Kerley (1965:154):

> The fields were selected in such a manner that they touched upon the outer edge of the bone and were fairly representative of the particular anatomic area of the section being examined. Each microscopic field covered a circular area of the cross section that was 1.25 mm [actually about 1.62 mm] in diameter. That was the area covered by a ten power objective lens combined with ten power wide-field oculars.
>
> The total number of recognizable osteons, including those that were partly obscured by the periphery of the field, was counted in each field. Then this number was totalled for all four areas to arrive at a single figure representing the osteon number for the outer zone of that particular section. The same procedure was used for counting osteon fragments and non-Haversian canals. The percentage of each field that was composed of circumferential lamellar bone was estimated, and the four fields were averaged—rather than totalled—to arrive at a figure that was representative of the outer zone lamellar bone.

In 1978, Kerley and I reexamined his original data and produced improved regression equations (Table 23). Note that the estimates calculated from the frequency of fragmentary (disrupted) osteons in the fibula are the most accurate. We also discovered that the size

Table 23. New regression formulas for estimating age (Y) from microscopic structures in long-bone cortex. X = factor value. (Kerley and Ubelaker 1978:546).

Factor	Regression	Standard Error of Estimate
Femoral osteones	$Y = 2.278 + 0.187X + 0.00226X^2$	9.19
Femoral fragments	$Y = 5.241 + 0.509X + 0.017X^2 - 0.00015X^3$	6.98
Femoral lamellar	$Y = 75.017 - 1.790X + 0.0114X^2$	12.52
Femoral non-Haversian	$Y = 58.390 - 3.184X + 0.0628X^2 - 0.00036X^3$	12.12
Tibial osteones	$Y = -13.4218 + 0.660X$	10.53
Tibial fragments	$Y = -26.997 + 2.501X - 0.014X^2$	8.42
Tibial lamellar	$Y = 80.934 - 2.281X + 0.019X^2$	14.28
Tibial non-Haversian	$Y = 67.872 - 9.070X + 0.440X^2 - 0.0062X^3$	10.19
Fibular osteones	$Y = -23.59 + 0.74511X$	8.33
Fibular fragments	$Y = -9.89 + 1.064X$	3.66
Fibular lamellar	$Y = 124.09 - 10.92X + 0.3723X^2 - 0.00412X^3$	10.74
Fibular non-Haversian	$Y = 62.33 - 9.776X + 0.5502X^2 - 0.00704X^3$	14.62

of the field used as a standard for counting should be about 1.62 mm, not 1.25 mm as originally published. This point is important because field size can vary on different microscopes, even when the same combination of lenses and occulars is employed. For accuracy, field size should be measured on each microscope used, and all counts should be adjusted to make them compatible with a field size of 1.62 mm.

Modifications of Kerley's method have been proposed by Ahlqvist and Damsten (1969) and by Singh and Gunberg (1970). Ahlqvist and Damsten contend that (1) reliance on several structures overly complicates the procedure without increasing its accuracy significantly; (2) use of circular fields makes it difficult to examine structures on the borders of a field, and (3) selection of the posterior location may introduce variations unrelated to age because it falls on the linea aspera, a site of muscle attachment. Their modified version employs only the midshaft of the femur and only the total percentage of remodeled bone (osteons and osteon fragments) in four square fields adjacent to the outer edge of the cross section. These fields are spaced around the circumference so they fall between Kerley's sites and avoid the linea aspera (Fig. 105). A ruled ocular micrometer (grid) containing 100 squares is inserted into the eyepiece so that one side (or 10 squares) measures one millimeter at the level of the section. The squares that are more than half-filled with either osteons or osteon fragments are counted and the result is coverted into a percentage. The percentages obtained from the four fields are averaged and transformed into an age estimate using the formula:

$$\text{Age} = 0.991x - 4.96 \pm 6.71$$

(x is the percentage of remodeled bone). This method may be easier to use than Kerley's, but is less accurate because the standard error is larger and because the sample on which it is based consisted of only 20 microscopic sections, most of them representing individuals over 50 years old.

The version developed by Singh and Gunberg (1970) employs: (1) the total number of osteons, (2) the average number of lamellae per osteon, and (3) the average shortest diameter of the Haversian canals in two fields randomly selected from the periosteal third of the cortex. They examined 139 sections taken from the mandible, femur, and tibia of 59 individuals with a 10x objective and a 10x widefield ocular to produce the regression equations on Table 24. The standard errors of their estimates are smaller than those of Kerley or Ahlqvist and Damsten. However, the ages of the individuals in the samples on which the equations are based range from 40 to 88 years, making their reliability for younger individuals uncertain. Their formulas consequently should not be applied unless a skeleton is obviously over 40 years of age (judging from macroscopic evidence). Anyone wishing to apply their method should consult the original publication by Singh and Gunberg (1970) for explicit instructions.

Table 24. Regression equations for estimating age from microscopic measurements on the mandible, femur, and tibia (after Singh and Gunberg 1970:377). X_1 = total number of osteons in two fields; X_2 = average number of lamellae per osteon; X_3 = average diameter of Haversion canals.

Regression Equations	Standard Error of Estimate
Mandible	
$20.82 + 0.85X_1 + 0.87X_2 - 0.22X_3$	2.55
$-18.99 + 1.13X_1 + 1.76X_2$	2.69
$32.23 + 0.92X_1 - 0.30X_3$	2.58
$74.73 + 1.52X_2 - 0.45X_3$	3.04
$-28.24 + 1.68X_1$	3.02
$5.31 + 5.00X_2$	3.83
$103.99 - 0.63X_3$	3.16
Femur	
$27.65 + 0.65X_1 + 0.78X_2 - 0.26X_3$	3.24
$-14.69 + 1.13X_1 + 1.11X_2$	3.55
$29.59 + 0.79X_1 - 0.28X_3$	3.25
$61.25 + 1.74X_2 - 0.44X_3$	3.52
$16.10 + 1.38X_1$	3.60
$2.00 + 5.16X_2$	5.01
$89.01 - 0.62X_3$	3.82
Tibia	
$43.52 + 0.291X_1 + 1.47X_2 - 0.34X_3$	3.02
$-3.40 + 0.67X_1 + 2.27X_2$	3.93
$48.61 + 0.53X_1 - 0.38X_3$	3.22
$54.79 + 2.19X_2 - 0.4X_3$	3.12
$-4.76 + 1.15X_1$	4.33
$5.10 + 4.88X_2$	4.59
$91.32 - 0.64X_3$	3.88

Thin sections must be prepared from the bone for examination under a microscope, regardless of the method employed. The Kerley and the Ahlqvist and Damsten procedures require undecalcified, ground thin sections; Singh and Gunberg use both decalcified, stained sections and ground thin sections. Detailed instructions for preparing ground thin sections from archeological samples of bone are provided in Appendix 2.

A disadvantage of these methods is the necessity of severing the bone to obtain a complete cross section for microscopic examination. Addressing this problem, Thompson (1979) developed an alternative that uses cores 4 mm. in diameter taken from the anterior midshaft of the femur, the medial midshaft of the tibia, the medial midshaft to the deltoid tuberosity of the humerus, and the lateral distal third of the ulna. Employing 19 variables in a sample of 116 human adults, he produced regression equations with standard errors as low as 6.2 years.

Prieto Carrero (1993) examined microscopic features of the iliac crest in a Spanish autopsy sample of 73 individuals with ages at death of 20 years or less and demonstrated growth-related changes useful for assessing age at death within this range.

Dental Microstructure. Based upon examination of 41 teeth with documented ages ranging from 2 to 69 years, Gustafson (1950) proposed a method of determining age at death using seven features of dental microstructure: attrition, cementum apposition, root resorption, periodontosis, secondary dentin apposition, transparency of the root, and closure of the root orifice. When considered together, changes in these features yield age estimates that have a standard error of only 3.6 years.

The age correlations were reexamined by Nalbandian (1959) and by Nalbandian and Sognnaes (1960), who established similar regression coefficients from samples of Swedish and United States Whites. Dechaume, Dérobert, and Payen (1960) agree that Gustafson's method is accurate when employed by an experienced histologist with proper equipment. Based on reassessment of the Gustafson age changes in a sample of 355 teeth from a contemporary Florida dental clinic, Maples (1978) developed multiple regressions using the Gustafson scoring procedure that show lower standard errors.

Although the Gustafson method appears to be reliable, it has several limitations for application to archeological material (Maples and Rice 1979, Burns and Maples 1976). First, a tooth must be destroyed to prepare the cross section, a loss that must be weighed against the value of the tooth for other types of anthropological research. Second, the method requires a detailed knowledge of dental anatomy, as well as expensive equipment for making cross sections. Third, one of the criteria employed by Gustafson is attrition, a process that occurs at highly variable rates in different populations. Variance in attrition also affects several other age changes observed by Gustafson. In summary, this method should not be attempted without a thorough review of the appropriate literature, acquisition of the necessary equipment, and understanding of the variability in the rates of change of the features within and among different populations.

A variation developed by Lamendin et al (1992) from a French autopsy sample of known sex and age at death employs two dental features: (1) periodontosis height multiplied by 100/root height and (2) root transparency. Periodontosis height is measured on the labial surface of single-rooted teeth between the cementoenamel junction and the line of soft tissue attachment. Transparency is measured on the labial surface from the apex of the root to the maximum expression of transparency. Age at death is estimated from the equation: $A=(0.18xP) + (0.42xT) + 25.53$, in which A equals age at death, P equals (periodontosis height x 100)/root height, and T equals (transparency height x 100)/root height. The mean error of this technique is only 8.4 years; accuracy is greater above 40 years in age.

Aging the Skeleton

In summary, the methods selected to estimate age at death depend upon which bones are available, the population believed to be represented, and the relative accuracy of the results. The greater the number of procedures employed, the more accurate the estimate is likely to be. Lovejoy et al (1985b) offer a multifactorial method that integrates data derived from the pubic symphyseal face, auricular surface, radiographs of the proximal femur, dental wear, and suture closure. Their application of this approach to the Hamann-Todd Collection indicates that multifactorial aging is substantially more accurate than reliance on any single age indicator. Variable results from an independent test of seven methods of estimating skeletal age led Saunders et al (1992) to the same conclusion.

Baccino and Zerilli (1997) recommend a two-step approach. A skeleton is first examined using the Suchey-Brooks pubic symphysis method. If the pubis is judged to be within the first three phases, it is used to generate the age estimate. If the pubis assessment falls above the first three phases, the Lamendin dental technique is used. This approach recognizes that the pubis is more accurate for young adults and the Lamendin method provides greater accuracy for older adults.

Subsequently, Baccino et al (1999) tested seven methods of estimating adult age using a French autopsy sample of known age and sex. Methods tested included evaluation of the sternal end of the fourth rib (Iscan method), single-rooted teeth (Lamendin method), symphyseal face of the pubis (Suchey-Brooks method), and femoral cortical remodeling (Kerley method), as well as three comprehensive methods consisting of: (1) the dental two-step strategy discussed above, (2) the mean derived from the results of the four individual methods, and (3) a "global" approach in which an investigator was allowed to examine all the evidence and use experience to produce an overall age estimate. The evaluation made independently by two investigators without knowing actual age demonstrated that the comprehensive approach achieved greater accuracy. Experience was important in the accuracy of estimates using a single technique.

4 Cultural and Pathological Alterations

In addition to the changes resulting from growth, age, sex, heredity, and other normal biological processes, modifications of the bones and teeth can be produced by (and therefore indicative of) cultural practices and pathological conditions. Considerable indirect as well as direct information on the way of life of a population may be provided by these conditions, when they are carefully described and correctly interpreted. Descriptions of the most common kinds of alterations will serve to illustrate their variety and the problems associated with their interpretation.

CULTURAL INFLUENCES ON THE SKELETON

Some kinds of alterations are intentional and produced for cultural or aesthetic reasons. The best examples are deformations of the skull and cuts, inlays, or other modifications of the teeth. Other abnormal conditions, such as squatting facets, may be unintentional consequences of specific behavior. A third type of modification is accomplished after death, either in connection with burial or other types of ritual, or as a consequence of natural disturbance.

Cranial Deformation

Cranial deformation or flattening is the alteration of the normal contour of the skull by applying external forces. These forces may be intentional, to produce a specific shape, or unintentional, the by-product of other behavior. Deformed skulls have been reported from all over the world. In the Americas, they are known from the northwestern, southwestern, and eastern parts of the United States, Mexico, the Antilles, and northern, western, and southern South America. As might be expected, this distribution has led to the proliferation of descriptive terms and classifications.

The types of deformation employed in the New World have been grouped into five types by Stewart (1973), which are defined morphologically and functionally as follows:

1. Vertico-occipital. This is the simplest form and consists only of flattening of the lower part of the occipital (Fig. 106). It seems to be the unintentional consequence of wrapping the infant with its head pressed against a hard cradle-board. The angle and severity of the flattening probably reflects the shape of the board and the length of time the infant was kept on it. This type has also been called "cradle-board deformity" and "tabular-erect."
2. Lambdoid. This term applies to flattening that occurs higher on the occipital, near the junction with the parietals (Fig. 107). It is not clear whether it is the intentional result of tying a flat object against the back of the head or the unintentional consequence of other behavior, such as use of a cradle-board.
3. Frontal. Occasionally, only the forehead (frontal) is flattened, implying that counter pressure was not exerted against the back of the skull. This kind of deformation is generally an unintentional result of using a band (tumpline) across the forehead to support weight carried on the back. Of course, the practice would have to begin in childhood to produce frontal flattening.
4. Fronto-occipital. When pressure is applied to the front and back of the skull simultaneously, flattening occurs on both the frontal and occipital (Fig. 108). Morphologically, this type is a combination of types 1 and 3 or types 2 and 3. The terms "fronto-vertico-occipital," "fronto-parallelo-occipital," and "tabular erect" apply to the first combination.

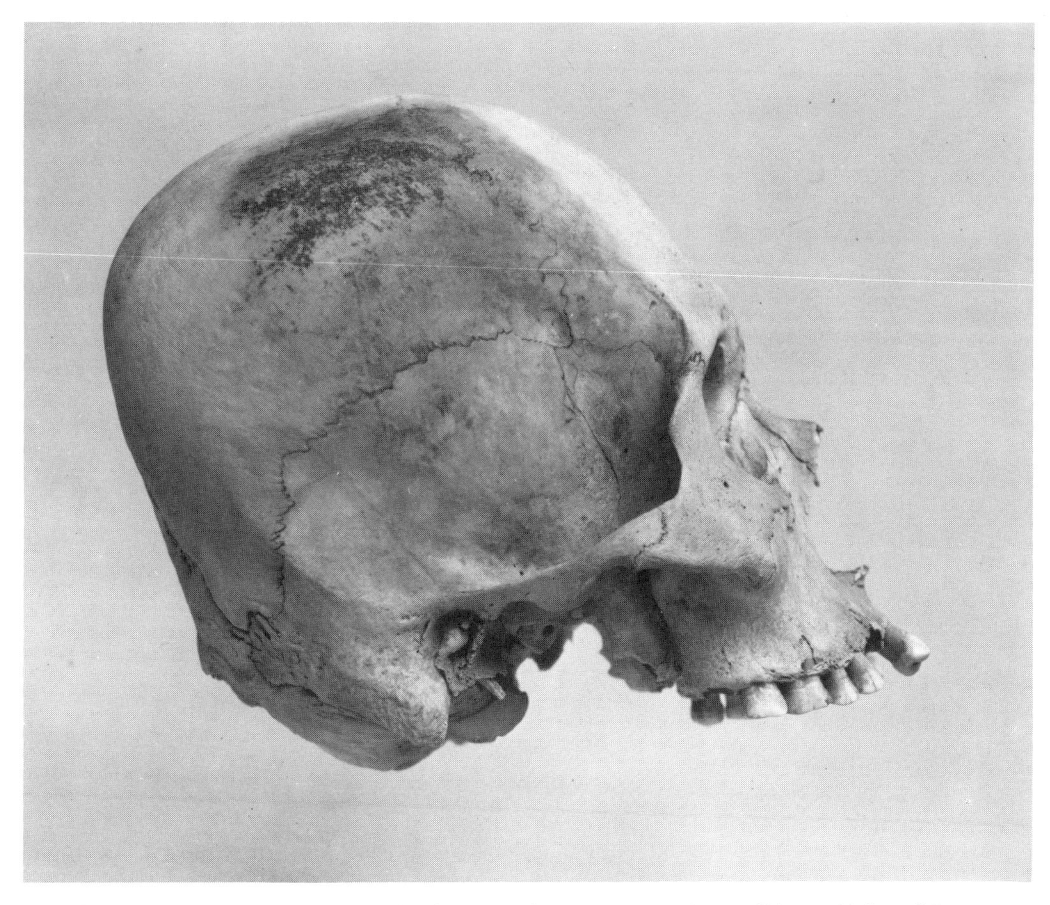

Fig. 106. A skull showing vertico-occipital deformation from a cemetery in the Chicama Valley of Peru.

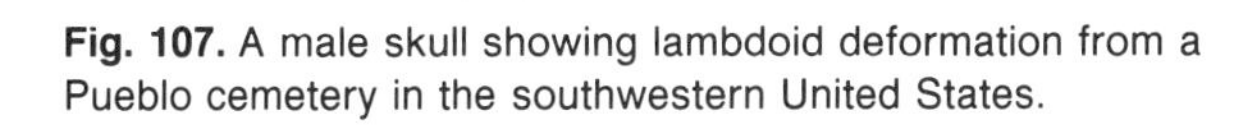

Fig. 107. A male skull showing lambdoid deformation from a Pueblo cemetery in the southwestern United States.

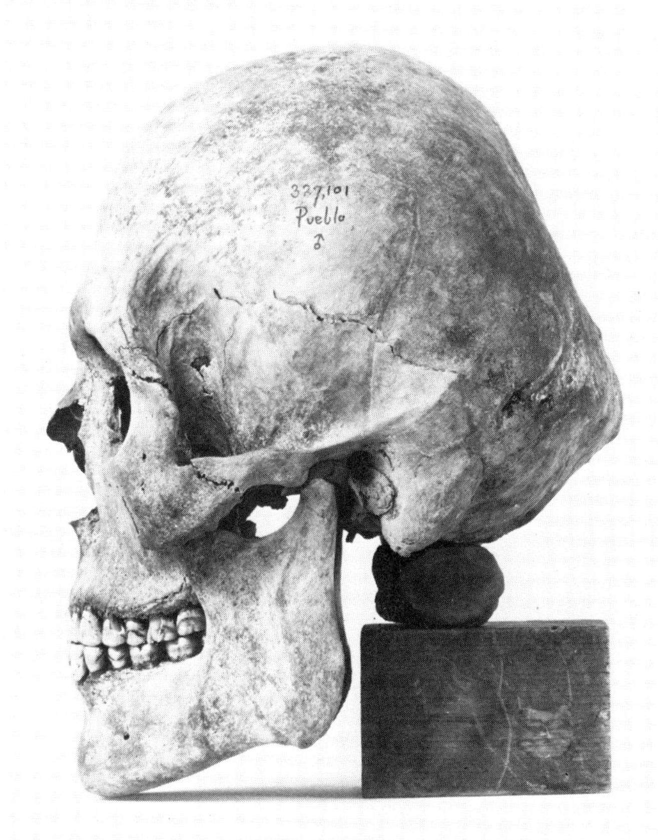

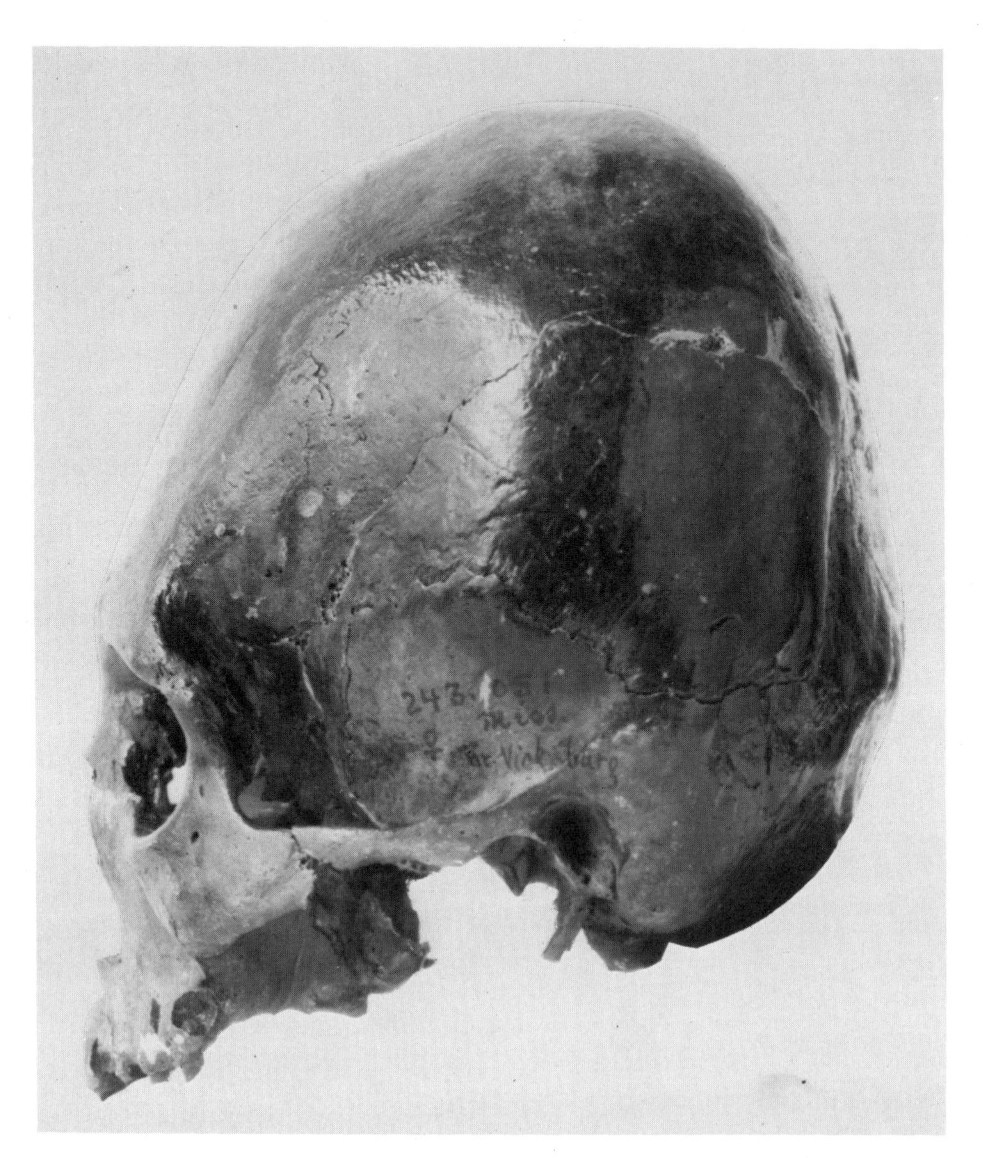

Fig. 108. A female skull showing fronto-occipital deformation from a precolumbian cemetery in Mississippi.

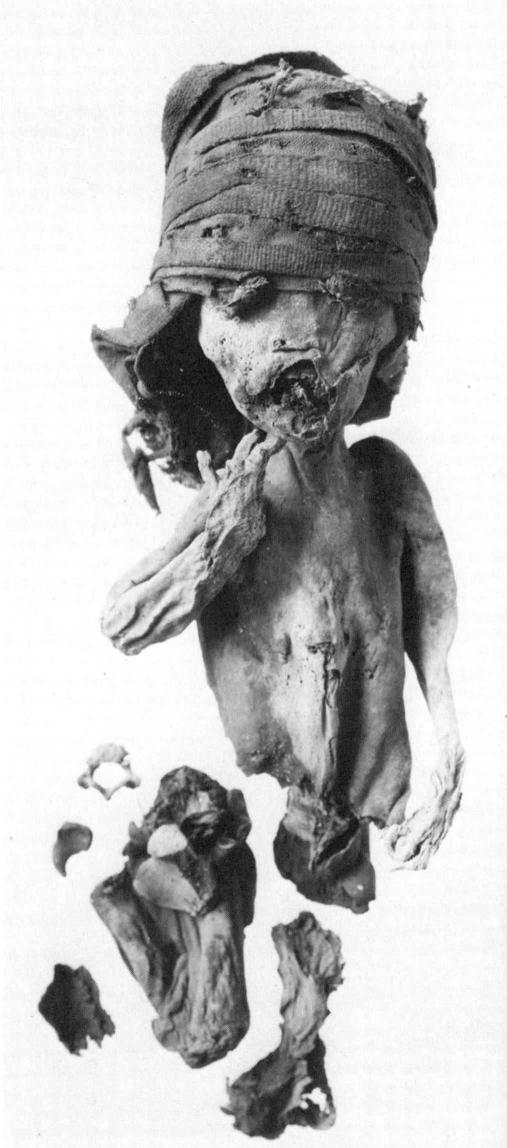

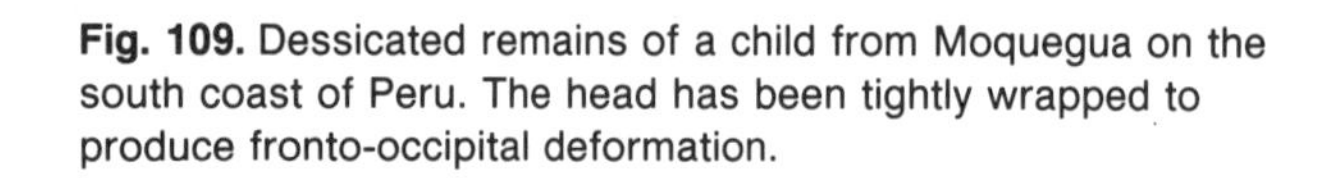

Fig. 109. Dessicated remains of a child from Moquegua on the south coast of Peru. The head has been tightly wrapped to produce fronto-occipital deformation.

"Fronto-lambdoid" or "tabular oblique" refer to the second. These types of flattening represent intentional modifications during early childhood by selective application of pressure. The most common methods are tight wrapping with cloth bands (Fig. 109) and tying small boards or similar hard, flat objects to the appropriate parts of the skull.

5. Circular. If a band (or perhaps a series of small boards) is wound tightly around the skull, it produces a circular depression around the entire circumference. This type of deformation has been called "circular," "pseudo-circular," (Stewart 1941), or "annular" (Shapiro 1928).

Dental Mutilation

People all over the world have modified their teeth by filing or inlaying them, usually for aesthetic effect. In the Americas, these practices are best known from Mexico, where filing was used to produce a variety of pointed, notched, grooved, and even asymmetrical shapes. These treatments were frequently accompanied by inlay with pyrite, jadeite or turquoise. Gold inlay has been reported from the coast of Ecuador. Most alternations occur on the incisors and canines because these teeth are the most visible.

Intentional Filing. Romero has classified the dental mutilations practiced by prehistoric Americans into 59 types, only five of which are known outside of Mesoamerica. Three of these occur in North America and two in South America. Romero recognizes three general loci of mutilation: (1) the edge of the dental crown (Fig. 110 A-C), (2) the labial (outer) surface of the crown (Fig. 110 D-E), and (3) both the edge and the labial surface (Fig. 110 F-G). Each incorporates several types and varieties of treatment (Romero 1970:52):

> In type A alterations of the occlusal edge alone are included; in B one of the angles of the crown is mutilated; in C both angles are symmetrically mutilated. Type D consists of all those cases in which straight filed lines appear on the labial surface of the crown. Type E is characterized by the presence of incrustations of pyrite, jadeite, turquoise or gold, or by beveling of nearly the whole labial surface (type E-5). Type F consists of forms in which both the occlusal edge and the angles of the crown, or both the occlusal edge and the labial face of the crown, are altered, mostly in asymmetrical form. Type G takes in all those cases in which there are incrustations, combined with alteration of the occlusal edge or of the angles, in either symmetrical or asymmetrical form.

In America north of Mexico, filed anterior teeth have been reported from several late prehistoric or early historic sites in Arizona, Illinois, and Georgia. All pertain to cultures known to have been influenced from Mesoamerica. Those from Texas (Fig. 111) may date back to the Archaic Period, however, several thousand years ago. Similarly, rare examples of incrustations and unusual patterns produced by filing from the coast of Ecuador support other cultural evidence for precolumbian contacts with Mesoamerica (Fig. 112; Ubelaker 1987a).

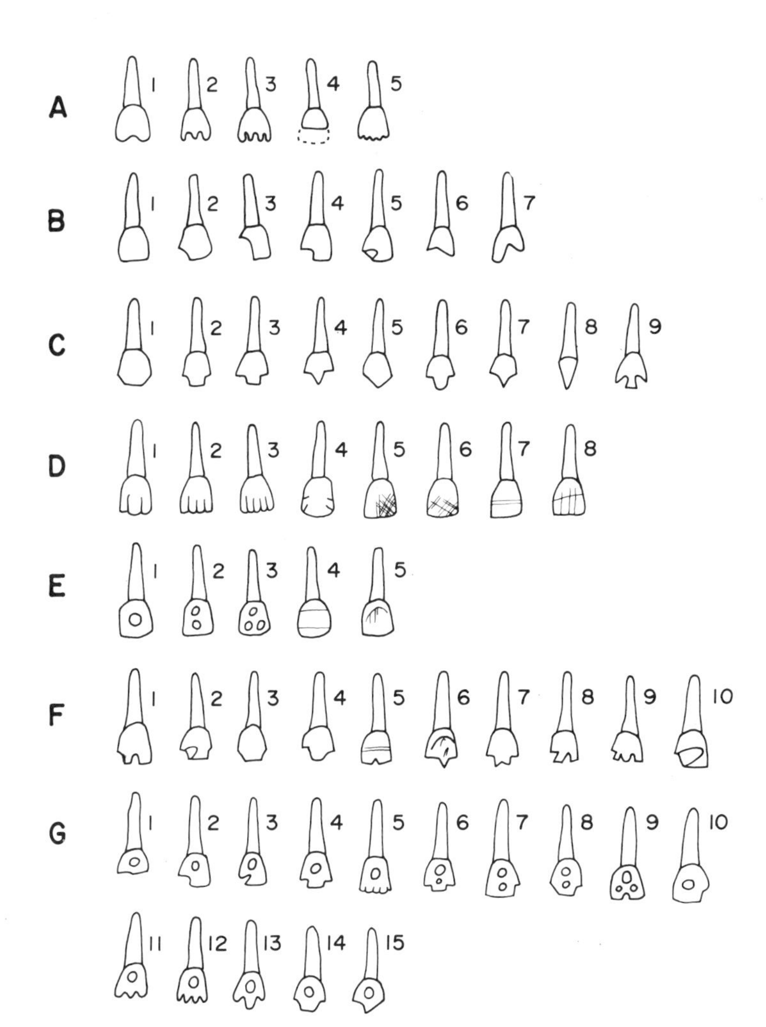

Fig. 110. Classification of types of dental mutilation employed for ornamental effect in precolumbian America. The principal classes are: A-C, modification of the edge; D-E, modification of the outer (labial) surface, and F-G, modification of both the edge and the labial surface (After Romero 1970:51).

Attrition. Intentional mutilations should not be confused with alterations of tooth morphology resulting from normal attrition, accidental fracture, disease or other factors. Mastication produces flattening of the occlusal surfaces. When attrition is extensive, and es-

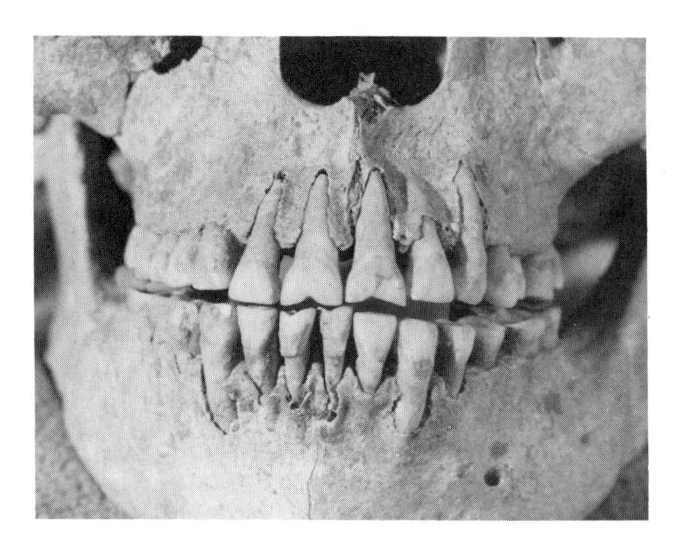

Fig. 111. Filed decoration of teeth from the Archaic Period in Texas.

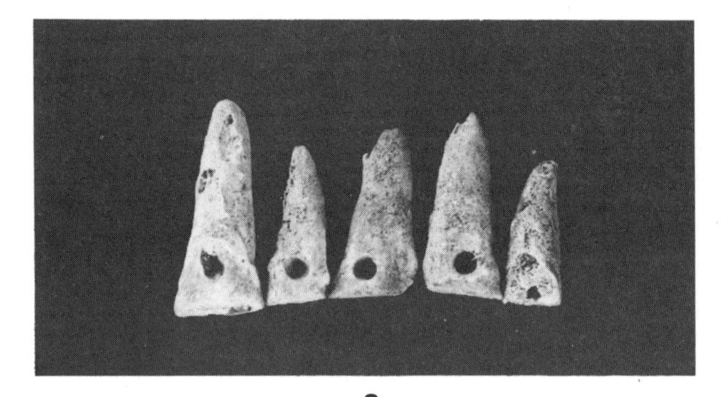

a

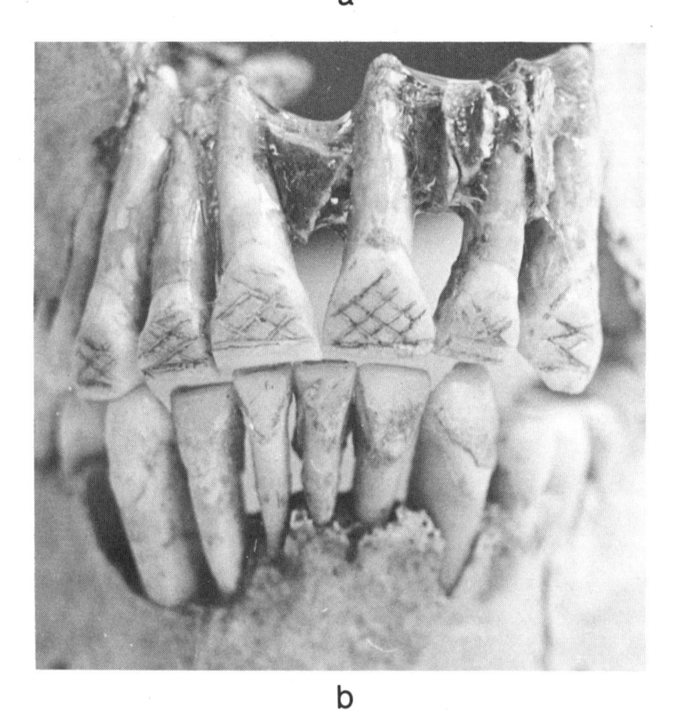

b

Fig. 112. Tooth decoration from precolumbian coastal Ecuador suggesting Mesoamerican influence. **a,** Perforations for inlay. **b,** Crosshatch.

pecially if it is accompanied by malocclusion, the wear may be distributed unevenly on the crown and resemble some of the alterations in Romero's types A-C. Examination of the occlusion permits differentiation of such attritional alterations from intentional mutilations.

Culturally Induced Alterations. Cybulski (1974) and Schultz (1977) have suggested that the occlusal grooves observed in some prehistoric dentitions may have been produced by pulling thin strings or threads through the teeth, perhaps during sewing or weaving. Cybulski also cites alterations on the anterior-occlusal surfaces of the incisors as possible indications that labrets were inserted through perforations in the lower lip.

Chipping. Turner and Cadien (1969) have called attention to natural chipping caused by abusive use of the teeth. They term the condition "pressure chipping," since it resembles morphologically the flake patterns on pressure-chipped stone artifacts. The highest frequencies occur among Eskimos, who frequently use their teeth to crush bones, chew leather, and perform other "heavy duty." Natural chipping differs from intentional mutilation in being more irregular and rough-edged.

Interproximal Grooves. Ubelaker, Phenice, and Bass (1969) observed shallow, polished grooves, usually occurring between the molars at the junction between the crown and root. The appearance and position of the grooves suggest they were produced by repeated insertion of an instrument between the teeth. The high association between grooves, carious lesions (Fig. 113), and alveolar resorption resulting from peridontal disease implies the instrument was inserted in an attempt to relieve discomfort. Wallace (1974) has proposed the more natural but less likely interpretation that the grooves result from the flow of grit particles suspended in saliva. Whatever their origin, the grooves represent modifications that should be distinguished from intentional mutilations produced for aesthetic effect.

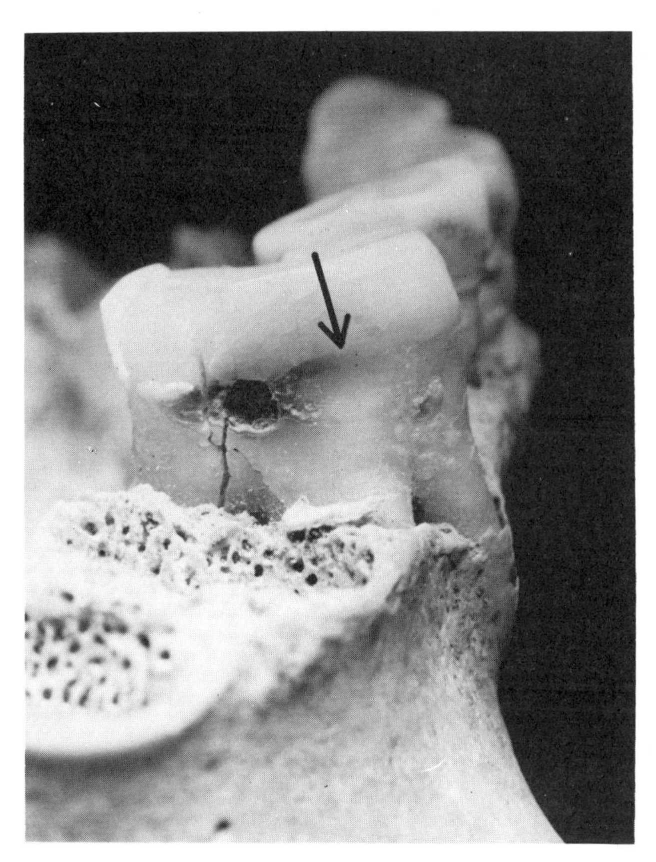

Fig. 113. A molar with a groove worn on the side at the junction between the root and the crown, associated with a carious lesion. The groove may have been produced by repeated insertion of a hard object, perhaps to alleviate pain caused by the decay.

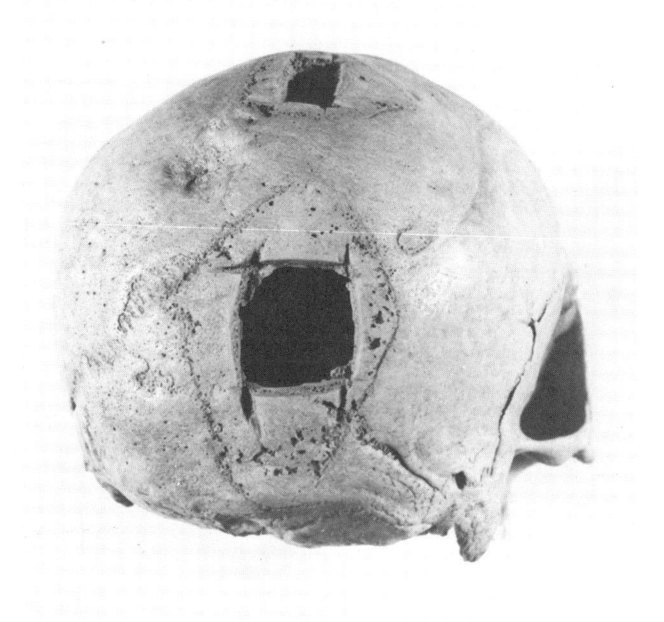

Fig. 114. Rear view of a skull from Cinco Cerros, Peru, showing two trephinations. The cut marks are clearly visible around the lower perforation.

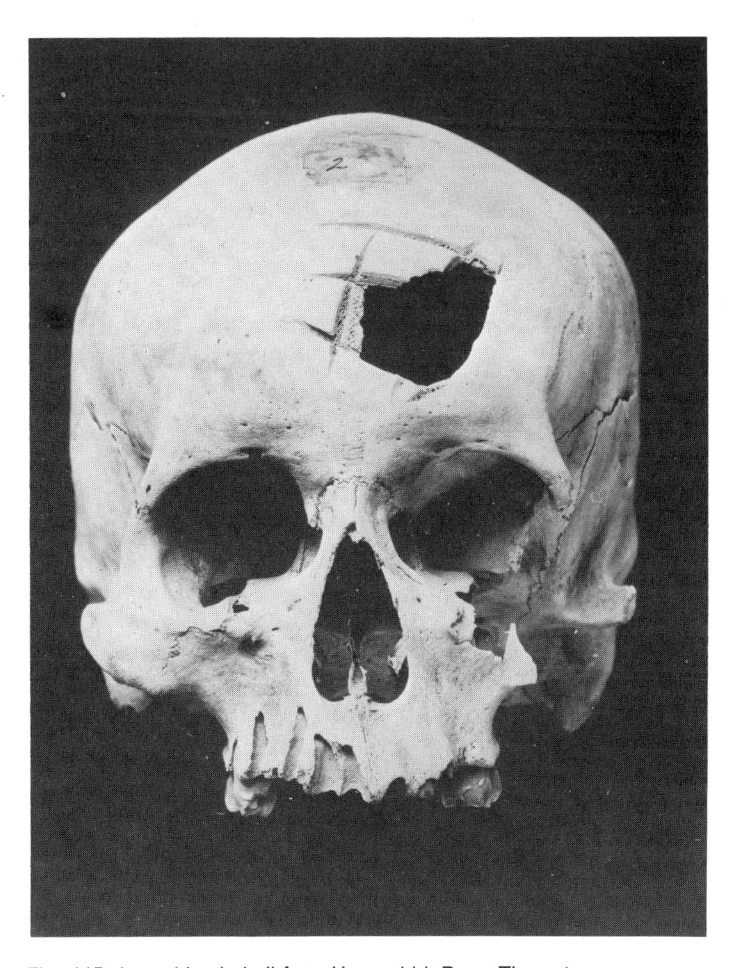

Fig. 115. A trephined skull from Huarochiri, Peru. The cut marks suggest that a larger portion was intended to be removed. Absence of regrowth suggests the patient died during the operation or soon after.

Trephination

A cultural practice affecting the vault is trephination, a surgical procedure that produces an opening through the living bone. This dramatic operation is known to have been performed in several parts of the world, including North America and Mexico, but was especially common in the south-central highlands of Peru. If completed, it produces a hole through the skull. The association between scars of trephination and cranial fractures or pathologies suggests a therapeutic purpose. If the patient died during or soon after surgery, the borders of the perforation remain sharp (Figs. 114–115). If the patient survived, however, the borders are masked by new growth (Figs. 116–117). With sufficient time, even large perforations may be completely healed over, leaving a shallow, scooped-out area marking the

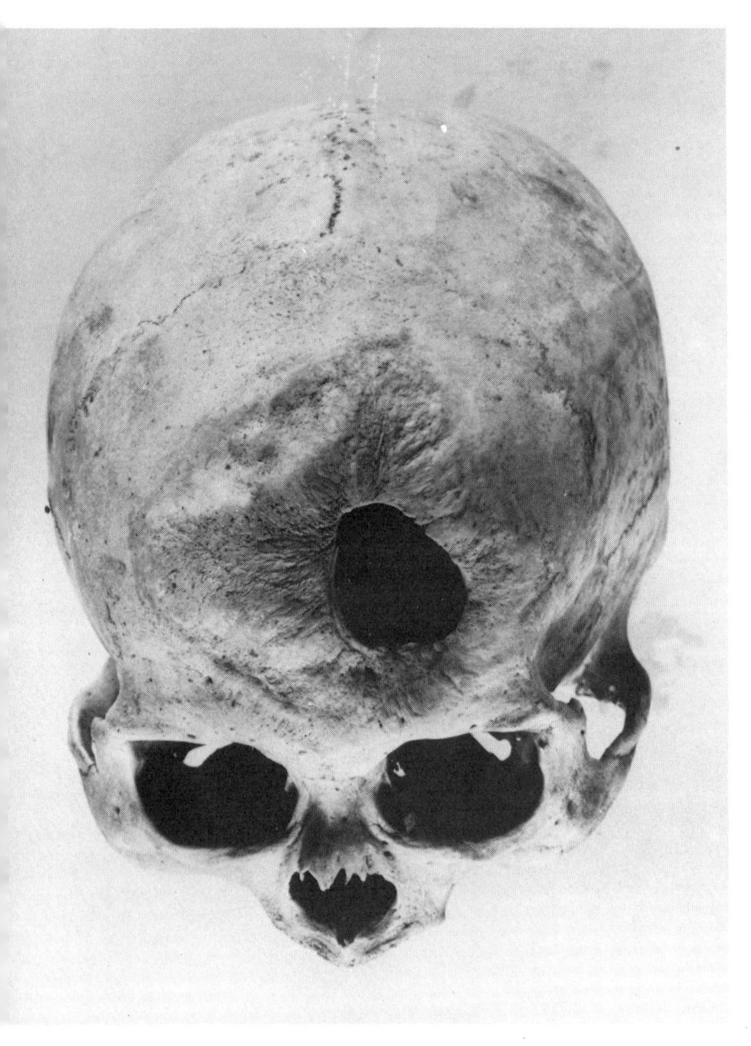

Fig. 116. A partially healed trephination on a skull from Cinco Cerros, Peru. The new growth indicates the individual survived for a considerable time after the operation was performed.

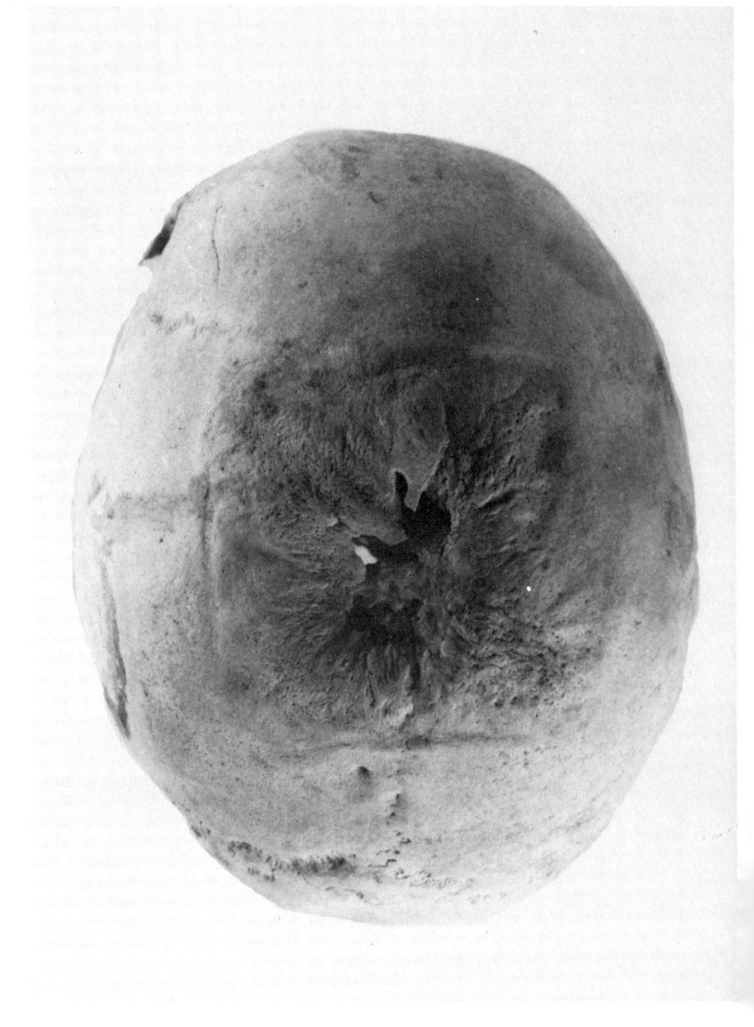

Fig. 117. A trephined skull from Cinco Cerros, Peru. The faint straight lines show where the scalp was cut to expose the bone. The perforation was produced by scraping and has nearly been closed by regrowth.

site of the trephination. Some skulls show evidence of two or more trephinations, accomplished either by cutting or scraping.

A problem in recognizing trephinations is the similarity of some types to perforations and lesions of different origins. A hole cut into a skull after death is virtually indistinguishable from one made during life. Both have sharp borders and exhibit signs of cutting or scraping around the margins. Since some peoples occasionally perforate the skulls of the dead for ceremonial purposes, mistakes in identification can occur. In addition, certain pathological processes can produce openings with sharp margins, although these lack the signs of cutting or scraping. Also, as Stewart (1975) has demonstrated, the depression at bregma for a congenital encephalocele can easily be mistaken for a partially healed trephination.

Positional Indicators

Squatting. Numerous investigators (Barnett 1954, Das 1959, Kate and Robert 1965, Morimoto 1960, Singh 1959, Thomson 1889) have suggested that modifications on certain articular surfaces of bones in the legs and feet are evidence of habitual squatting. These features appear in high frequencies on fetal bones, perhaps reflecting the flexed position of the normal fetus. In adults, they are common only in populations that spend considerable time squatting. The trait has also been observed in Rhesus monkeys who favor a squatting position (Singh 1963).

This interpretation has been contested by Trinkaus (1975), who contends that only facets on the posterior-superior femoral condyles and a groove on the femoral intercondylar line are definitely correlated with

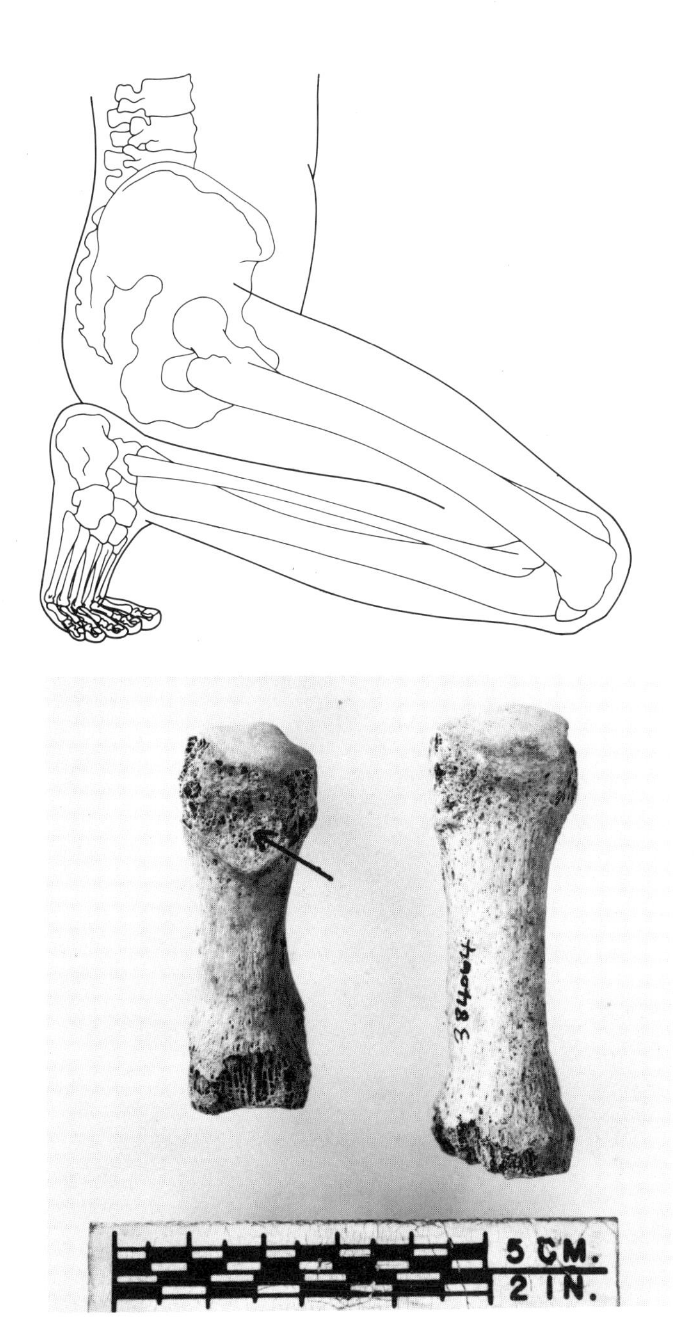

Fig. 118. Alterations of foot bones indicating habitual kneeling. **a,** Hyperdorsiflexion of the toes during kneeling. **b,** Morphology of normal (right) and modified (left) metatarsals. The modified portion is identified by the arrow.

hyperflexion of the hip during squatting. My own survey of modern and prehistoric samples showed the intercondylar area to be so variable that neither the grooves nor the shape of the line could be scored satisfactorily.

Kneeling. Alternations on the metatarsals and foot phalanges possibly caused by prolonged hyperdorsiflexion of the toes during kneeling were first noted in the sample from the Late Integration Period cemetery of Ayalán on coastal Ecuador, where they seemed to occur predominantly in females (Fig. 118). These features have been reported from both sexes in other New World populations, although in smaller frequencies. They are highly correlated with the presence of the squatting indicators (femoral condylar facets and osteochrondritic imprints) (Ubelaker 1979).

Post-mortem Modifications

Chemical Erosion. The primary cause of post-mortem modification of bone is chemical erosion. The rate and nature of the erosion depend on several factors, including temperature, soil type, soil acidity, moisture, method of burial, and even the structural-chemical condition of the bone at the time of death. Since decomposition can occur at different rates on different parts of a skeleton, it may produce effects that simulate pathological or cultural changes. In interpreting such alterations, it is important to consider the archeological context, especially the position of the skeleton.

Differences in kinds and degrees of modifications of parts of a single skeleton may assist in reconstructing differential exposure. The cranium of a forensic case from Palmara Island was so well preserved that charred deposits retained a fresh appearance years after death (Fig. 55), perhaps because of the formation of adipocere, a white soap-like material resistent to erosion. The long bones, by contrast, had extremely eroded surfaces suggesting exposure to water and sand erosion.

Mechanical Erosion. The left side of the face of the Palmara cranium shows a type of flattening sometimes termed "coffin wear," because it develops over long periods of time from small movements of the bone against a hard surface, such as the floor or side of a coffin (Fig. 119).

Sun Exposure. Bones exposed to sun for prolonged periods of time exhibit whitening and a slight increase in brittleness. The effect is usually diffuse, without sharp margins, but the difference between exposed and unexposed surfaces can be dramatic. The pattern may permit reconstructing the position of a bone during the period of exposure.

Unusual situations sometimes produce interesting patterns of sun-induced alterations. The cranium

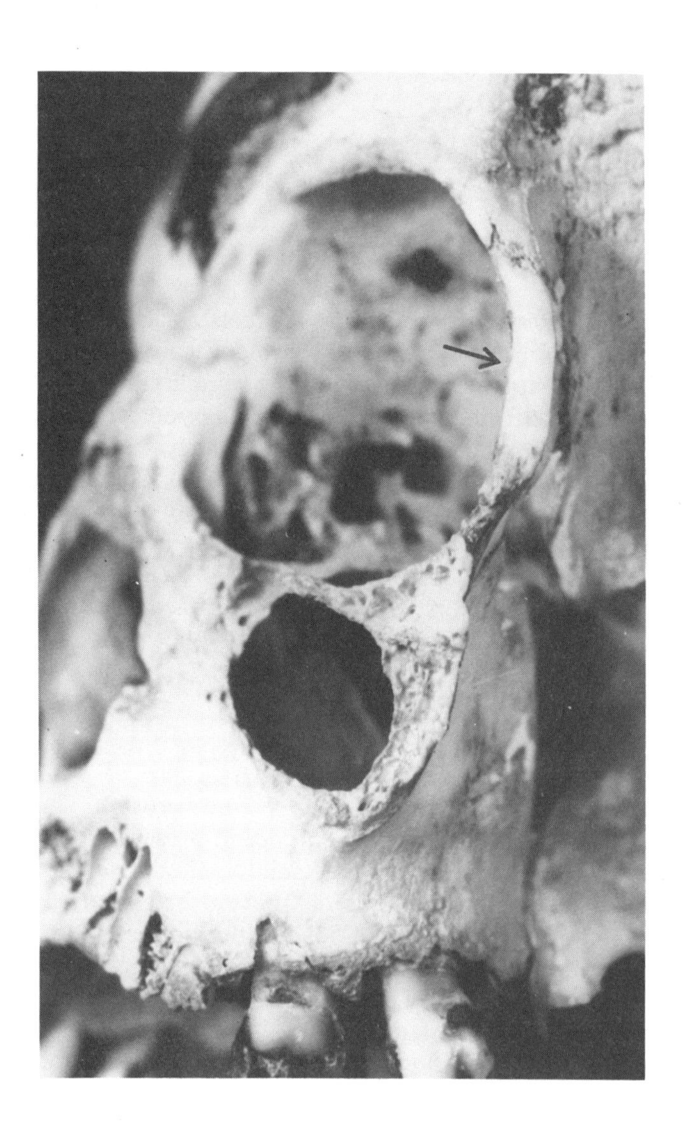

Fig. 119. Mechanical erosion on the left margin of the orbit produced by rubbing against a hard surface.

of a forensic case from Omaha NB exhibited large white spots on the upper frontal and parietal areas (Fig. 120). These contrasted dramatically with the surrounding darker bone. The vault showed no erosion, although the facial area was severely damaged. The remains had lain in an unused cistern for nearly nine years prior to discovery. When found, the cranium was resting on its base directly below a manhole cover that had several circular perforations. On sunny days, these allowed small beams of light to strike the cranium. Because of the thickness of the cover, the light was diffused except for a brief period around midday, when the sun was directly overhead. During the nine-year interval, this exposure produced the color modifications.

Marine Exposure. Salt water can cause a bleaching effect. Marine exposure usually also leaves other indicators, such as deposits of algae or barnacles.

Animal Activity. Burrowing insects and rodents may leave misleading marks on skeletal remains. The internal surface of a skull from the Dominican Republic shows lesions that could easily be mistaken as pathological (Fig. 121). However, the specimen was removed from sandy soil, where insects and mollusks frequently produce such modifications. Fragments of

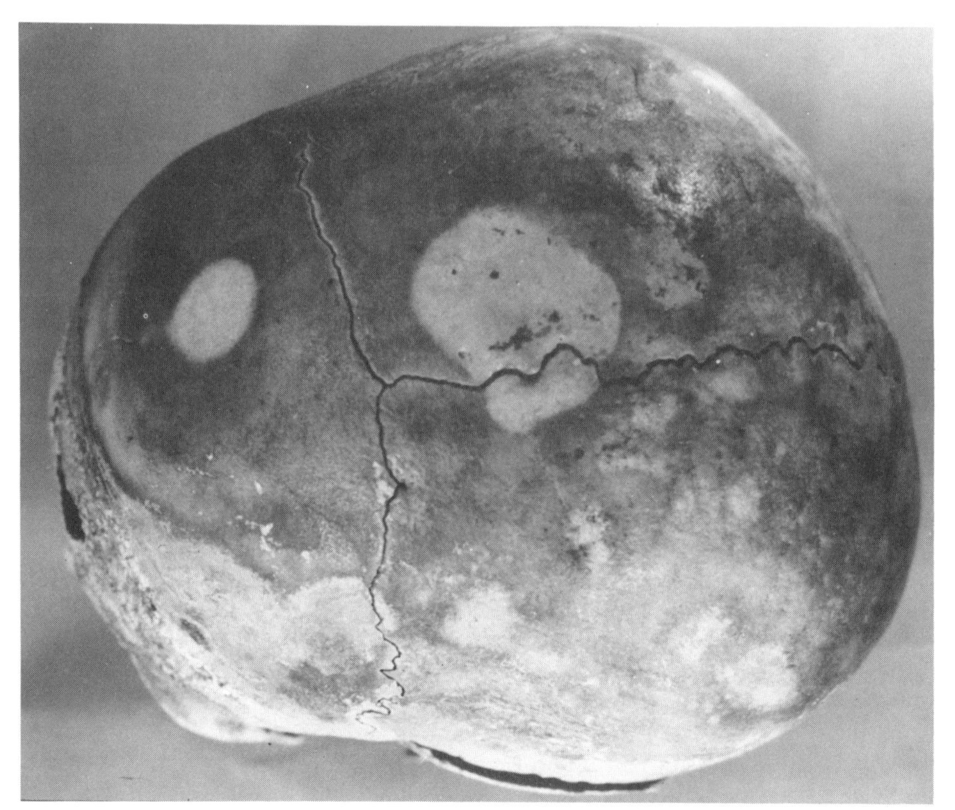

Fig. 120. Bleached spots on the vault of a cranium produced by localized, long-term exposure to sunlight.

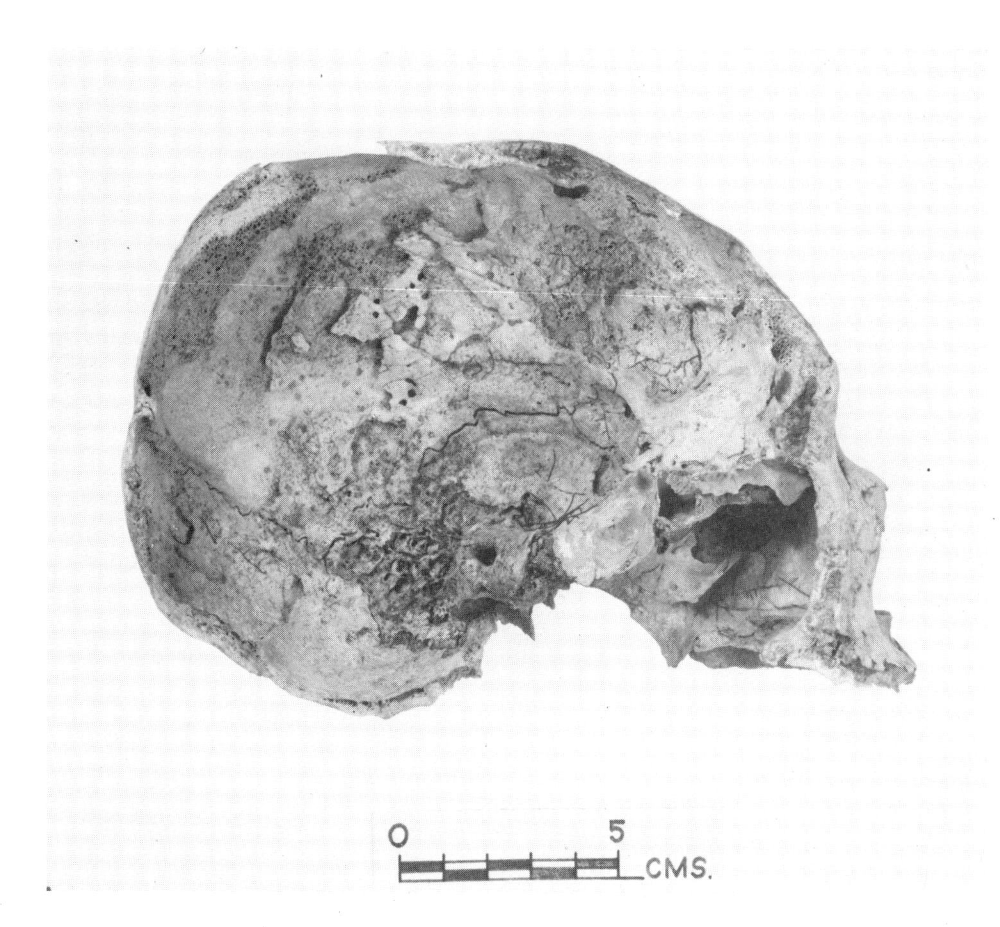

Fig. 121. Interior of the vault of a cranium exhibiting post-mortem deterioration that might be mistaken for pathology. Observations of the soil and other aspects of the context of a burial are important for explaining conditions of this kind.

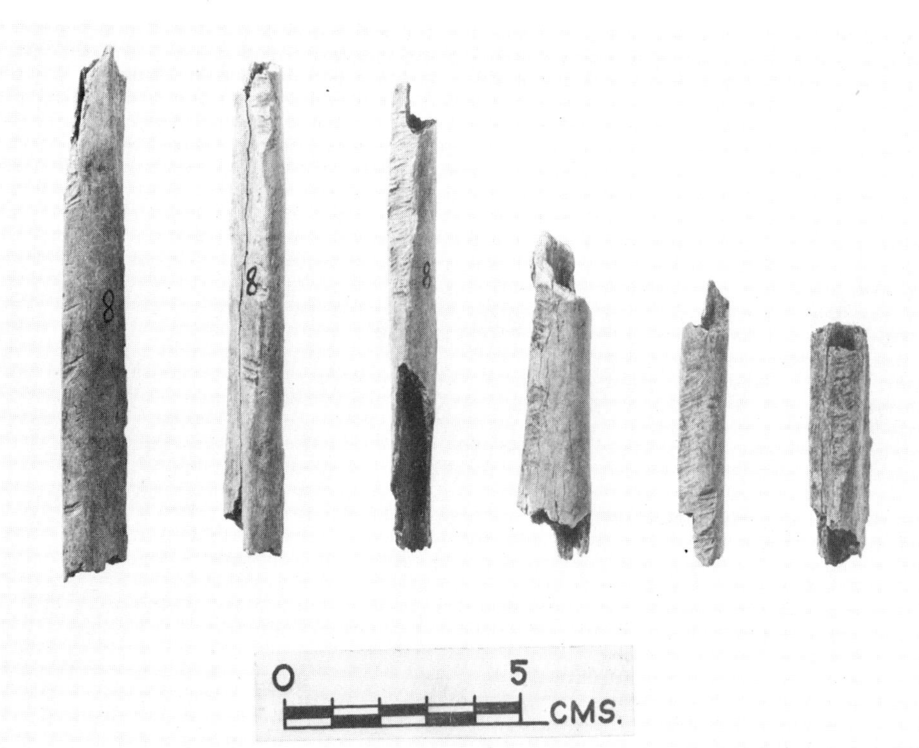

Fig. 122. Tooth marks left by rodents, a post-mortem modification that should not be confused with marks produced as a by-product of cultural practices.

the shafts of long bones in Ecuadorian burial urns have irregular, linear notches cut by rodents seeking calcium or protein residues in the bone (Fig. 122). Such marks, which commonly occur on skeletal remains that were either exposed on the surface or shallowly buried, should not be confused with knife marks. Cuts made by a knife or other sharp instrument are usually located near the joints, since they were produced during intentional disarticulation or defleshing. They have sharper borders than marks left by rodent teeth, and are narrower, straighter, and more widely spaced (Figs. 123–124).

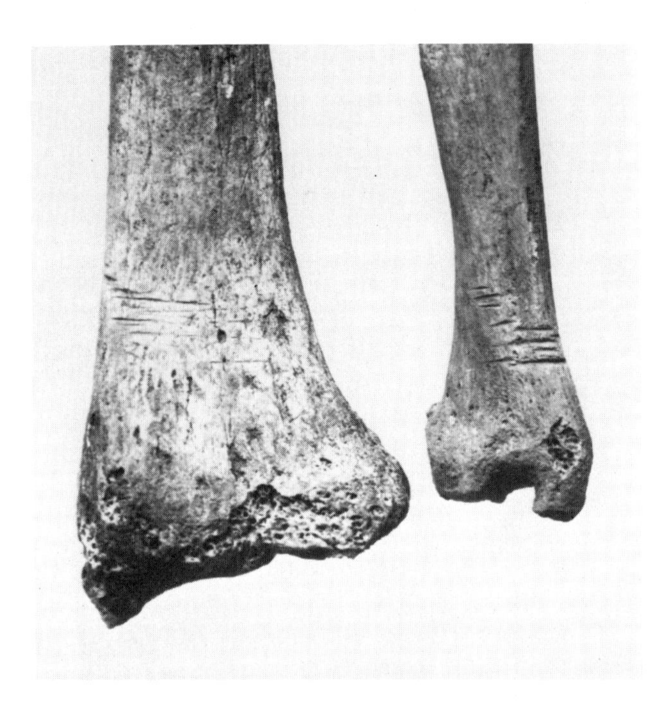

Large mammals, such as dogs, coyotes, and wolves, tend to gnaw on the ends of long bones, destroying the articular surfaces and epiphyses. Bears tend to break the diaphyses of long bones and may perforate bones with their canines, leaving diagnostic marks (Fig. 125; Murad and Boddy 1987).

Ceremonial Alteration. A less frequent but widespread cause of post-mortem bone modification is for ceremonial use. Skulls have been mounted on sticks (Vignati 1930), hung in houses, painted (Hrdlicka 1905), and even employed as containers. When a skull has been altered intentionally, either for such a purpose or as a result of use, the consequences may resemble trephinations and natural erosion. For exam-

Fig. 123. Cut marks on the distal ends of a radius and an ulna from the Potomac Creek site in Virginia, probably produced during intentional disarticulation or defleshing of the body.

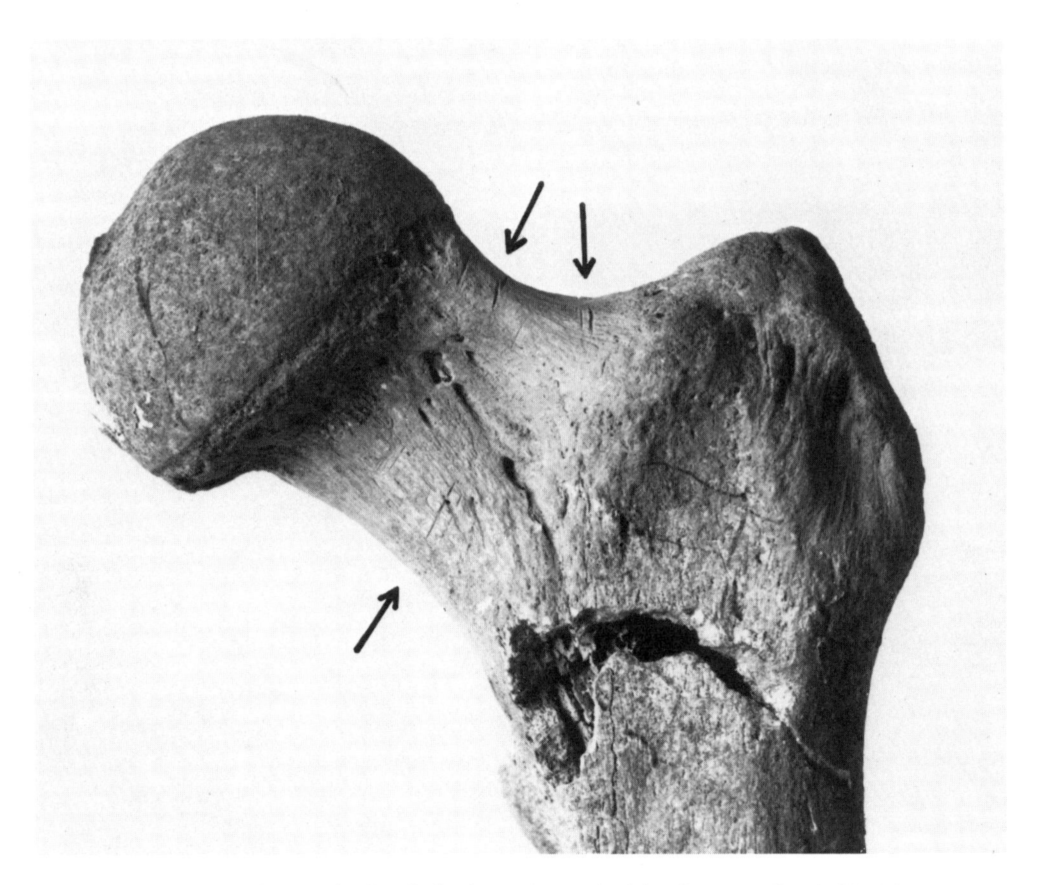

Fig. 124. Cut marks on the proximal end of a femur from a burial at Potomac Creek, Virginia suggesting purposeful disarticulation of the body as part of the mortuary procedure.

ple, Phenice (1969b) found eight skulls in a large series from Kansas that exhibited perforations in the occipital. The well defined margins argued against healed trephination as the explanation, and this was confirmed by x-rays showing absence of a dense margin around the holes. The presence of polishing on the edge eliminated the possibility of unsuccessful trephination and suggested that the skulls were used ceremonially. Phenice inferred that the perforations were made so the skulls could be placed on pegs in the houses for display.

Gillman (1875) reported several crania from Michigan sites with perforations located on the sagittal suture, between the parietals. This central location and the clean-cut margins suggested the cuts were made after death, perhaps to permit insertion of a string or rope by which the skulls could be hung. He referred to the following note from a correspondant: "It is an interesting coincidence that the head hunting Dyaks of Borneo have a house in the center of their village, in an upper story of which they keep the heads which they capture suspended by a string which passes through a perforation in the top of the skull" (Gillman 1875:238–9).

Hooton (1922:124) described six skulls from Ohio displaying nearly circular holes on the upper portion of the cranial vault. His interpretation that they were drilled post-mortem, probably with a stone drill, was supported by Willoughby (1922:61), who wrote: "the position of the holes seems to indicate that at least a part of them were intended for the passage of a suspending cord. Others may have been used for the insertion of feathers or other decorations."

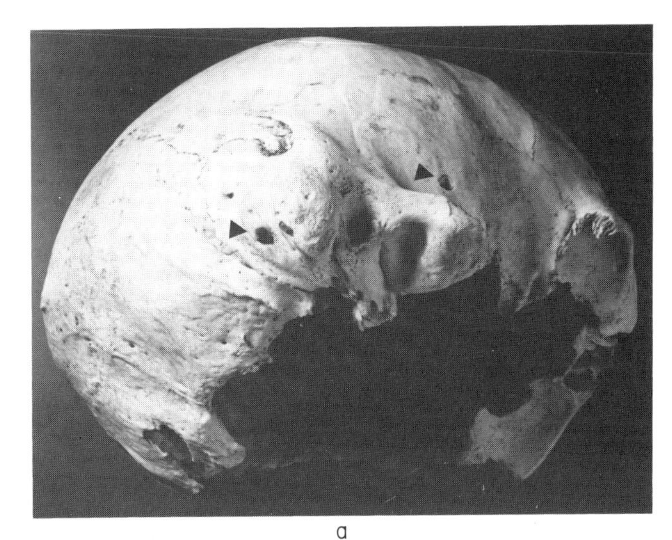

a

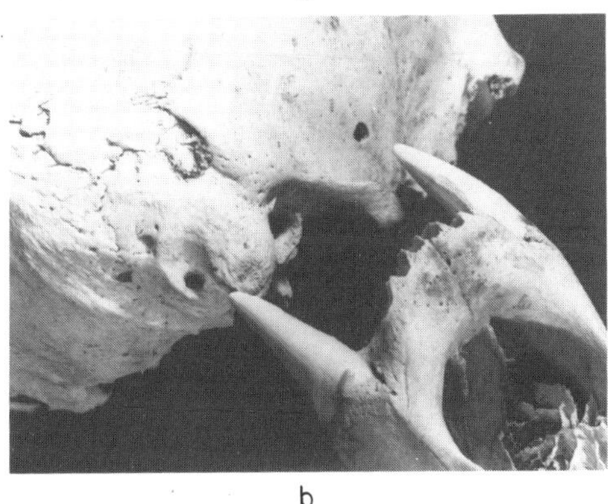

b

Fig. 125. Damage by carnivores. **a,** Paired perforations in a cranium. **b,** Matching separation of the upper canines of a bear (Courtesy of Mark Skinner).

PATHOLOGICAL ALTERATIONS

Numerous disease processes leave their mark on bone and thus provide the opportunity to learn something about the illnesses and health problems of prehistoric peoples. Unfortunately, most diseases do not affect the skeleton, so the knowledge we can acquire is incomplete. Identification of the surviving evidence is complicated by the fact that different pathological conditions may affect bone in a similar manner. Changes in the joints that appear to represent simple osteoarthritis may result from trauma, osteomyelitis, gout, hemophilia, psoriasis or other causes. A tibial lesion may exhibit the pattern expected of syphilis, but actually be the effect of *Staphylococcus aureus* osteomyelitis, Paget's disease or some other condition. Even simple peridontal disease with accompanying alveolar resorption may be a secondary expression of dietary deficiencies rather than a direct result of poor hygiene. Much of the confusion in the literature on paleopathology is the consequence of attempts at diagnosis by non-specialists. We should reserve such judgments to the relatively few persons qualified to make them because of their thorough knowledge of the disease process and its effect on bone. These individuals do not include all pathologists, and certainly not all physical anthropologists.

Although speculation on the causes should be avoided, it is important to recognize an abnormal condition and bring it to the attention of a specialist. Ten categories of disorders affect bone and may be encountered in archeological remains. In the order of their frequency of occurrence, they are: (1) arthritis, (2) fractures, (3) infections, (4) congenital disorders, (5) circulatory disturbances, (6) tumors, (7) metabolic disorders, (8) endocrine disorders, (9) diseases of blood-forming tissue, and (10) miscellaneous diseases.

Arthritis

The most common pathological condition is arthritis, produced by a gradual breakdown of the cartilage between the adjoining bones of a joint so that the articular surfaces come into direct contact. A 1975 classification of arthritic disorders (Table 25) includes many varieties that cannot be diagnosed reliably from skeletal remains. Their diversity serves to emphasize the complexity of the problem of identification and the probability of erroneous diagnoses by an untrained observer.

The most frequent form of arthritis is osteoarthritis, the term applied to degenerative changes during the normal process of aging. Osteoarthritis is a gradual alteration of the articular cartilage and articular surfaces of the bone as a consequence of long-term mechanical stress, repeated minor irritation of the cartilage, or disruption of circulation of the blood to the area. Its most common expression, a build-up of osteophytes (lipping) along the margins of the vertebral centra, was discussed as a method of estimating age at death (Fig. 94). Although usually considered as pathology, osteophytic development is normal unless it occurs prematurely.

Similar kinds of changes occur on most other joint surfaces as normal aging phenomena. Arthritic changes of the elbow begin as small bony nodules or pits (Fig. 98 b-c). Later, the nodules and pits enlarge and some lipping develops on the margins of the articular surfaces. Eventually, the cartilage in the joint is destroyed, allowing the bones to rub together. The friction produced by movement rapidly causes polishing of the areas in contact. This effect, called "eburnation" (Fig. 98 d), is usually accompanied by advanced stages of lipping and pitting (Ortner 1968).

The same progression from slight pitting and lipping to marked lipping and eburnation occurs in most of the other joints, although the rates are different. The greater severity of the condition at one joint may indicate differences in stress, which can serve as clues to occupations or other activities. The age when general arthritic changes begin can be an indicator of the intensity of physical activity. For example, Ortner (1968) found that degenerative arthritis of the elbow is more frequent among Eskimos than Peruvian Indians, perhaps reflecting different subsistence practices. Roche (1957) has shown that osteophytosis is more common in Whites than in Blacks, and Stewart (1947) reported more lipping in the cervical region in Whites than in Eskimos and Pueblo Indians, but the significance of these differences has not been ascertained.

Although most of the arthritic changes observed in archeological remains represent osteoarthritis, other causes can sometimes be identified. Excessive trauma to joints, especially lacerations and dislocations, can produce disruption of circulation, destruction of cartilage, and ultimately changes in the bone that are virtually identical to those attributed to osteoarthritis. They are usually confined to one joint, whereas osteoarthritic symptoms affect the whole skeleton. Isolated traumatic events may also result in fractures, infections or other problems, which can be detected

Table 25. Classification of the varieties of arthritis (after Aegerter and Kirkpatrick 1975:623).

I. Secondary arthritis
  - Gout and pseudogout
  - Alkaptonuric arthritis
  - Hemophilic joint disease
  - Psoriatic arthritis

II. Degenerative joint disease
  - Osteoarthritis and arthritis of chronic trauma
  - Chondramalacia
  - Neurogenic arthropathy and congenital indifference to pain
  - Cortisone arthropathy

III. Villonodular synovitis

IV. Infectious arthritis
  - Suppurative arthritis
  - Nonsuppurative arthritis

V. Hypersensitivity (collagen disease) arthritides
  - Rheumatoid arthritis
  - Arthritis of the collagen diseases

VI. Miscellaneous types of arthritis
  - Intermittent hydrarthrosis
  - Palindromic rheumatism
  - Arthritis mutilans
  - Cysts of the menisci
  - Baker's cyst

VII. Diseases of periarticular tissues
  - Fibrositis
  - Bursitis
  - Ganglia
  - Synovial chondrometaplasia

with radiograms if they are not evident from macroscopic examination.

Other causes of arthritic change have rarely been identified in archeological samples and can be diagnosed reliably only by experts. Secondary arthritis is a reflection of several generalized diseases. In the case of gout, a congenital imbalance in uric-acid metabolism leads to the deposition of uric-acid crystals in the joints. These crystals cause inflammation and ultimately destruction of the bones. Pseudogout (Chondro calcinosis) is calcification of the articular plate and associated cartilage by the deposition of calcium salts. Hemophilia can produce arthritis as a by-product of defects in the mechanism of clotting of the blood. Aegerter and Kirkpatrick (1975) suggest that a minor trauma invokes bleeding into the joint cavity, affecting the maintenance of the cartilage plate. The initial result is pitting on the articular surface of the bone, but destruction grows progressively worse. Although psoriasis is usually thought of as a skin disease, it too can create arthritic lesions when the inflammatory reaction occurs in the joints.

In summary, any condition that disrupts the normal metabolism of the joint can produce arthritic changes in the bones. These effects can be produced by normal activity (osteoarthritis), trauma, infection (tuberculosis, fungus, virus, syphilis, bacteria), diseases of collagen (rheumatoid arthritis), and a number of other causes. Although diagnosis should be left to experts, it is essential to describe the location and severity of arthritic changes throughout the skeleton.

Fractures

A fracture is a disruption of the normal structure of a bone, usually resulting from trauma. This disruption can be restricted to part of the cortex or it can completely sever the bone. In either case, fractures are usually accompanied by laceration of the adjacent soft tissue and blood vessels. Infection frequently results and complicates the process of repair. In addition, disruption of the supply of blood can lead to necrosis (physiological death) of all or part of the bone tissue surrounding the fracture.

Repair Process. The process of repair or healing varies with the complexity of the fracture, the constitution of the individual, and several other factors. Generally, however, the following sequence occurs:

1. The supply of blood is disrupted, causing immediate death of the bone adjacent to the fracture. Simultaneously, blood coagulates in the surrounding tissues.
2. A fracture callus forms around the site of the injury, bridging the separated parts and affording support while permanent bone is being replaced. This callus begins to develop about 16 days after a fracture occurs and is complete after about 30 days.
3. The callus is slowly converted to permanent bone by a process of remodeling. Years may be required to restore the original appearance.

Recent fractures can usually be recognized immediately by anyone familiar with the morphology of normal bone. Healed fractures may not be obvious, since the bone attempts to reassume its original shape; a radiogram will usually show the characteristic disruption or at least an irregularity in the configuration of the cortex, however. The most obvious fractures are those in which the parts were not alligned properly during healing. This is most likely to occur if the break is complete because the muscles tend to pull the ends out of place (Figs. 126–127). This not only retards healing, but shortens and disfigures the limb considerably.

Pathological Fractures. Pathological fractures are produced when normal stress is applied to bones weakened by disease, malnutrition or other causes. The repair process is about the same as in traumatically induced fractures.

Traumatic Fractures. Most fractures result from trauma and thus offer information about accidental injuries or perhaps intentional violence. A high incidence of fractures of the distal ends of the radius and ulna, known as Colles fractures (Fig. 128), may indicate frequent falls. Broken forearms are often a sign of warfare, being produced when the arms were raised for defense against blows. Stress fractures, in which separation is incomplete, are associated with excessive physical activity, especially during childhood and adolescence.

Bony alterations induced by impact usually reflect the instrument that produced them. A blunt object, such as a rock, causes general fractures and a "pushed-in" effect (Fig. 129).

Spondylolysis. A defect in the neural arch of a vertebra, this condition can lead to separation from the vertebral body on one side or a completely unattached arch. It usually occurs in the lumbar region and varies in frequency among different populations (Stewart 1931, 1935).

Sharp Objects. Knife cuts are usually encountered on the border of a rib or adjacent bone in forensic cases. The incisions are often slight and can be over-

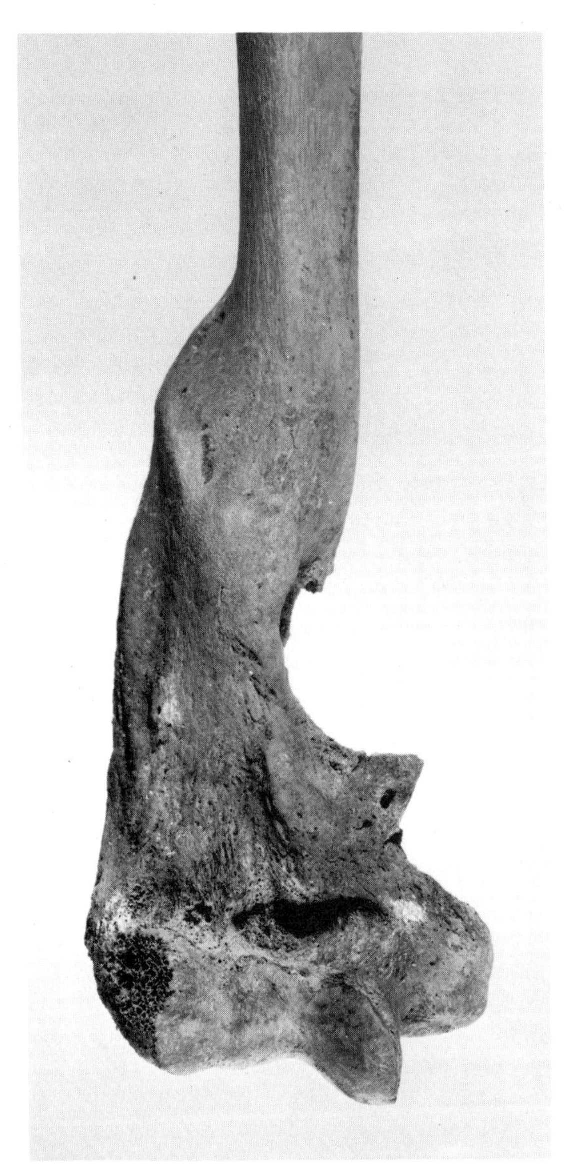

Fig. 126. Healed fracture of a humerus from a skeleton encountered at Pueblo Bonito, New Mexico. The broken ends were pulled by the muscles so they slightly overlapped, shortening the upper arm.

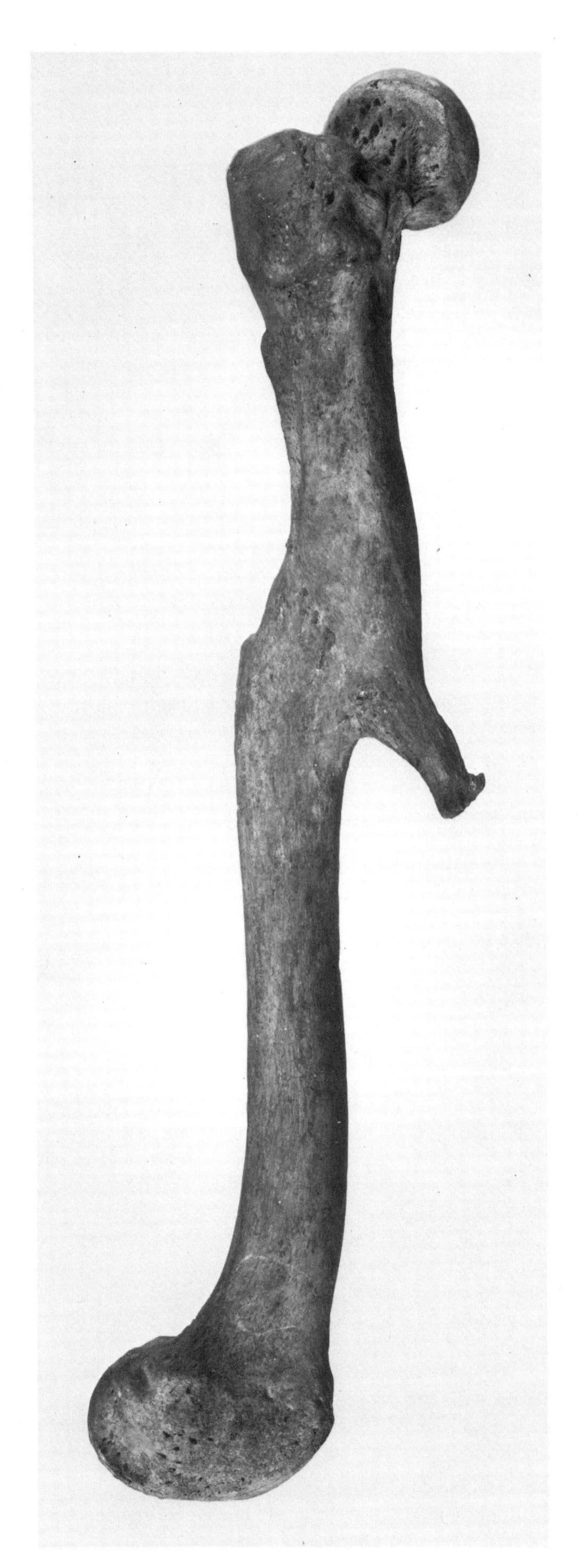

Fig. 127. Healed fracture of a femur from a skeleton encountered at Pueblo Bonito, New Mexico. The severity of the displacement resulting from the break is indicated by the projection, which was not incorporated into the reunited shaft. The leg was not only shortened considerably, but probably also disfigured.

Fig. 128. Healed Colles fracture (dark band) at the distal end of the radius from a skeleton from the cemetery of Ayalan on the coast of Ecuador. This type of fracture often results from a fall.

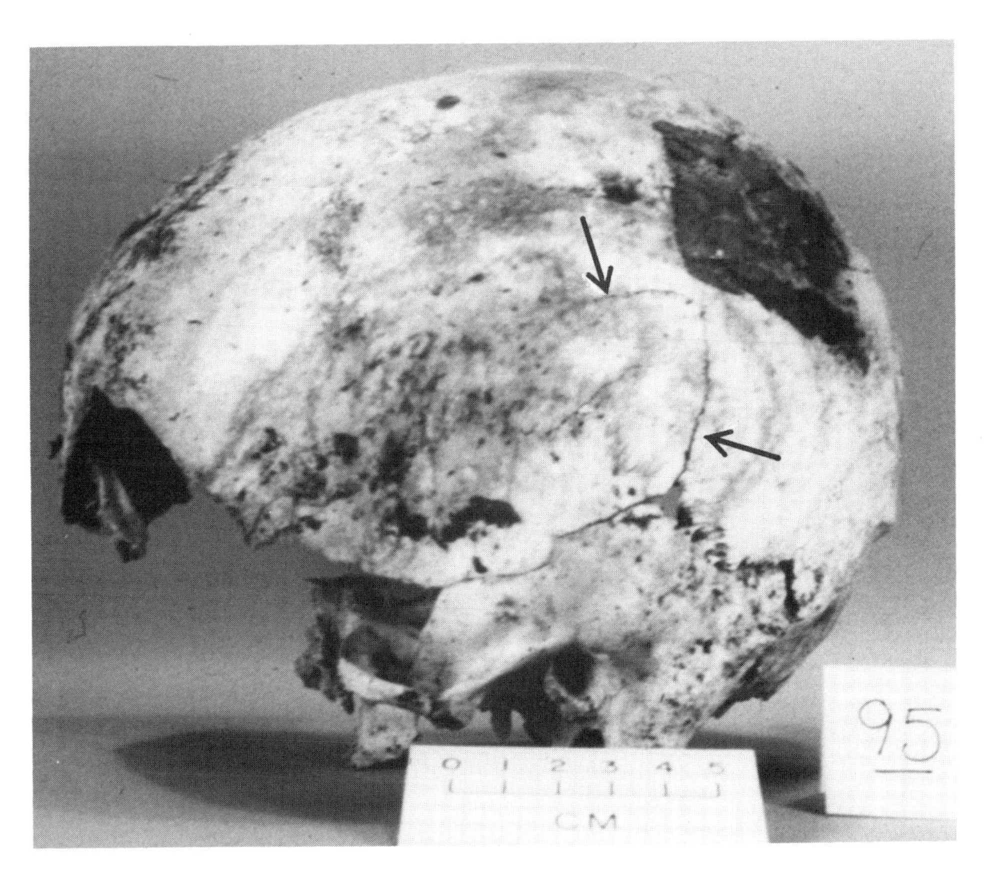

Fig. 129. "Pushed-in" area on the left side of a cranium, diagnostic of impact with a blunt instrument.

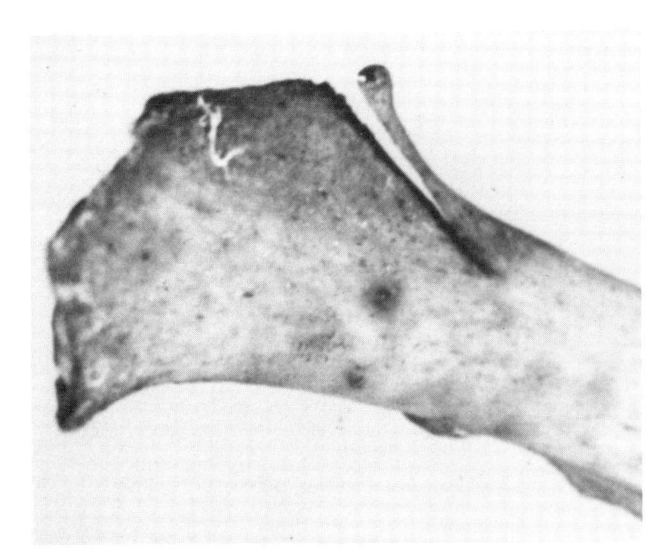

Fig. 130. Outward bending of a sliver of bone, signifying the cut was inflicted at or about time of death.

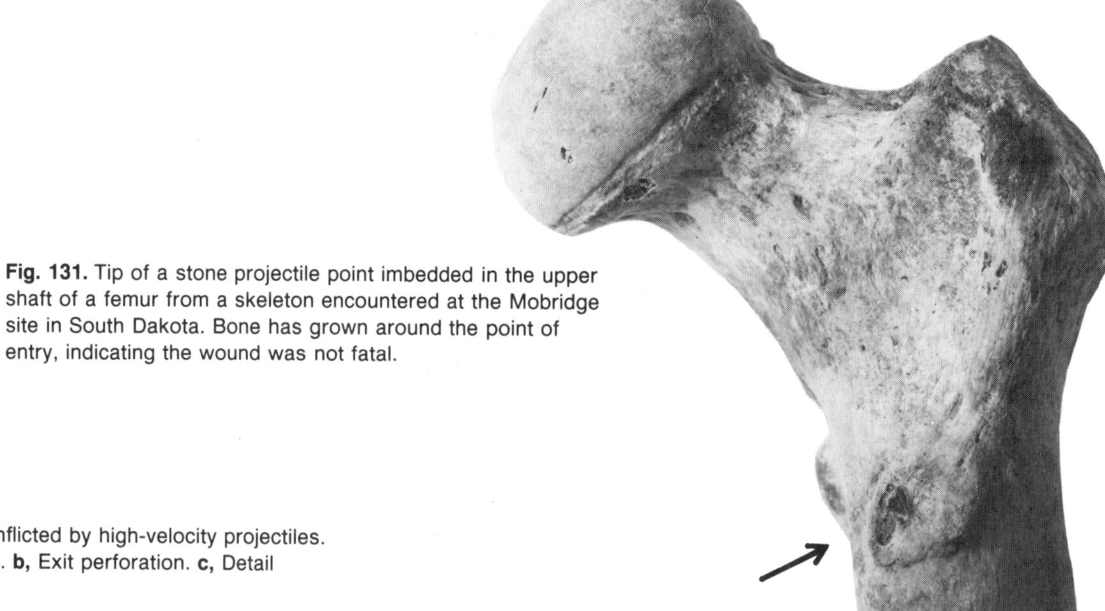

Fig. 131. Tip of a stone projectile point imbedded in the upper shaft of a femur from a skeleton encountered at the Mobridge site in South Dakota. Bone has grown around the point of entry, indicating the wound was not fatal.

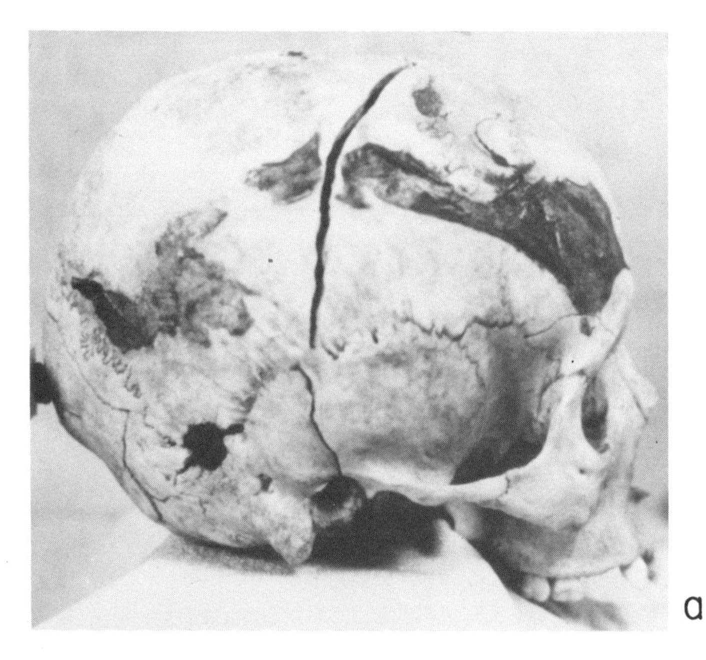

Fig. 132. Damage inflicted by high-velocity projectiles. **a,** Entry perforation. **b,** Exit perforation. **c,** Detail of exit perforation.

looked if the soft tissue has not been removed completely. Living bone has elastic properties and outward bending of a sliver partially severed by a knife usually indicates the trauma was inflicted at or about the time of death (Fig. 130). The same force and instrument applied to dry bone would cause it to fracture.

Projectiles. Occasionally, traumatic effects of cultural events take the form of projectiles imbedded in bone. Such features are direct evidence of violence and sometimes of cause of death. If the projectile is distinctive, its origin may be recognizable. The tip of a projectile point was observed in the upper shaft of a femur from the Mobridge site in South Dakota (Fig.

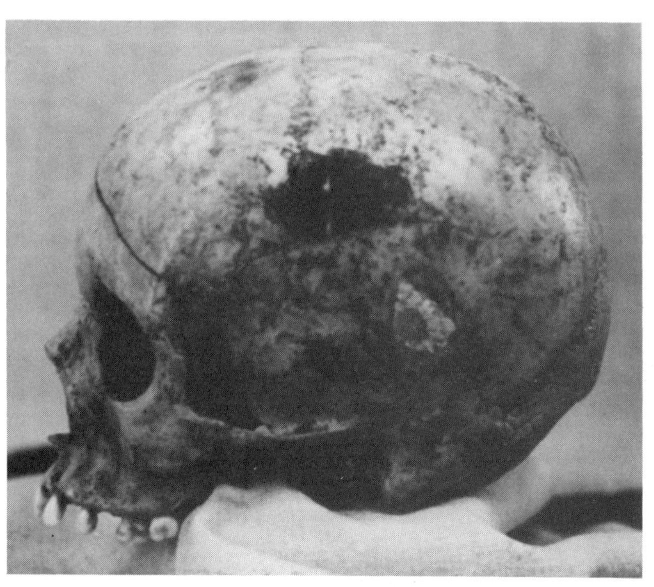

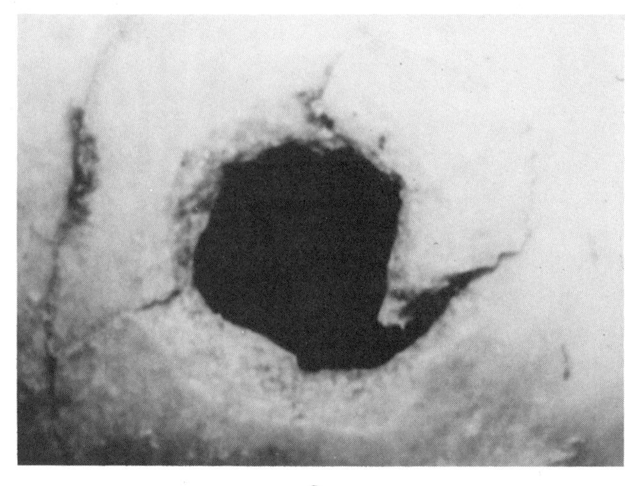

131). The bone displays considerable remodeling around the zone of penetration, indicating that the individual did not die immediately from the wound.

High-velocity projectiles may create perforations and bone loss, as well as traumatic fractures. The cranium of a young woman shot at close range has an entrance wound that approximates the diameter of the bullet and associated extensive fracture lines (Fig. 132a). The edges of the entrance perforation are beveled, making the opening slightly larger on the interior than on the exterior. The exit performation is usually much larger and shows the opposite beveling, the exterior opening being larger than the interior opening (Fig. 132b-c).

Dislocations. Trauma may force a bone out of normal articulation in a joint. If it remains out of place long enough, characteristic changes may occur in the morphology or new articular surfaces may be formed. In one skeleton from the Mobridge site in South Dakota, the head of the femur was displaced above its normal articulation with the acetabulum (Fig. 133). Changes in the configuration of the associated ilium

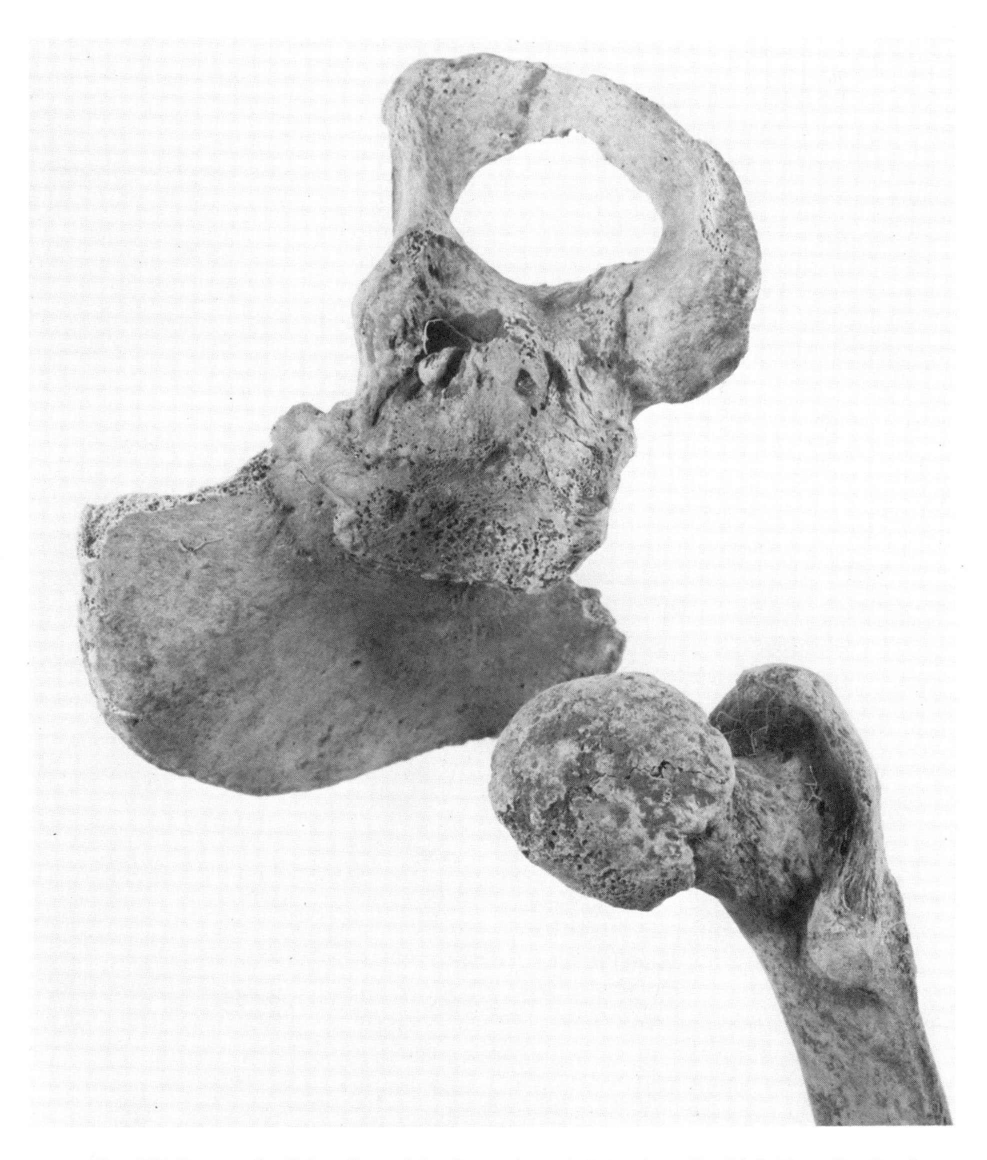

Fig. 133. Traumatic dislocation of the femur in a skeleton from the Mobridge site, South Dakota. Displacement has produced an alteration in the configuration of the socket, as well as flattening of the head of the femur (Compare the shape of this head with that in Figure 131).

Fig. 134. Infection on the shaft of a tibia from a skeleton from Pueblo Bonito, New Mexico.

suggest a new, irregular acetabulum was being formed. Flattening or mushroom-shaping of the head of the femur may also occur with a congenital dislocation. Traumatic dislocations are distinguishable from congenital ones by the presence of remnants of the surfaces of the original joints.

Infections

Bones are normally protected by flesh and are exposed only at the time of fracture or when objects penetrate the soft tissue. Except in such cases, infective organisms must be introduced from the circulatory or lymphatic systems. They generally enter the medullary cavity and quickly work their way outward to the subperiosteal area of the cortex. The organisms then spread along the shaft beneath the periosteum and eventually penetrate the shaft. Wherever the blood supply is disrupted, the bone dies. This dead "necrotic" bone is removed by the formation of pus, which is discharged along with small particles of bone through perforations (known as cloaca) in the living bone tissue (Fig. 134). Advanced bone infection of this type is called "osteomyelitis." Since the organisms responsible are carried by the circulatory systems, they may attack several bones (Fig. 135). If the periosteum becomes inflamed or if blood-borne organisms are stopped by the body defenses, the infection may be limited to the outer layer or surface of a bone. The most common infective organism is *Staphylococcus aureus,* although *Salmonella*, fungi, viruses, tuberculosis, and syphilis also produce osteomyelitic lesions.

Tuberculosis osteomyelitis is a secondary manifestation of general tubercular infection. It most frequently occurs in the spine, especially in the lower thoracic or lumbar region. Since the anterior portion of the affected vertebra collapses first, the vertebral column usually exhibits twisting (kyphosis). In long bones, both the metaphysis of the shaft and the adjacent joint are affected. Thus, tuberculosis can produce both arthritis and osteomyelitis.

Bone syphilis may be acquired in utero (congenital syphilis) from an infected mother or later in life (acquired syphilis) by contact with an infected person. In acquired syphilis, the spirochetes are transported in the blood stream and usually attack the tibia, fibula, and skull, although other bones may be involved. The effects are similar to those produced by most other kinds of osteomyelitis infections. In congenital syphilis, the spirochetes attack the epiphyseal areas of any or all bones. If the child survives, the infection prevents normal development of the skeleton. A "saber shin" tibia, displaying unusual anterior bowing and sharp anterior margins, is most characteristic. Congenital syphilis also produces distinctive deformation of the permanent teeth and occasionally of the deciduous molars. There is often a marked reduction in size (hypoplasia) and considerable disfigurement of the crowns as well. Teeth so affected are known as Hutchinson's incisors and Mulberry molars.

Fig. 135. Bones from Ossuary II in Maryland showing evidence of infection in the form of perforations (cloaca), localized enlargement, and generalized disfiguration on the surface of the shaft.

Congenital Disorders

Congenital disorders comprise a variety of inherited abnormalities, most of which are rarely found in skeletal populations. One condition that has occasionally been reported is Klippel-Feil, a congenital defect manifested in the fusion of two or more vertebrae, usually in the cervical region. A possible example was encountered at the Mobridge site in South Dakota (Fig. 136).

Another rare congenital disorder is fusion of the proximal ends of the radius and ulna. Examples have been encountered in prehistoric cemeteries in South Dakota (Fig. 137a) and southern Maryland (Fig. 137b). The latter individual was a new-born infant. This type of congenital fusion could easily be confused with fusion resulting from a healed fracture.

Circulatory Disturbances

A circulatory disturbance is any condition that disrupts the normal flow of blood to living bone. The result is usually necrosis, or death of the tissue. A variety of factors can affect circulation. The most common is trauma, in the form of fracture or laceration of the soft tissue. Infections, degenerative diseases, and congenital disorders can all cause circulatory disturbances.

Tumors

Tumors are abnormal growths that either originate in bone tissue (primary) or spread to bone from other places of origin (secondary). Primary bone tumors may develop in the cartilage and expand to the bone, or they may form in the bone itself. In either case, the result is a structural defect in the cortex. Most tumors are inactive (benign) and a radiogram shows them surrounded by a dense layer of bone. Malignant (active) tumors are rare, but can be recognized by the absence of a dense margin.

Metabolic Disorders

In general, metabolic diseases reduce the mass of bones, either by interfering with normal development or by causing excessive deterioration. The literature is complex and only a few conditions that may be detected in archeological remains will be described.

A number of diseases in this category inhibit the ability of the body to form osteoid, a material essential to the construction of bone. This general problem is termed "osteopenia." One cause is scurvy, which results from a deficiency of Vitamin C or ascorbic acid. It produces a "ground glass" appearance of the trabeculae and causes the formation of a band of dense bone at the metaphysis. These bands are retained as the bone increases in length and are known as Harris lines in radiograms (Fig. 138).

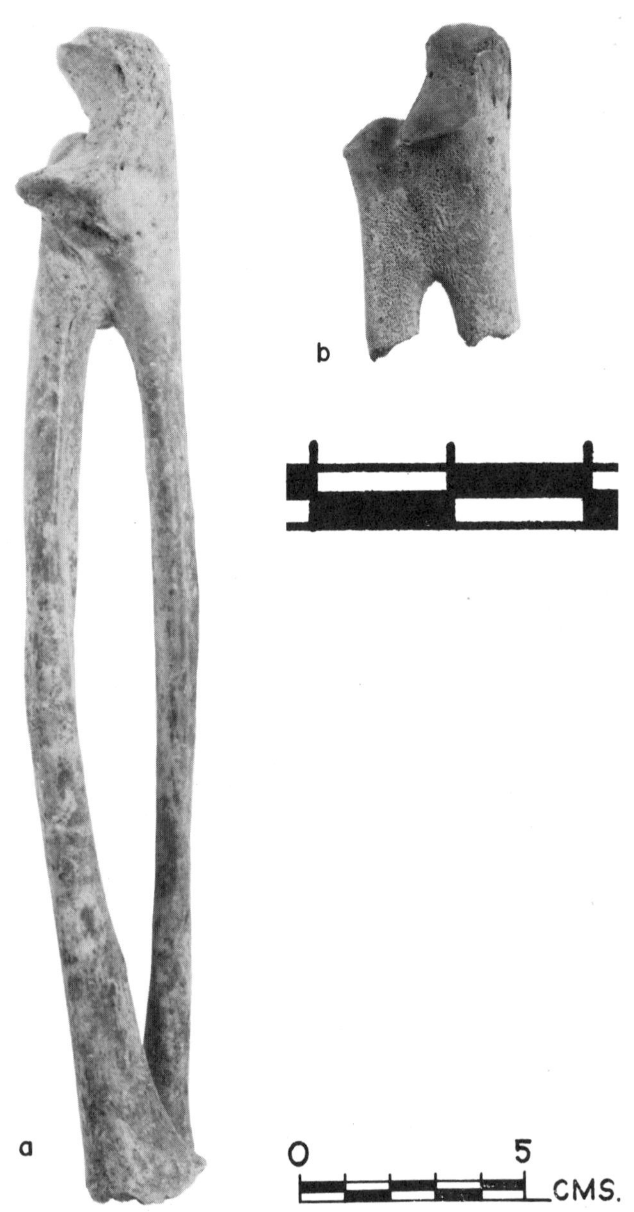

Fig. 137. A rare congenital disorder expressed in fusion of the proximal ends of the radius and ulna. **a,** An adolescent from the Mobridge site, South Dakota. **b,** A new-born infant from Ossuary II, Maryland. (Both scales are in centimeters).

Fig. 136. Fused cervical vertebrae from a skeleton from the Mobridge site, South Dakota. This is characteristic of a congenital condition known as Klippel-Feil.

Congenital osteopenia is known as "Osteogenis Imperfecta." The inherited inability of the body to produce osteoid effectively is manifested in a weakened skeleton, with bones that fracture easily and heal slowly. A group of related diseases represents the inability of the body to mineralize osteoid. This general condition, termed "Osteomalacia," also creates a structurally imperfect skeleton. It may result from a deficiency of Vitamin D, an excessive demand by the body for Vitamin D because of renal disorder or other problems, or the inability of newly formed osteoid to mineralize (Hypophosphotasia). Regardless of the cause, these diseases are usually referred to as rickets in subadults and osteomalacia in adults. Because of its structural weakness, bone affected by osteomalacia bends and fractures easily when subjected to physical stress.

Loss of cortical bone is also a normal phenomenon of aging. This type, usually referred to as osteoporosis, involves both a reduction in the thickness of the cortex and an increase in the porosity of the bone. It should not be confused with the effects of metabolic disorders.

Endocrine Disturbances

Endocrine disturbances are glandular malfunctions that affect bone. The glands most often involved are the thyroid, parathyroid, and pituitary.

Thyroid deficiency during the early period of growth causes cretinous dwarfism. This condition results from the inability of the body to convert cartilage to bone. The long bones increase normally in width, but not in length. Growth of the cranial vault is not inhibited, but growth of the base is restricted. The over-all effect is a skull of normal size (but with a constricted base) on a very short post-cranial skeleton.

Overactive parathyroids produce overactive osteoclasts, which rapidly destroy bone. Underactive parathyroids result in subnormal stature and structurally deficient bone.

Prolonged or excessive pituitary secretion causes the excessive growth characteristic of giantism and acromegaly. Inadequate pituitary activity has the opposite effect, manifested in a type of dwarfism known as hereditary achondroplasia.

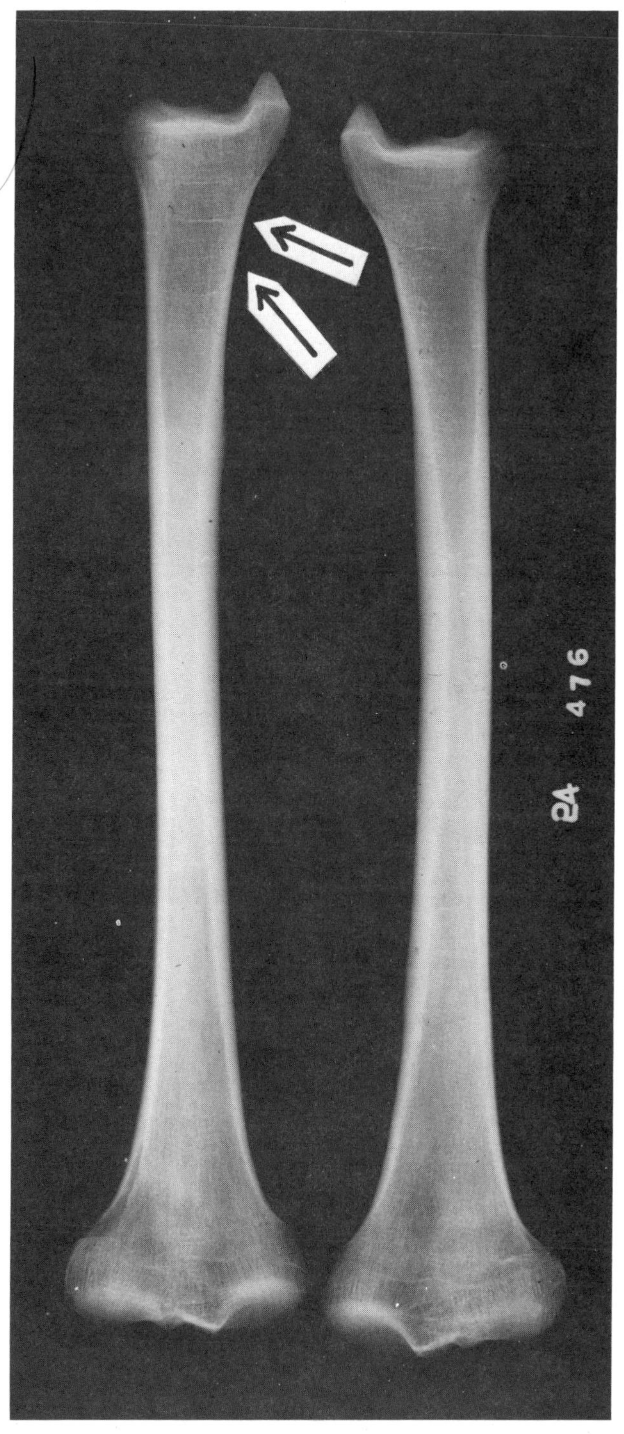

Fig. 138. Radiogram of the left and right tibiae of an adult, showing Harris lines indicating periods of interrupted growth (Courtesy of Claire Cassidy).

Diseases of Blood-forming Tissue

A number of diseases inhibit the production of normal blood. Examples include Hand-Schüler-Christian disease, eosinophilic granuloma, and Letterer-Siwe disease. Although different in etiology, these diseases all cause "punched-out" circular lesions on the skull. They manifest themselves in the long bones by enlargement of the medullary cavity at the expense of the inner surface of the cortex. The periosteum may respond by producing new bone on the outer margin.

Anemias also produce distinctive bony responses. As in the diseases listed above, the defective blood-forming tissue in the medullary cavities expands at the expense of cortical bone. Several types of anemia (sickle cell anemia and thalassemia, for example) are identifiable by a "hair-on-end" appearance in a radiograph of the cranial vault.

Miscellaneous

This category includes a variety of diseases, such as the dysplasias, Paget's disease, and dental disease. The dysplasias are tumor-like lesions in the cortex that contain misplaced fibrous structures and other tissues. In Paget's disease, the rate of remodeling is accelerated, producing a defective, deformed bone.

Dental pathologies are probably the most common diseases and the easiest to diagnose in archeological remains. They include caries, abscesses, effects of periodontal disease, hypoplasia, and malocclusion. In describing dental lesions, it is important to provide complete information and to avoid confusing lesions with normal pits and grooves in the teeth.

5 Ancestry, Identity, and Time Since Death

Variation among populations in differences between males and females, the nature and timing of age changes, and body proportions contributing to stature make identifying ancestry prerequisite to other kinds of skeletal analyses. Although estimating affiliation is difficult, a number of morphological indicators and mathematical formulas can be used.

Forensic research has produced techniques for identifying individuals from skeletal features and for reconstructing facial appearance that may lead to identification. Some of these procedures are applicable to prehistoric remains.

ANCESTRY

A question almost always asked of forensic anthropologists, and one relevant to historical archeologists, is "What race is it?" Despite its importance, ancestral affiliation is difficult to assess. Individuals classified socially as members of a particular "race" vary greatly in physical appearance. Some population differences are apparent in the skeleton, but variation within groups and overlap among groups reduce the accuracy of identifications from an individual skeleton. The presence of concentrations of extreme expressions of some skeletal traits, however, can suggest affiliation with one of the major groups.

Cranium (Fig. 139)

Asians and American Indians. Crania have very forward-projecting malar bones and comparatively flat faces. The nasal aperture is moderate in width and has a slightly pointed lower margin. The orbits tend to be more circular than those of other groups and the palate is moderately wide. At least among American Indians, the suture between the maxilla and malar tends to be straight.

Blacks. Crania usually show relatively little projection of the malars, more rectangular orbits, and wider interorbital distances. The nasal aperture is very broad and lacks a sharp lower border. The palate tends to be very wide and somewhat rectangular. The anterior alveolus of both the maxilla and mandible tends to be very projecting (prognathism). Many crania exhibit slight coronal depression just posterior to the coronal suture.

Whites. Crania are typically characterized by very receding malar bones giving the face a pointed appearance. The nasal aperture is very narrow and has a prominent sharp lower border. The palate is relatively narrow and trianguloid. The suture between the maxilla and malar tends to be curved. The frequency of metopism (suture from nasion to bregma) is higher than among other groups.

Post-cranial Skeleton

Biological differences are expressed in the curvature of the long bones. In particular, Blacks tend to have relatively straight femora, with very little torsion or twisting of the neck and head. In contrast, Asian femora tend to be quite curved, with considerable torsion at the neck. Whites are intermediate in both curvature and torsion. American Indian femora frequently show marked flattening or platymery on the anterior upper shaft.

Fig. 139. Morphological differences in the facial bones. a, Asian/American Indian. b, Black. c, White.

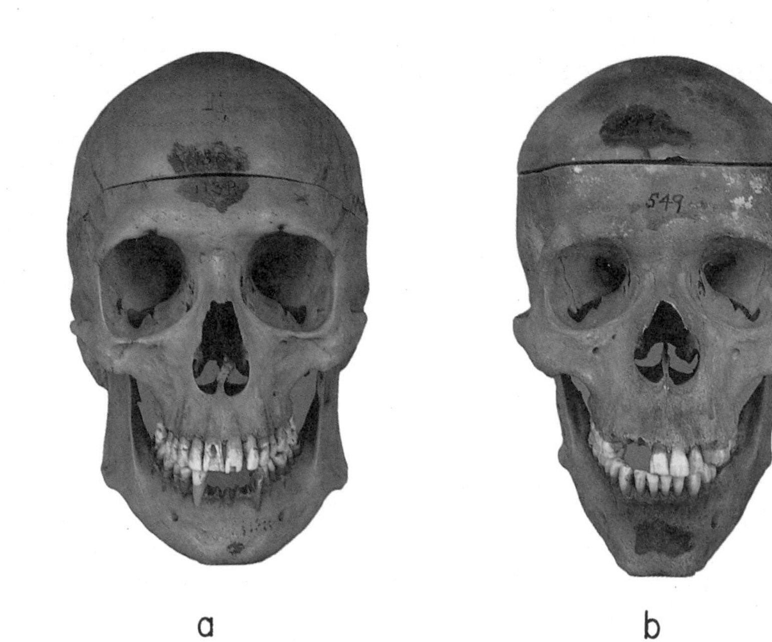

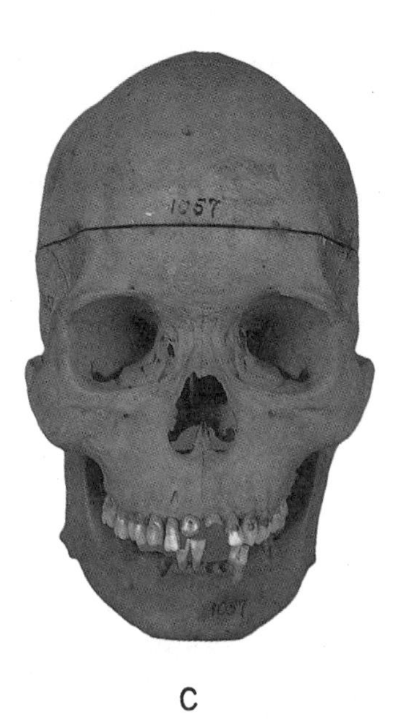

Dentition

Although important differences exist at the population level, they may not be diagnostic for individuals.

Occlusion. In general, Asian and American Indian populations have excellent dental occlusion, with a near perfect edge-to-edge bite. Most Whites untreated by orthodontics show an overbite, where the maxillary teeth project over the mandibular teeth. Occlusion among Blacks is typically intermediate and variable.

Attrition. Ancient American Indians usually display marked occlusal attrition (wear on the biting surfaces). Among some groups, especially those with hunter-gatherer subsistence, the teeth may be worn down very rapidly. This condition contrasts markedly with most modern populations, where occlusal wear is much more gradual.

Size. In general, teeth of aboriginal Australians, Melanesians, American Indians, and Eskimos are among the largest. Teeth of Lapps and Bushmen tend to be among the smallest, with other populations presenting intermediate values (Lasker and Lee 1957).

Shovel Shape. Many Asians have maxillary incisors with prominent marginal ridges on the lingual surface, giving them a "shovel-shaped" appearance. In extreme cases, the ridges also occur on the buccal surface in a pattern known as "double shoveling" Occasionally, the ridges extend so far lingually that they meet, forming a "barrel-shaped" incisor. American Indians and Eskimos have particularly high frequencies of this trait, but shoveling is absent in some individuals and may occur in Blacks and Whites.

Carabelli's Cusp. An extra cusp may be present on the mesiolingual aspect of maxillary molars. The trait is most pronounced on the first molars, but variations sometimes occur on other molars. Frequency is highest among Whites and comparatively low among Blacks, Asians, and American Indians.

Protostylid. An extra cusp or tubercle sometimes occurs on the mesiobuccal surface of mandibular molars. The frequency of protostylids is highest among Asians including American Indians and Eskimos.

Anomalies. There is some evidence suggesting that maxillary lateral incisors of diminished size and variable form are more common among Whites, such as peg-shaped forms and miniature versions of normal teeth.

Mathematical Approaches

In 1962, Giles and Eliot introduced a scoring method to estimate ancestral affiliation based on their study of Blacks and Whites from the Hamann-Todd and Terry Collections and American Indians from Kentucky, the Gulf states, and the southwestern United

Table 26. Variables and multiplication factors for identifying ancestry (after Giles and Eliot 1962).

Variables	Males: White vs. Black	Males: White vs. Indian	Females: White vs. Black	Females: White vs. Indian	Male vs. Female
Basion-prosthion	3.06	0.10	1.74	3.05	−1.00
Glabella-occipital l.	1.60	−0.25	1.28	−1.04	1.16
Maximum cranial br.	−1.90	−1.56	−1.18	−5.41	
Basion-bregma ht.	−1.79	0.73	−0.14	4.29	
Basion-nasion length	−4.41	−0.29	−2.34	−4.02	1.66
Max. bizygomatic br.	−0.10	1.75	0.38	5.62	3.98
Prosthion-nasion ht.	2.59	−0.16	−0.01	−1.00	1.54
Nasal breadth	10.56	−0.88	2.45	−2.19	
Sectioning point	89.27	22.28	92.20	130.10	891.12

Table 27. Variables and multiplication factors for differentiating Blacks and Whites (after Jantz and Moore-Jansen 1987:62).

Variable	Male	Female
Max. cranial breadth	−0.070103	−0.063754
Basion-bregma ht.	−0.066245	−0.056871
Basion-nasion	−0.122604	−0.127035
Basion-prosthion l.	0.152699	0.198088
Min. frontal breadth	0.077145	—
Nasal breadth	0.156295	0.249499
Orbit height	0.205818	0.227995
Black mean	3.031190	6.024326
White mean	0.780894	2.469136
Sectioning point	1.90604	4.246731

States. Eight measurements of the cranium are multiplied by a determined factor, and the results are added or subtracted to produce a score that can be assessed for racial affiliation. The measurements and discriminant function coefficients are specified on Table 26.

To establish whether a particular male cranium is White or Black, the coefficients in the first column should be used. Basion-prosthion length should be multiplied by 3.06, the product added to the product of glabella-occipital length multiplied by 1.60, and so on. The product of each multiplication should either be added or (when preceded by a minus sign) subtracted from the previous total. The grand total should be compared with the sectioning point. In this comparison, a score exceeding 89.27 indicates Black and less than 89.27 indicates White affiliation.

These discriminant functions compute affiliation of an unknown male to the samples from which the functions were derived, whereas the population represented by an archeological skeleton or a forensic case may be quite distinct from those used by Giles and Eliot. A more recent study at the University of Tennessee, Knoxville of a large forensic sample of Blacks and Whites from all parts of the United States has established a new sectioning point of 62.89 for males. The sectioning point for females is unchanged. Measurements and multiplication factors are the same as on Table 26.

The University of Tennessee Project has proposed a slightly different set of measurements for application to forensic cases (Table 27; Jantz and Moore-Jansen 1987). In this system, a value above the sectioning point indicates Black and one below the sectioning point indicates White affiliation.

Discriminant function approaches to establish differences using bones of the post-cranial skeleton are summarized by Krogman and Iscan (1986:268-301). The computer program Fordisc 2.0 (Ousley and Janz 1996) offers customized discriminant functions to estimate likely ancestry from those measurements available.

FACIAL REPRODUCTION

Special circumstances may warrant attempting to reproduce facial appearance. Anthropologists want to compare fossil hominids with our species and police want to establish the identity of an individual in a forensic case. Many details of soft tissue are not directly reflected in the cranium; thus, facial reproduction represents only an educated guess or approximation.

There are two approaches: (1) rebuilding the face directly over the cranium and (2) sketching a likeness using information provided by an anthropologist. Both begin with an attempt to estimate the amount of muscle and other soft tissue, which is usually accomplished with the aid of small cylindrical markers applied at specific points (Fig. 140). The length corresponds to values derived from measuring males and females of the principal racial groups (Table 28). For American Whites, values have been refined according to body build (Table 29; Rhine, Moore, and Weston 1982). The thickness employed in reproduction should match

Table 28. Facial tissue thickness values for Black, White, and Japanese males and females (After Rhine and Campbell: Table 3).

Location	Black		White		Japanese	
Midline	Male	Female	Male	Female	Male	Female
1. Supraglabella	4.75	4.50	3.75	3.50	3.00	2.00
2. Glabella	6.25	6.25	4.75	4.25	3.80	3.20
3. Nasion	6.00	5.75	5.00	4.50	4.10	3.40
4. End of nasal	3.75	3.75	2.00	2.00	2.20	1.60
5. Mid-philtrum	12.25	11.25	11.50	10.00	—	—
6. Upper lip margin	14.00	13.00	9.50	8.25	—	—
7. Lower lip margin	15.00	15.50	—	—	—	—
8. Chin-lip fold	12.00	12.00	10.00	10.00	10.50	8.50
9. Mental eminence	12.25	12.25	10.25	10.00	6.20	5.30
10. Beneath chin	8.00	7.75	6.00	6.25	4.80	2.80
Lateral						
11. Frontal eminence, left	8.25	8.00	—	—	—	—
Frontal eminence, right	8.75	8.00	—	—	—	—
12. Supraorbital, left	4.75	4.50	5.75	5.25	—	—
Supraorbital, right	4.75	4.50	—	—	4.50	3.60
13. Suborbital, left	7.50	8.50	4.25	4.50	—	—
Suborbital, right	7.75	8.25	—	—	3.70	3.00
14. Inferior malar, left	16.25	17.25	—	—	—	—
Inferior malar, right	17.00	17.75	—	—	—	—
15. Lateral orbits, left	13.00	14.25	6.75	7.75	—	—
Lateral orbits, right	13.25	12.75	—	—	5.40	4.70
16. Zygomatic arch, left	8.75	9.25	4.25	5.25	—	—
Zygomatic arch, right	8.50	9.00	—	—	4.40	2.90
17. Supraglenoid, left	11.75	12.00	6.75	7.00	—	—
Supraglenoid, right	11.75	12.25	—	—	—	—
18. Occlusal line, left	19.50	18.25	—	—	—	—
Occlusal line, right	19.00	19.25	—	—	—	—
19. Gonion, left	14.25	14.25	10.50	9.50	—	—
Gonion, right	14.75	14.25	—	—	6.80	4.00
20. Sub-M, left	15.75	16.75	—	—	—	—
Sub-M, right	16.50	17.25	—	—	10.20	9.70
21. Supra-M, left	22.25	20.75	—	—	—	—
Supra-M, right	22.00	21.25	—	—	14.50	12.30

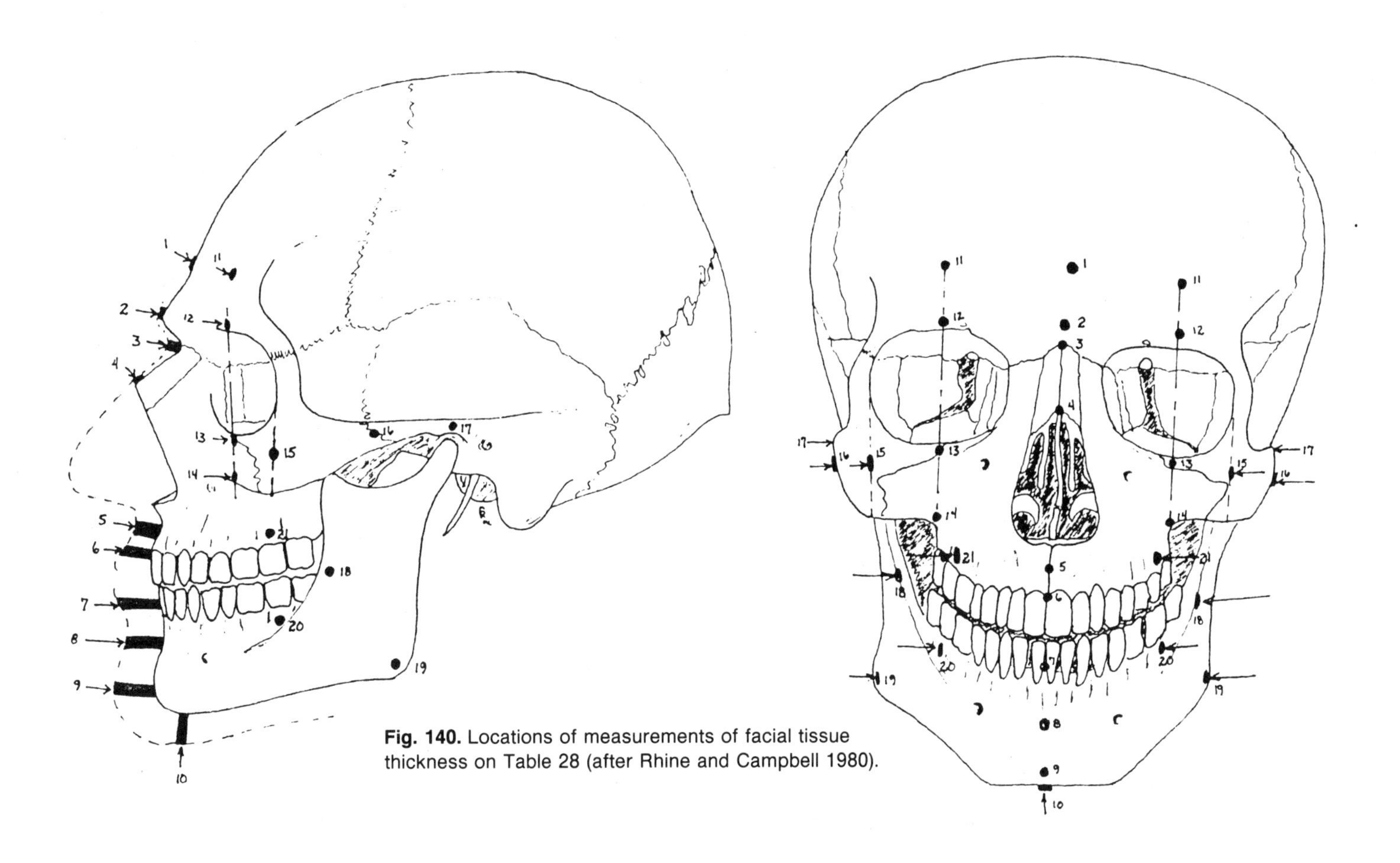

Fig. 140. Locations of measurements of facial tissue thickness on Table 28 (after Rhine and Campbell 1980).

Table 29. Facial tissue thickness variations according to body build (after Rhine, Moore and Weston 1982).

Measurement	Emaciated		Normal		Obese	
Midline	Male (3)	Female (3)	Male (67)	Female (19)	Male (8)	Female (3)
Supraglabella	2.50	2.50	4.25	3.50	5.50	4.25
Glabella	3.00	4.00	5.25	4.75	7.50	7.50
Nasion	4.25	5.25	6.50	5.50	7.50	7.00
End of Nasals	3.00	2.25	3.00	2.75	3.50	4.25
Mid Philtrum	7.75	5.00	10.00	8.50	11.00	9.00
Upper Lip Margin	7.25	6.25	9.75	8.50	11.00	11.00
Lower Lip Margin	8.25	8.50	11.00	10.00	12.75	12.25
Chin-Lip Fold	10.00	9.25	10.75	9.50	12.25	13.75
Mental Eminence	8.25	8.50	11.25	10.00	14.00	14.25
Beneath Chin	5.0	3.75	7.25	5.75	10.75	9.00
Bilateral						
Frontal Eminence	3.25	2.75	4.25	3.50	5.50	5.00
Supraorbital	6.50	5.25	8.25	6.75	10.25	10.00
Suborbital	4.50	4.00	5.75	5.75	8.25	8.50
Inferior Malar	8.50	7.00	13.50	12.50	15.25	14.00
Lateral Orbit	6.75	6.00	9.75	10.50	13.75	13.25
Zygomatic Arch, midway	3.50	3.50	7.00	7.00	11.75	9.50
Supraglenoid	5.00	4.25	8.25	7.75	11.25	8.25
Gonion	6.50	5.00	11.00	9.75	17.50	17.50
Supra M^2	8.50	12.00	18.50	17.75	25.00	23.75
Occlusal Line	9.25	11.00	17.75	17.00	23.50	20.25
Sub M_2	7.00	8.50	15.25	15.25	19.75	18.75

the sex, race, and body build of the unidentified individual.

Both approaches to facial reproduction have merit. Rebuilding probably provides more accurate proportions when accomplished by a highly skilled artist, but requires much more time and effort than sketching and does not produce as life-like an appearance.

Rebuilding

After the appropriate tissue-thickness markers have been set in place, clay or other suitable material is used to fill the intervening spaces (Fig. 141). Artificial eyes are inserted into the orbits; lips and eyelids are fashioned using strips of clay. The process may end when the features have been refined or may be elaborated by adding hair and clothing.

The reproduction illustrated is the work of Lewis L. Sadler, Head of the Department of Biocommunication Arts, University of Illinois at Chicago. The cranium used is in the Terry Collection at the Smithsonian Institution and is accompanied by a mold of the face made immediately after death, although Sadler was not aware of its existence when he made his reproduction. The cast made from the death mask incorporates postmorten changes in the soft tissue: even so, it strikingly resembles the modeled face. The

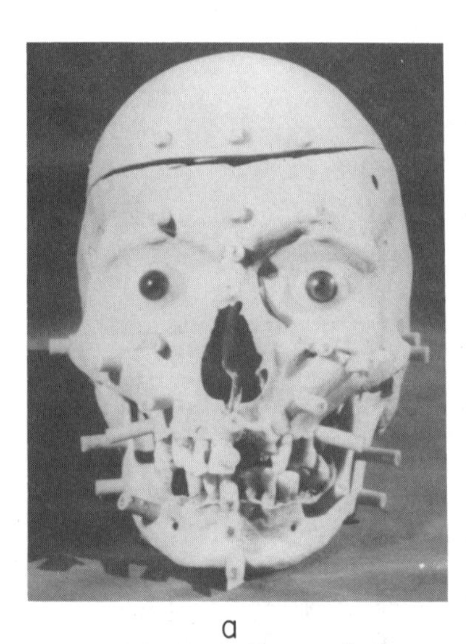

a

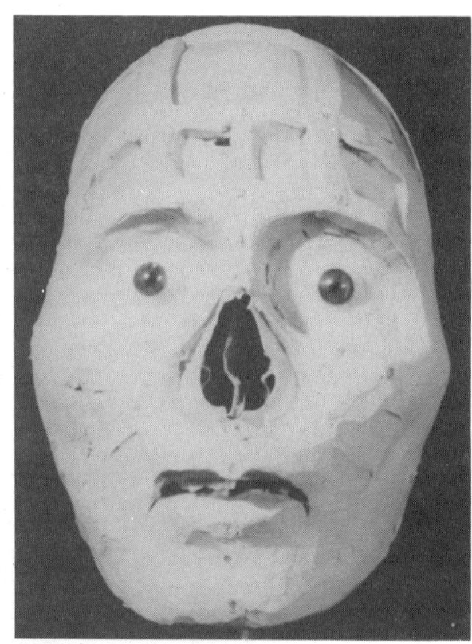

b

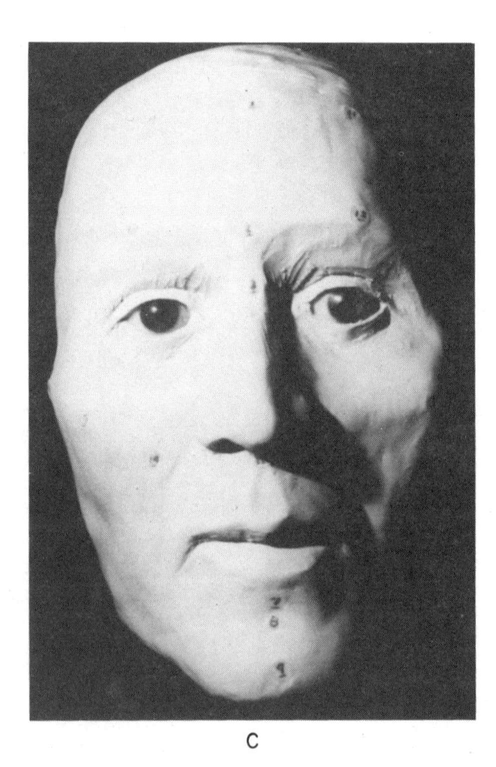

c

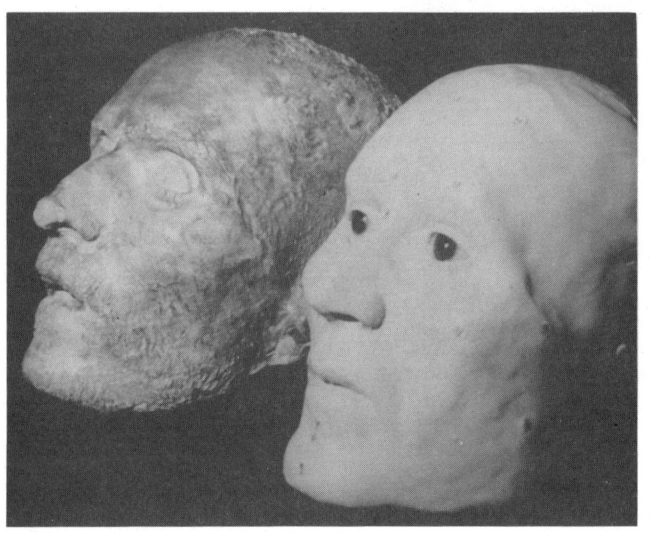

d

Fig. 141. Rebuilding soft tissue of the face. **a,** Tissue thickness markers in place, artificial eyes in orbits. **b,** Intervening spaces filled on the face. **c,** Reproduction completed. **d,** Result compared with the death mask.

close similarity shows how successful this technique can be in the hands of a gifted artist with a sound understanding of facial anatomy.

Sketching

This approach also begins by placing the appropriate tissue-thickness markers on the cranium. The anthropologist then describes to the artist the aspects that should be highlighted. The facial components in kits used by police departments for obtaining information from eye witnesses can be useful. The anthropologist can select the appropriate chin type, nose form, eye separation, etc., which the artist can use to guide production of a composite drawing based on the contours established from the cranium and the tissue-thickness markers.

A cranium and mandible, and the reproduced facial image are shown in Figure 142. Anthropological examination indicated the individual was a young White male of medium build. Facial structure showed slight asymmetry in the nasal-orbital area, as well as a slight overbite. Using the composite approach, the artist sketched a face with slight asymmetry and a bulbous upper lip reflecting the overbite. Hair form was not emphasized since its nature could not be established. Unfortunately, there were no unusual characteristics that might have facilitated identification of the individual.

By contrast, the cranium and mandible in Figure 143 present unique features that enhance the probability of identification. They represent a young female of mixed ancestral affiliation, who had suffered major cranio-facial trauma long before death. This trauma

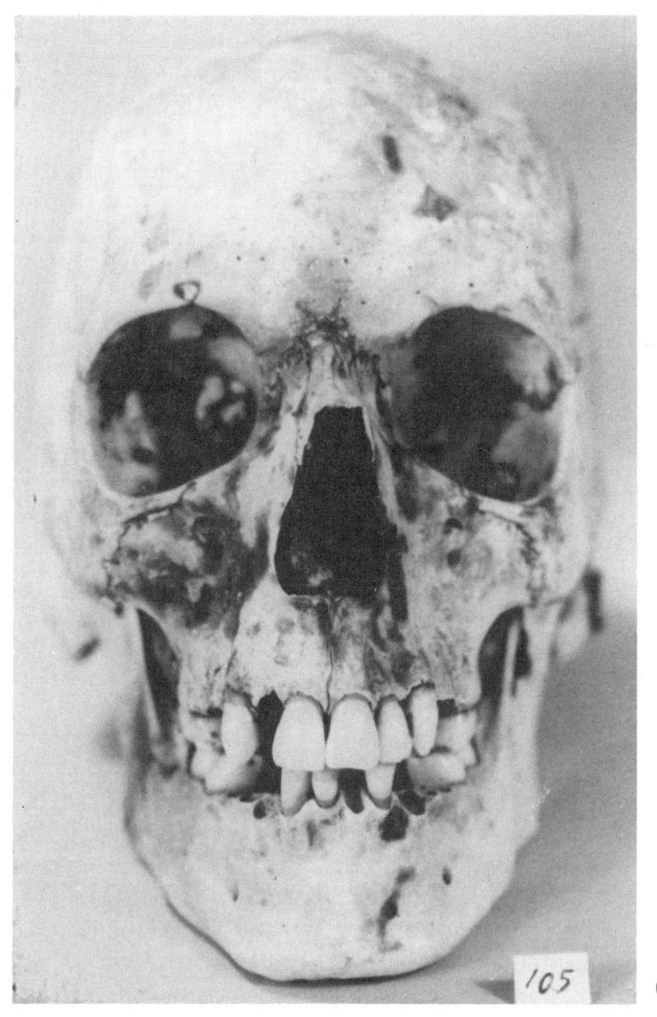

a

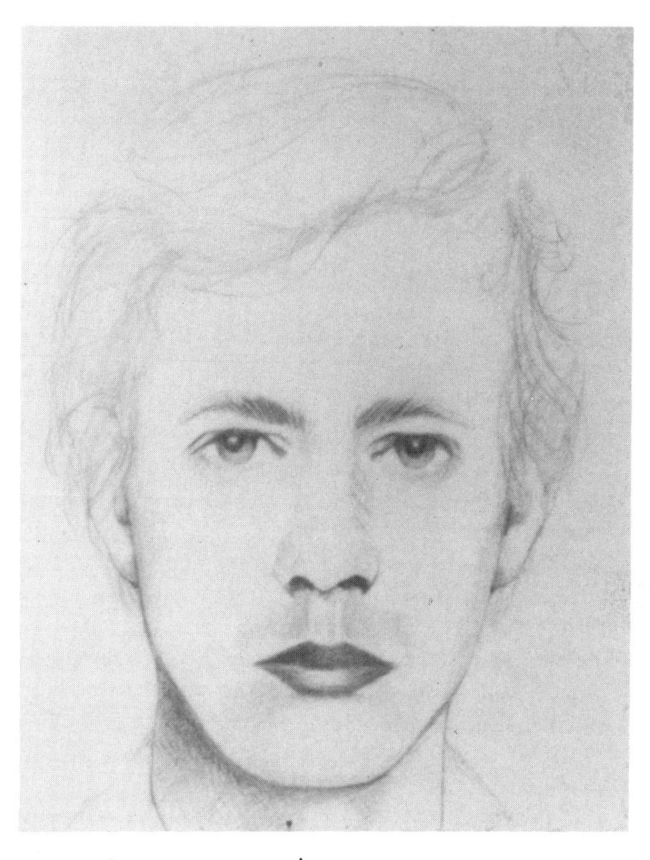

b

Fig. 142. Sketch reproducing living appearance from a cranium with no unique features.

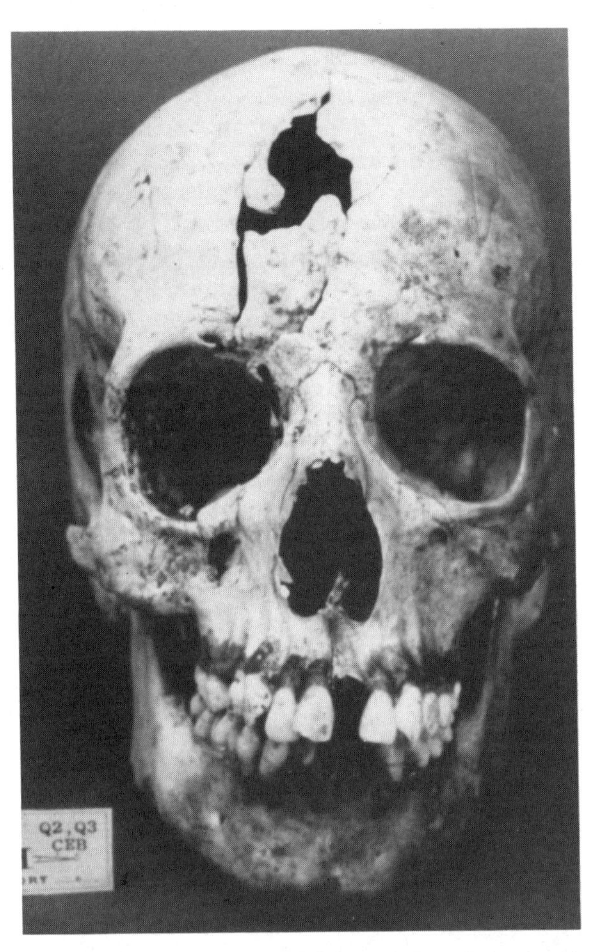

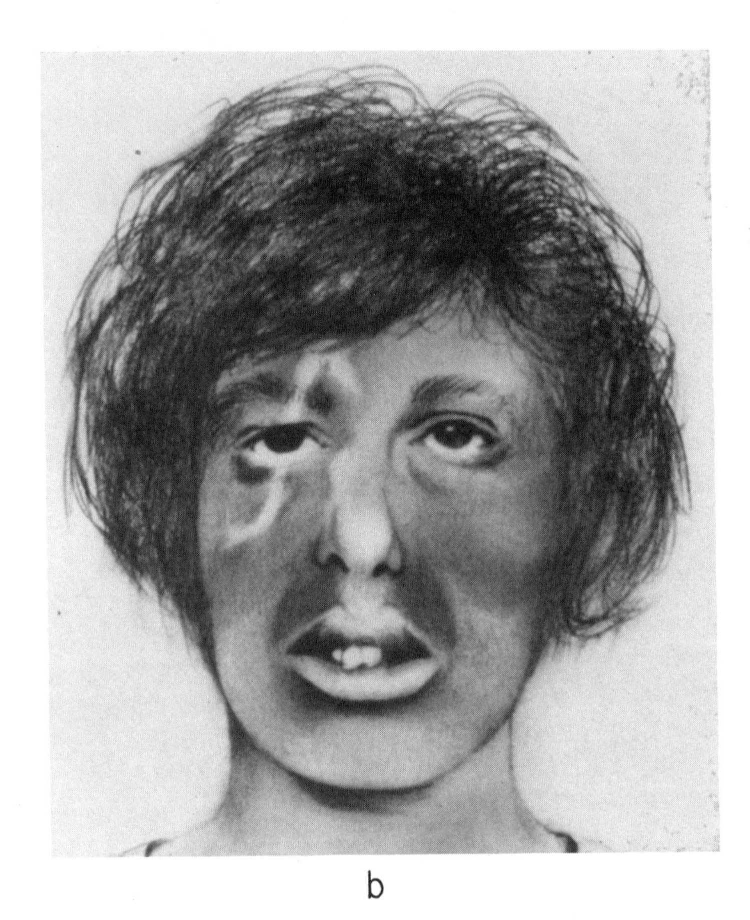

Fig. 143. Sketch reproducing living appearance from a cranium with major cranio-facial trauma and other unusual features.

produced a massive fracture of the right frontal and maxilla. A glass eye associated with the remains indicated that restorative surgery had been performed. A surgical wire attached to the right lacrimal provided another clue to the type of surgery. The lower margin of the right orbit is markedly lower than that of the left orbit and the nasal aperture is very asymmetrical. The dentition shows a pronounced overbite causing considerable prognathism and the anterior teeth are crowded. The anthropologist-artist team reproduction highlights these unique features.

The cranium and mandible of a middle-aged Black male in Figure 144 were recovered while considerable soft tissue and hair remained, enhancing the accuracy of the reproduction of hair form and length. The upper dentures were associated and were unusual in having gold crowns on the maxillary lateral incisors. Presumably, the individual had wanted to preserve his previous appearance and had the dentures made accordingly. Public display of the sketch led to identification in this case. Comparison of the reproduction with a photograph taken during life shows that most facial features match. Nasal width is an exception, illustrating the difficulty in accurately inferring aspects of facial anatomy.

POSITIVE IDENTIFICATION

Identification of a specific individual is the goal of forensic analysis and may be a concern of archeologists who encounter tombs potentially assignable to monarchs and other historical figures. The physical anthropologist contributes to an identification by supplying enough information about the person (age, sex, stature, ancestral affiliation, body build, time since death, and other distinguishing characteristics) that the pos-

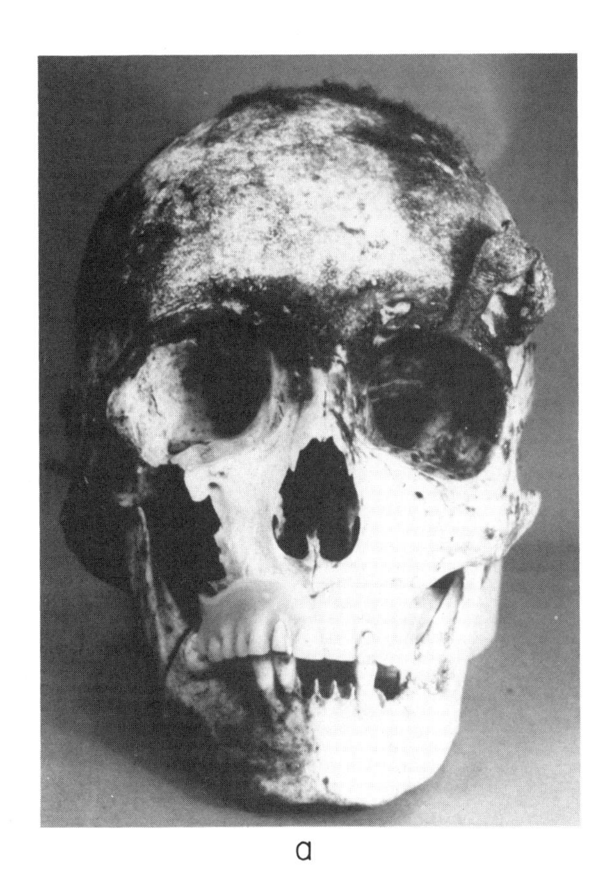

a

b

c

Fig. 144. Sketch reproducing living appearance from a cranium retaining soft tissue and hair. **a,** Cranium. **b,** Reproduction. **c.** Photograph taken before death.

sibilities can be narrowed. Positive identification requires recognizing one or more unique features, usually documented by medical and/or dental records.

Dental Identification

Dental records are the principal resource in forensic cases because most people receive dental treatment and their files are accessible from local dentists. Such records generally include radiographs, which provide unique information about the structure of the teeth and surrounding bone, and the appearance of dental restorations. Comparisons require skilled interpretation and should be done by forensic odontologists. Many months or even years may have elapsed between the last available radiograph and the death of the individual and teeth may have been lost or otherwise altered during this interval. "Explainable differences" do not detract from identification, but a a single unexplainable difference is decisive.

An excellent example of dental radiograph comparison that would lead to positive identification is illustrated in Figure 145. The radiograph on the top was taken from the remains, whereas the one on the bottom was taken during life. Note that aspects of restorations on the mandibular second premolar and the maxillary molars are identical. The additional restorations visible on the top radiograph were completed prior to death but after the radiograph on the bottom was taken.

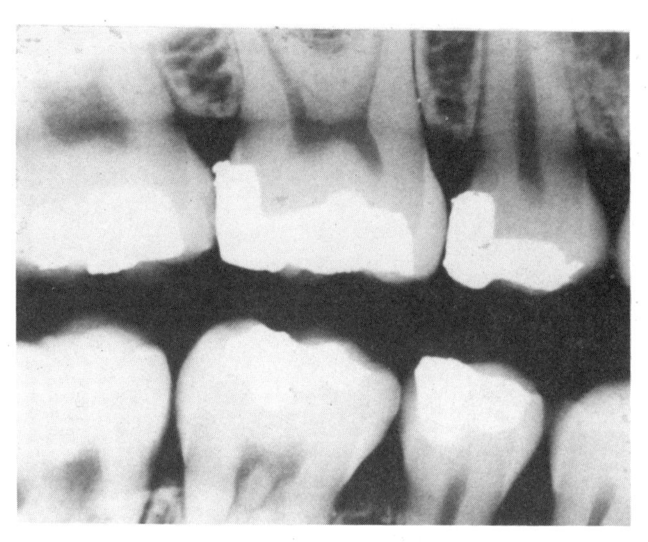

a

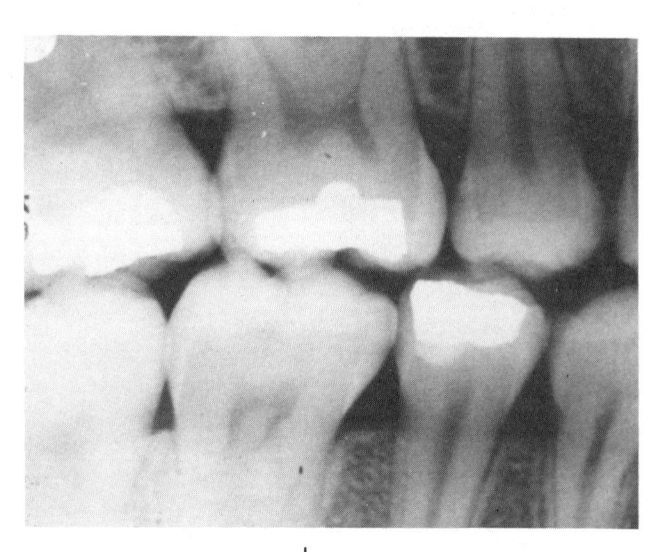

b

Fig. 145. Positive identification using dental radiographs. **a,** Restorations at time of death. **b,** Dentist's radiograph showing identical restorations on several teeth.

Frontal Sinus

The frontal sinus consists of a complex of lacunae within the frontal bone between the eyes and just above the nasal bridge. The sinus usually appears during the first few years and becomes radiographically visible between the ages of seven and nine. Short of trauma, disease or surgery, it remains relatively unchanged throughout life. Its utility for forensic identification using radiographs has been recognized since 1921 (Schuller 1921).

Radiographic comparison of frontal sinus patterns was the crucial evidence identifying a murder victim in 1980. FBI agents brought to my laboratory a well preserved human calvarium believed to be that of a Massachusetts prostitute who belonged to a group practicing Satanism. Circumstances suggested she had witnessed the murder of another prostitute and that the leader of the group, fearing she would inform the police, killed her and disposed of all of the body except the calvarium. An important aspect of the case was proving the calvarium belonged to the victim.

Morphological examination indicated that the calvarium represented a young adult female, perhaps of White ancestry. Although this information was consistent with the description, it was insufficient for positive identification. A search of local hospitals revealed that shortly before her death, the victim had suffered a severe headache and that frontal and lateral

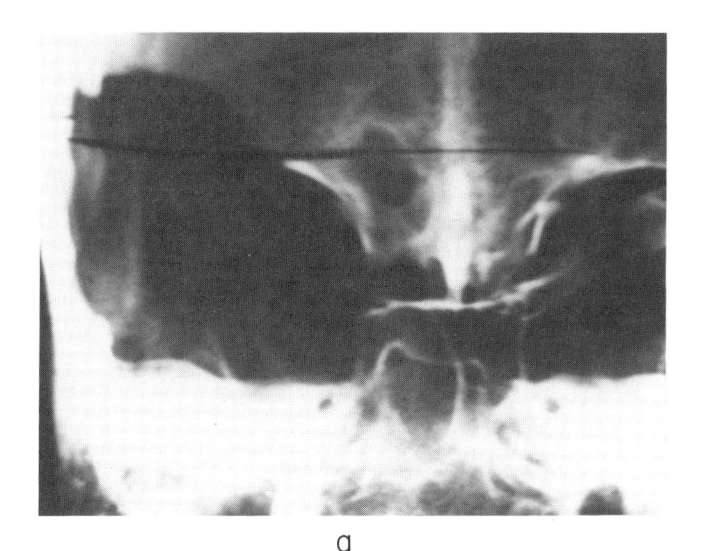

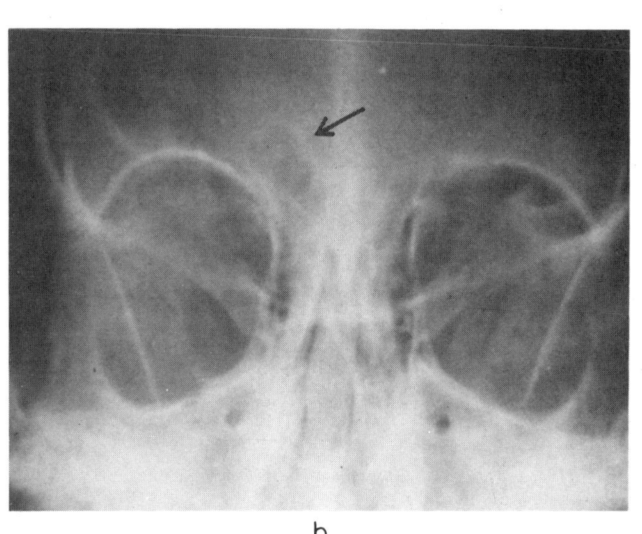

Fig. 146. Positive identification using radiographic comparison of the frontal sinus. **a,** The unknown individual. **b,** Matching hospital radiograph taken shortly before death.

radiographs had been taken of her head for diagnosis. Comparing these with the unknown calvarium revealed an exact match of details of the frontal sinus and associated structures (Fig. 146), as well as details of the sella turcica on the lateral radiograph.

To establish the uniqueness of the frontal sinus and its potential for positive identification, I compared 35 radiographs for a total of 595 comparisons (Ubelaker 1984b). This established that no two crania were alike. The number of differences between individuals averaged about 8, with a range from 3 to 15. I and a forensic pathologist both testified that the two sets of radiographs had to be from the same person. The Court permitted this expert testimony to be heard and evaluated by the jury, which convicted the defendent of murder.

Post-cranial Bones

Although details of other bones are unique, they are seldom documented radiographically. Furthermore, most parts of the skeleton undergo changes with age and the process of remodeling. Trauma, surgical procedures, and disease can also alter radiographic appearance in a short time. Thus, although an exact match can allow positive identification, the existence of differences does not rule it out.

Radiographs permitted identification of an American Indian from South Dakota from remains submitted by the FBI in 1984. The cranium of a 30-year-old male showed blunt-force trauma inflicted at or about the time of death. Other evidence suggested the remains might be those of an individual reported missing 10 months earlier. Although the teeth showed many restorations, neither dental records nor dental radiographs could be located. Search did produce several radiographs taken of the missing person's shoulder and ankle. Their careful comparison revealed an unusual pattern on the lateral border of the right scapula that matched perfectly (Fig. 147). To verify that this feature was unique, I examined 100 right scapulas and failed to find a match. The comparison was used in court to establish positive identification.

ESTIMATING TIME SINCE DEATH

Estimating the time elapsed between death and discovery is an important but frequently elusive aspect in forensic analysis. Immediately after death, the body cools, develops varying degrees of rigor mortis, and then undergoes tissue decay that culminates in skeletonization. The rates of these processes, especially tissue decay, are dramatically affected by diverse environmental factors. If the corpse was buried, rate of decay will be influenced by soil acidity, groundwater retention, and type of container. If exposed above ground, major roles will be played by foraging mam-

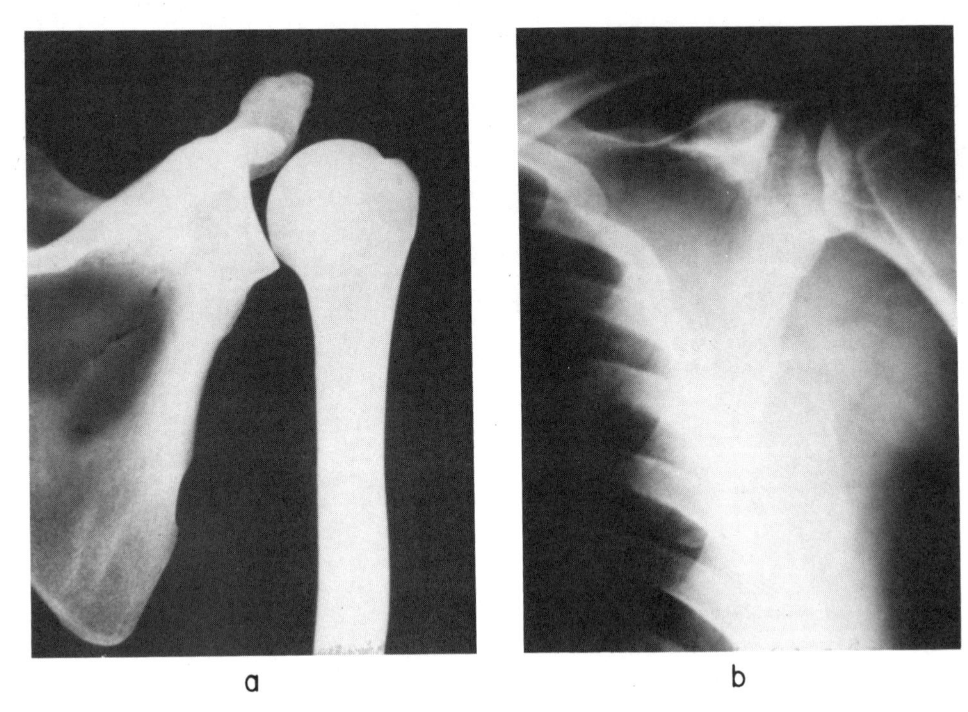

Fig. 147. Positive identification using a unique pattern along the lateral border of the right scapula. **a,** The unknown individual. **b,** Matching hospital radiograph taken during life.

mals, birds, and arthropods. All variables differ by environment, season, and method of treatment of the body. In a hot and humid climate, with exposure to scavengers, a body can become skeletonized in two weeks. By contrast, in extremely dry regions, such as the desert coasts of Chile and Peru, dessicated soft tissue can be preserved for thousands of years. Extreme variations may exist within a single mortuary area used in the recent past.

Experiments by Rodriguez and Bass (1983) indicate that a direct correlation exists between stage of decomposition and species of arthropod present. Decomposition occurs most rapidly during spring and summer, when arthropod populations are most active. Rate of decay may be slowed if the soft tissue is protected by clothing, burial, etc.

Additional clues may come from the state of preservation of associated materials. Morse (1983) presents experimental data on rates of decomposition of various types of clothing and other grave goods under different conditions. Rayon disintegrated most rapidly. Other materials tested showed the following order of increasing resistence to decomposition: paper, untreated cotton, treated cotton, silk, wool, human hair, polyester, triacetate, nylon, leather, plastic, and acrylic.

In 1987, Willey and Heilman suggested plants might provide evidence. Stems and woody roots of perennials have annual growth rings, which can be counted to establish the age of a plant growing through clothing or having some other relationship with the remains that could only have been established after deposition at the site of discovery.

Additional discussion of approaches for estimating time since death and the variables involved is provided by Krogman and Iscan (1986:21-36) and Haglund and Sorg (1997).

6 Prehistoric Population Dynamics

The availability of more fully documented skeletal remains; more precise methods of establishing age and sex, and more accurate diagnoses of nutritional, pathological, and other kinds of external influences has enabled skeletal biologists to make broader inferences about past populations than they could when anthropometric data were the primary focus of research. Sometimes they can corroborate hypotheses based on cultural evidence; sometimes they can suggest interpretations for archeologists or linguists to evaluate. In this chapter, I will describe a few examples. More extensive discussions are provided elsewhere (for example, Brothwell and Sandison 1967; Blakely 1977).

The two approaches to be discussed in the following pages are: (1) biological distance, the measurement of the morphological affinities between two or more populations, and (2) demography, the reconstruction of population size and composition. They are end products of the kinds of observations and analyses set forth in the preceding chapters. Their validity depends on the care with which the samples have been collected and described, but it cannot be repeated too often that even samples appearing to satisfy all the requirements may not be representative of a prehistoric population. We must work with the best materials and the best methods available, but maintain a critical attitude about the results.

ESTIMATING BIOLOGICAL DISTANCE

Biological distance is the expression of the morphological affinities between two or more populations. The assumptions are the same as those employed for assessing cultural relationships; that is, specimens displaying the most similarities are considered most closely related. Traditionally, physical anthropologists have reconstructed biological relationships by comparing standardized measurements, observations, and indices derived from individuals. In recent years, this field of research has been aided considerably by the development of computers and application of new statistical methods of analysis, which permit many measurements and observations to be evaluated simultaneously. For example, it is now possible to compare 35 measurements per skull from several hundred individuals representing three populations and to establish their general degree of morphological similarity. Such a comparison might show that populations A and C were more alike (and therefore presumably more closely related) than either was to population B.

The extent to which the relationships inferred from these kinds of comparisons reflect reality depends mainly on two factors: (1) the adequacy of the sample, and (2) the selection of traits that are genetically rather than environmentally determined. The former is more easy to control than the latter. For most purposes, a sample of 100 or more adults from each group being compared is sufficient, provided the specimens have not been selected in any way that affects their representativeness.

Choice of the criteria to be compared is a more difficult problem. Since the object of the study is determining genetic relationship, only traits not susceptible to alteration by environmental or nutritional variables should be used. Unfortunately, none of the morphological features on the human skeleton are completely free from non-genetic influence. Inheritance patterns have been suggested for several dental traits, but even these are not immune to distortion by nutrition, disease, and other environmental factors. Experiments on rats indicate that some skeletal details may be largely genetic, but similar studies have not been conducted with sufficient thoroughness on humans to identify such features, if they exist. Thus, the "biological distance" measured by comparing human populations probably incorporates a combination of genetic and environmental factors. In spite of this limitation, the results are useful for evaluating biological hypotheses generated from cultural data, such as continuity from prehistoric to historic groups, presence of individuals captured from another population, miscegenation, and even social practices. Does the appearance of new vessel shapes and decorative techniques in a local ceramic tradition indicate incorporation of alien individuals into the population, local innovation, or cultural diffusion without the introduction of new genes? To resolve this kind of problem, the biological data should be analyzed independently and the results compared with those obtained from other evidence to achieve the most logical interpretation.

A word of warning should be injected concerning misuse of biological data because it occurs all too commonly in popular literature. Many of the errors stem from adherence to the discredited "typological" definition of a population. According to this view, a certain set of traits identify a "type" and all individuals possessing these traits belong to the same class. Among biologists, this concept has been replaced by "populational thinking," which recognizes that all levels of biological difference (communities, varieties, species, genera, etc.) incorporate a range of variation, and no single individual or set of individuals is likely to possess the entire complement of traits. In practical terms, this means that a sample of 100 adult skeletons from a cemetery will exhibit a range of variation in characteristics. This range will generally have the configuration of a bell-shaped curve, with the largest number of individuals near the middle of the range and a few at each extreme. Curves for different populations will have their peaks at different locations, but will generally overlap at the extremes. Once these curves have been defined, a large sample of specimens can be correctly identified by comparing the pattern of variation with those obtained from populations of known origin. Small samples, and especially single individuals, may be impossible to identify with certainty because their representativeness, and consequently their position on the curve is unknown. The overlap in geographical distribution is sufficient that a single skull from an Asian population could exhibit many traits normally present in highest frequencies in African populations. Adherence to the typological definition would cause it to be erroneously classified as African.

The potential of skeletal analysis for resolving archeological problems involving biological hypotheses cannot be realized until the genetics of bone development is better documented. When the traits or measurements useful as genetic population markers have been identified, analyses of the kind that follow should become routine. In the meantime, we can be building the large and well documented samples of skeletons required for such research.

Correlating Prehistoric and Historic Groups

Jantz (1974) provides an excellent example of the contribution detailed analysis of skeletons can make to the solution of an archeological problem. Archeologists have shown that the Redbird Focus in the upper Plains area of the central United States is similar culturally to the Lower Loup Focus. Since Lower Loup culminated in the historic Pawnee, they assumed the Redbird Focus was also ancestral to the Pawnee. Wood (1965) suggested, however, that the Redbird Focus was ancestral to the Ponca. Although most of the artifacts were not diagnostic, a few sherds were identified as Stanley Braced Ware, a type of pottery usually associated with the Arikara. Wood maintained that the basic population was Ponca and that the Arikara pottery was either imported or made by women captured from the Arikara.

Jantz examined skeletal samples representing the Arikara, Ponca, Pawnee, and Omaha in an effort to find evidence to resolve the disagreement. He made seven standard measurements on each skull and compared them by multivariate analysis. This showed a clear separation in the males between the Caddoan-speaking Pawnee and Arikara, and the Dehegila Siouan-speaking Omaha and Ponca (Fig. 148). The females exhibited a different pattern. The Arikara and Omaha were distinct, as among the males, but the Ponca and Pawnee were very similar (Fig. 149). According to Jantz, "The intermediate status of the Ponca sample is explicable in terms of Wood's hypothesis that Arikara females were present at Ponca Fort [from which the sample was obtained]. . . . It is also worth noting that, compared to the males, the females show less effective separation. This would result if female mo-

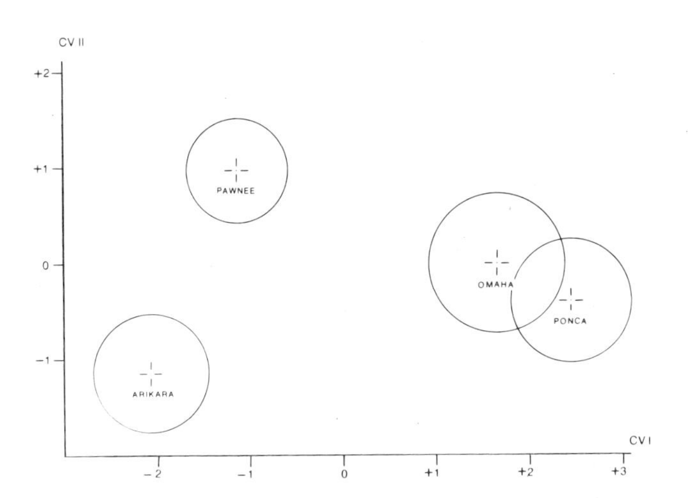

Fig. 148. Biological distance between four groups of northern Plains Indians, as indicated by statistical comparison of seven standard measurements on the skulls of males. These data show a clear separation between the Siouan-speaking Omaha and Ponca and the Caddoan-speaking Arikara and Pawnee (After Jantz 1974: Fig 1).

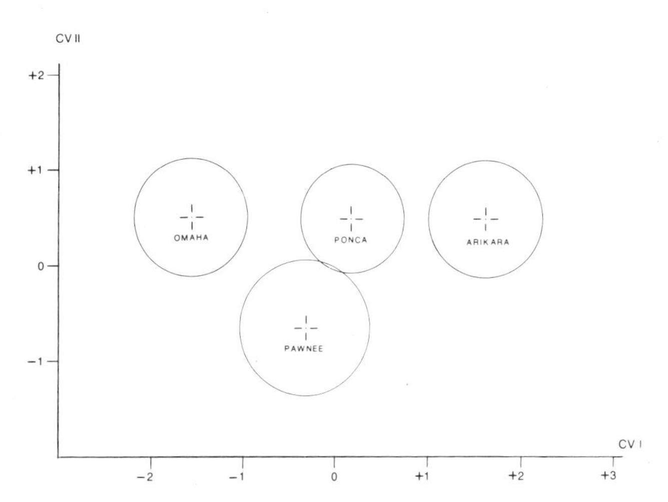

Fig. 149. Biological distance between four groups of northern Plains Indians, as indicated by statistical comparison of seven standard measurements on the skulls of females. The Omaha and Arikara separations are comparable to those obtained from the data on males, but the Ponca and Pawnee females are much more similar (After Jantz 1974: Fig 2).

bility between tribal groups were greater than male mobility. Thus, at any one time, there would be more foreign females than males in a particular sample, tending to minimize intertribal distances" (1974: 9).

When Jantz compared the crania associated with the Redbird Focus with the Arikara (Caddoan) and Omaha (Siouan) samples of known affiliation, he found that they were clearly within the Omaha range of variation. Since the Redbird data were based on only two skulls, the possibility that they are unrepresentative cannot be eliminated until a larger sample has been examined. With this reservation, the evidence from physical anthropology tends to favor Wood's interpretation.

Dental morphology can also assist in establishing relationships between past and present populations. Turner recorded the frequencies of nine commonly studied features on the crowns of the permanent teeth in four Asian populations to establish the ancestry of the Ainu. Two of the samples were archeological and the other two modern. The former consisted of 277 individuals from An-yang, a Chinese site of the late Shang Period dating about 1100 B.C., and 101 individuals from a Jomon Period site of comparable antiquity in Japan. These samples were compared with data obtained from living Ainu and Japanese. Turner concluded that "the dentition of recent Japanese are too much like that of the 3100-year-old An-yang Chinese and too dissimilar to recent Ainu to mean anything other than that modern Japanese could easily be descendants of migrants from north China. The Ainu are probably descended directly from the Jomon population" (1976: 912–3). Furthermore, the hypothesis that the Ainu are remnants of an ancient European-like stock was unsupported; their dentition places them clearly within the Asian group.

Recognizing Interbreeding

Biological data can supplement cultural evidence for interaction between populations. Archeological and ethnohistorical research indicate that the Arikara separated from the Pawnee in what is now Nebraska about A.D. 1500 and moved northward along the Missouri River (Fig. 150). They settled in north-central South Dakota, where they became neighbors of the Mandan and experienced increasingly intense contact with European traders and settlers. Large skeletal samples from five cemetery sites offered a basis for determining whether this cultural interaction was accompanied by physical intermixture. Jantz (1973) made 15 standard measurements on each of the crania and compared

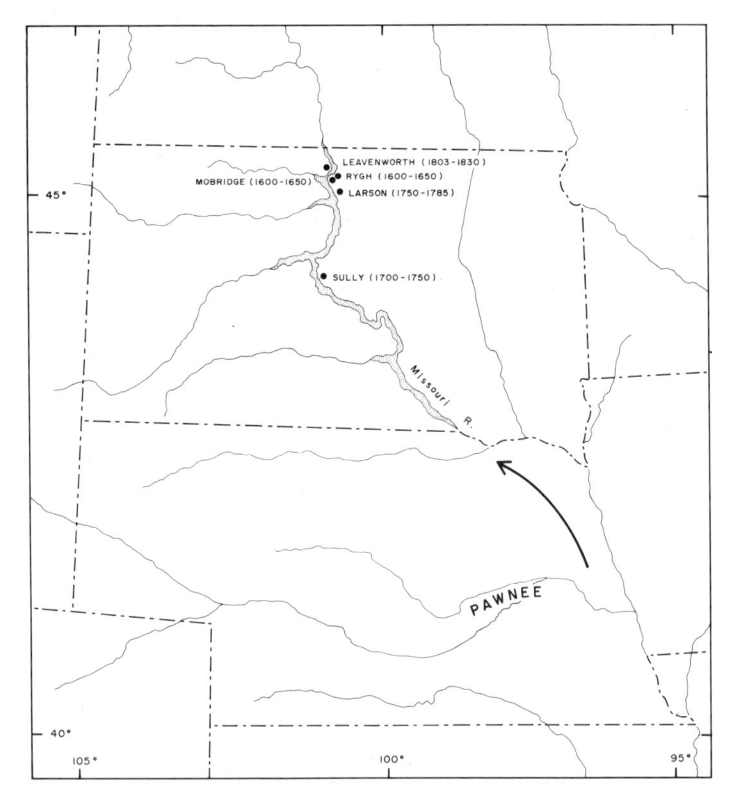

Fig. 150. Locations of five sites in South Dakota occupied by the Arikara between about 1600 and 1830. Analysis of the cranial morphology of skeletons from the cemeteries indicates increasing mixture with neighboring Indian groups and Whites during this period.

them using multivariate statistics. He had two goals: (1) to establish the nature and degree of morphological changes, and (2) to provide a basis for judging whether the changes (if they existed) could be attributed to gene flow resulting from interbreeding.

The five cemeteries were arranged in chronological order and given estimated durations based on the amount of European artifacts included with the burials. No trade objects were encountered in the Mobridge and Rygh cemeteries. In the Sully cemetery, 21 percent of the burials contained items of European origin. This percentage increased to 36 percent in the Larson cemetery and 85 percent in the Leavenworth cemetery. The estimated durations shown on the map are based on analysis of the kinds and quantities of these trade goods.

Jantz' study showed a consistent and directional change in the cranial morphology of the skeletal populations from the earliest to the latest site. Biological change can generally be explained by three factors acting alone or in combination: (1) genetic drift, (2) natural selection, and (3) gene flow. Genetic drift could be ruled out because it produces random variations rather than the kind of directional change observed. Natural selection was a doubtful cause because its effects are seldom evident in so short a time (only about 200 years). This left gene flow as the most likely explanation from the biological point of view. Further support came from a comparison of the Arikara measurements with similar data on Mandans and 17th century British, which showed (1) an increasing similarity in the number of Arikara skulls that could be classified as Mandan or White, and (2) an increasing similarity to both the Mandan and White population samples. Jantz' study thus not only documented a change in Arikara cranial morphology between about 1600 and 1850, but also identified the probable cause as gene flow, and established the process as biological mixture with Mandans and Whites.

Inferring Matrilocal and Patrilocal Residence

Skeletal remains incorporate evidence useful for reconstructing social organization. In their analysis of five Seneca cemeteries in New York, Lane and Sublett (1972) assumed that each cemetery reflected a "rural-neighborhood type settlement pattern of the exogamous units," and that the five cemeteries together represented an endogamous breeding population. They reasoned further that if residence was patrilocal, the skeletons of the males should show little morphological variability within a cemetery, but considerable variability from one cemetery to another. The females, being of external origin, should show greater variability within a cemetery but less between-cemetery variability than the males. If residence were matrilocal, the reverse situation should prevail.

To test these hypotheses, Lane and Sublett computed the frequencies of 33 traits on 290 crania from the five cemeteries. Comparison revealed the patterns of variability expected of patrilocal residence. Since the burials date from A.D. 1850 to 1930, this result corroborates ethnohistorical reports of a shift to patrilocal residence in post-European times.

Differentiating Diffusion and Migration

Observations of dental traits may correct interpretations based on cultural data. In the Wadi Halfa area of Nubia, archeologists reconstructed cultural continuity in a Meroitic population from about the 6th century B.C. to about A.D. 350. At that time, a set of new traits appeared and the resulting configuration dominated lower Nubia for the next two centuries. The cultural differences between the Meroitic and the "X-Group" complexes were sufficiently great that archeologists assigned the latter to an invading population.

To evaluate this hypothesis, Greene (1967) calculated the frequencies of 16 dental traits in large samples of Meroitic, X-Group, and Christian crania. No significant differences among the three groups emerged from the statistical comparisons, strongly suggesting biological continuity. Thus, the cultural differences observed by archeologists "must be explained on the basis of cultural facts and processes rather than on the basis of biological differences [indicating invasion]" (Greene 1967:57).

RECONSTRUCTING DEMOGRAPHY

In recent years, archeologists and physical anthropologists have expressed an expanding interest in problems of prehistoric demography. Archeologists are becoming increasingly concerned with interpreting the sizes, locations, and differential functions of sites and the changes in population concentration in time and space, and with inferring from these kinds of evidence the existence of fluctuations or trends in population size and distribution. Their reconstructions are based mainly on cultural remains, but the relevance of data potentially available from skeletal remains is recognized. These data fall into two principal categories: (1) vital statistics, such as life expectency, probability of death at specified ages, and crude mortality rates, and (2) population size and density. Similar kinds of data and assumptions are involved in both kinds of inferences, but the latter are more difficult to make. Rather than discuss the procedures and hazards in the abstract, I will describe the methodology I employed to reconstruct the demographic situation in southern Maryland about the time of European contact from information obtained by excavating and analyzing the contents of two ossuaries.

Reliability of the Data

Demographic reconstruction employs the procedures developed by demographers to study modern populations except that they derive data from censuses of the living, whereas paleodemographers use censuses of the dead. The reliability of a reconstruction depends on: (1) the accuracy of the estimates of age and sex of the skeletons in the sample, and (2) the extent to which the sample is representative of a population.

Sex and Age Estimates. Obviously, the most reliable methods available must be employed for estimating sex and age. When the skeletons are complete and well preserved, several criteria should be used, with greatest emphasis placed on the data obtained from the most reliable method (see Chapter 3). When the remains are incomplete or comingled, the choice will be dictated by the frequency of the diagnostic bones. For example, tabulating the bones of adults in the Maryland ossuaries revealed that the largest number of individuals were represented by femora. Consequently, I used that bone to estimate age at death. Fortunately, the technique for estimating age from the femur—microscopic cortical remodeling—is the most accurate known. Whatever the methodology, the exact procedures should be described to permit readers to judge the reliability of the reconstructions. If subjective methods are used, the name and qualifications of the person making the identifications should be given. Age estimates recorded on specimens or catalog cards should not be used unless it is certain they were made by a competent investigator.

Sample Validation. The greatest potential for error in demographic reconstructions based on skeletal remains lies in the representativeness of the sample. Any demographic statement about an extinct population rests on the assumption that the numbers, ages, and sexes of the skeletons accurately reflect the death rate in the population or that any bias can be recognized and taken into account. An inadequate sample can result in grossly inaccurate interpretations. Sources of error include: (1) undetected differential disposal of the dead, with the result that some categories of subadults or adults are not represented in the cemeteries; (2) inadequate archeological sampling of a cemetery, if locations of burials were not random, and (3) selection by the excavator of well preserved, complete, adult individuals.

The remarkable variability in aboriginal burial customs seriously limits the potential utility of many prehistoric skeletal samples for demographic reconstruction. Several methods were often employed by a single population. The discovery of a cremation demonstrates that this method was used, but does not rule out the possibility of other forms of burial. If only the young adults were cremated and individuals of other ages were treated differently, a demographic reconstruction based on the cremations would be very biased. Many peoples use different procedures for infants, whose consequent absence from an archeological sample would distort the demographic reconstruction. Ethnohistoric data on related groups may provide information useful for evaluating the adequacy of skeletal samples. In the absence of such data, a sample cannot be used with complete assurance. If both sexes are represented in all the expected age ranges, the sample is likely to be reliable but all uncertainty can never be eliminated.

A second handicap stems from the sampling procedure. Suppose that a population buried all its dead in one cemetery, but did not distribute them randomly. One section was allocated to children, another to older males, others to individuals in other categories. A small excavation in the area reserved for older males would produce a biased sample, which would give a distorted picture of the demographic situation. It is difficult to assess the existence of this practice without completely excavating a cemetery. If this is impossible, large samples should be collected from several parts.

Differential preservation is another source of sampling error. When conditions are unfavorable, the fragile bones of the infants, the children, and the elderly are usually destroyed. The skeletons of adults may survive, but erosion may make determination of ages unreliable. The problems are compounded when part of a cemetery has been destroyed by erosion, construction, or other sources of disturbance. In the absence of information to the contrary, it must be assumed that the missing skeletons had the same age and sex distributions as those remaining. Of course, if spatial segregation was practiced in interment, the surviving sample is likely to be biased.

A final limitation of skeletal samples is the bias introduced by the excavator or the museum curator. The necessity of saving all bones has not always been appreciated. Fragmentary remains, infants, and even post-cranial bones of well preserved adults have often been discarded, only the complete skulls being saved. No museum collection should be used for demographic reconstruction without investigating its history before and after its arrival at the museum.

Few archeological samples meet all the requirements. The Maryland ossuaries are notable exceptions for several reasons. First, the Indians collected the remains of all their dead periodically for burial in a communal grave. Second, all bones were discovered during excavation. Third, every scrap was saved for analysis. Nevertheless, we cannot be certain the sample is complete. Some individuals may have died away from their villages and their bodies not recovered, or there may have been another method of disposal for selected individuals that was not reported in the ethnohistoric literature. Although these possibilities must be kept in mind, the ossuaries still provide unusually well controlled samples.

It should be remembered that each ossuary contains the remains of those who died over a period of about three or four years. Any demographic data calculated from these samples thus represents an average over that length of time. In larger cemeteries, the period of use is usually much longer and the probability of demographic variation correspondingly greater. A demographic reconstruction based on such a sample still represents only an average for the duration of its use. Fluctuations can be detected only by subdividing the burials into shorter chronological intervals based on associated cultural materials.

Assembling Data on Sex and Age

The first step in demographic reconstruction is allocating the individuals to categories according to age at death. Five-year periods are usually used because they are long enough to encompass most of the probable error in the age estimates, yet short enough to allow patterns in death rate to be recognized easily. All the individuals in a sample should be assigned to

Table 30. Age distribution of individuals interred in Ossuaries I and II in Maryland.

	Ossuary I		Ossuary II	
Age Interval	No.	%	No.	%
0.0–4.9	36	29.03	56	32.37
5.0–9.9	14	11.29	12	6.94
10.0–14.9	6	4.84	7	4.05
15.0–19.9	2	1.61	14	8.09
20.0–24.9	5	4.03	8	4.62
25.0–29.9	14	11.29	10	5.78
30.0–34.9	16	12.90	18	10.40
35.0–39.9	13	10.48	17	9.83
40.0–44.9	9	7.26	11	6.36
45.0–49.9	7	5.65	9	5.20
50.0–54.9	2	1.61	7	4.05
55.0–59.9	0	0	3	1.73
60.0–64.9	0	0	0	0
65.0–69.9	0	0	1	0.58

an age category regardless of the state of preservation or the difficulty of determining age. More error is introduced by omitting a fragmentary skeleton that is difficult to age than by including it. When possible, age categories should be divided into males and females to permit observation of demographic differences between the sexes. The numbers and percentages of individuals in each age category constitute the basic data for all reconstructions. Table 30 provides these data for the two Maryland ossuaries.

Life Expectancy

Mortality Curve. The next step is to plot the percentage of individuals in each age category in the form of a mortality curve. This is a demographic profile of the population. The mortality curves for the Maryland ossuaries show several interesting features (Fig. 151). Note the high death rate between birth and five years of age, followed by a dramatic reduction during adolescence. Imagine how the shape of this curve would be affected if the archeologist had excluded infants from the sample or if they had not been placed in the ossuaries by the Indians.

The importance of using the most accurate method for determining age at death is obvious when the adult mortality curves calculated from different criteria are compared (Fig. 152). One pair of curves is based on ages estimated from macroscopic changes in the pubic symphysis; the other pair is derived from microscopic observation of femoral remodeling. The latter data suggest more individuals lived beyond the age of 45 and fewer died between 20 and 40. The curves obtained from the microscopic method are more accurate for two reasons: (1) the method itself is more reliable for establishing age in older adults, and (2) the number of individuals represented by femurs was larger than the number represented by pubic bones.

Survivorship. The survivorship curve is the reverse of the mortality curve. It indicates what percentage of a theoretical original population of 100 persons remains alive at the end of each five-year period. The survivorship curves for the ossuaries (Fig. 153) show that 71 percent of all individuals born in the population represented by Ossuary I were still alive after five years, whereas only 68 percent of those in the population represented by Ossuary II were living. The number of survivors was about equal in the two

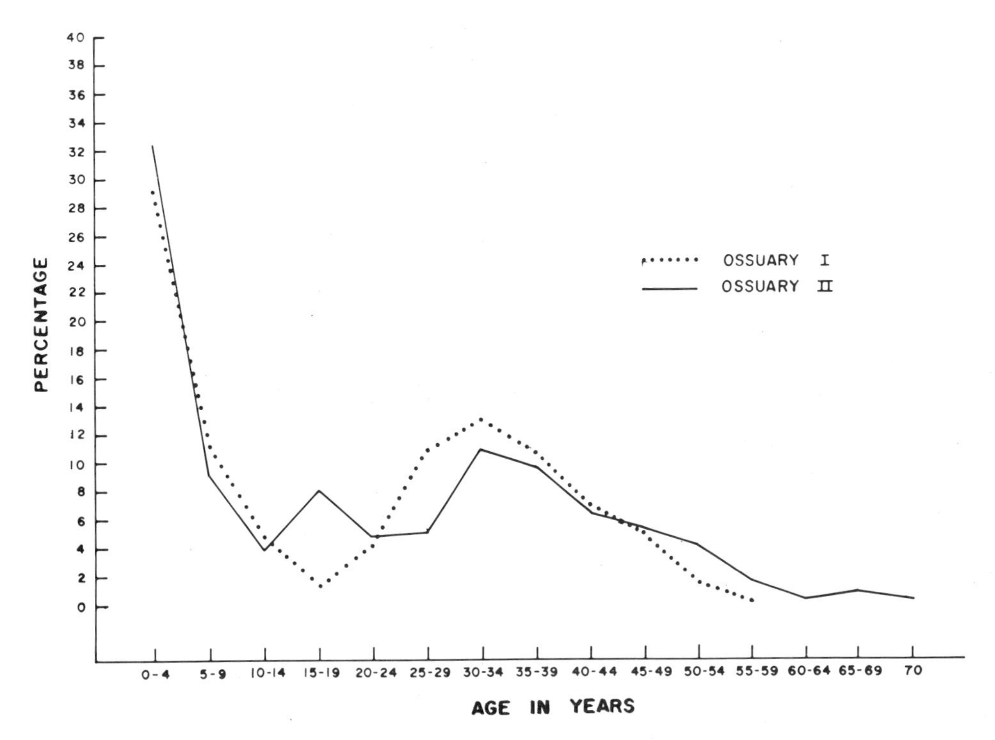

Fig. 151. Mortality curves reconstructed for the populations represented in two ossuaries in Maryland. There was a higher death rate in the population associated with Ossuary II between the ages of 10 and 20, but more individuals survived beyond the age of 59.

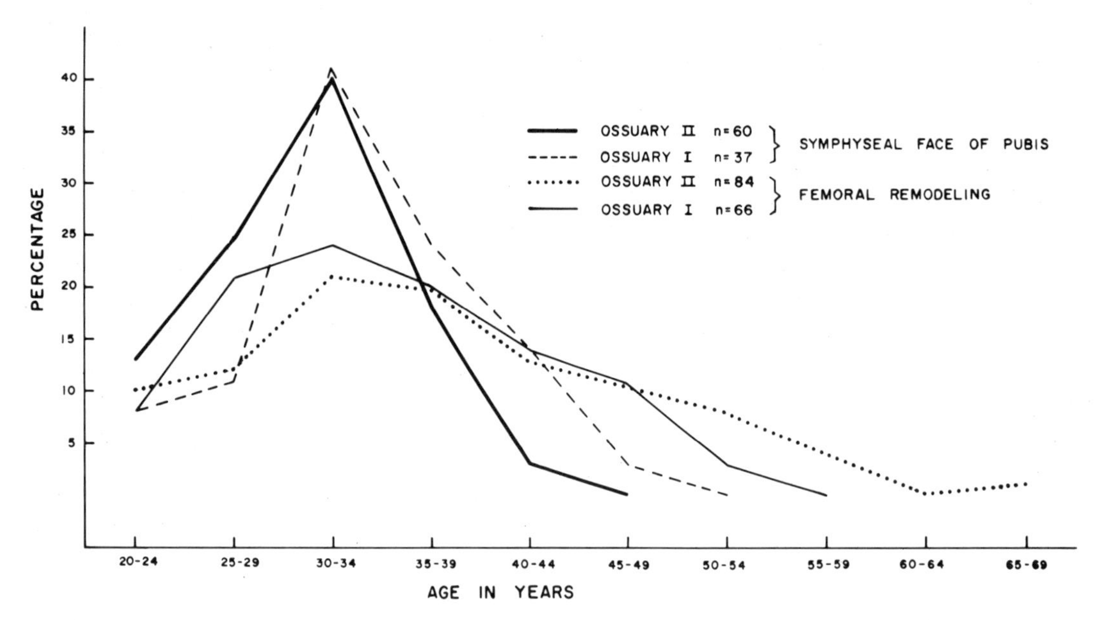

Fig. 152. Adult mortality curves calculated from ossuary remains using two criteria for estimating age at death. The estimates obtained from changes in the symphyseal face of the pubis imply a mortality between the ages of 30 and 34 almost double that estimated from femoral remodeling. The later method also indicates greater longevity.

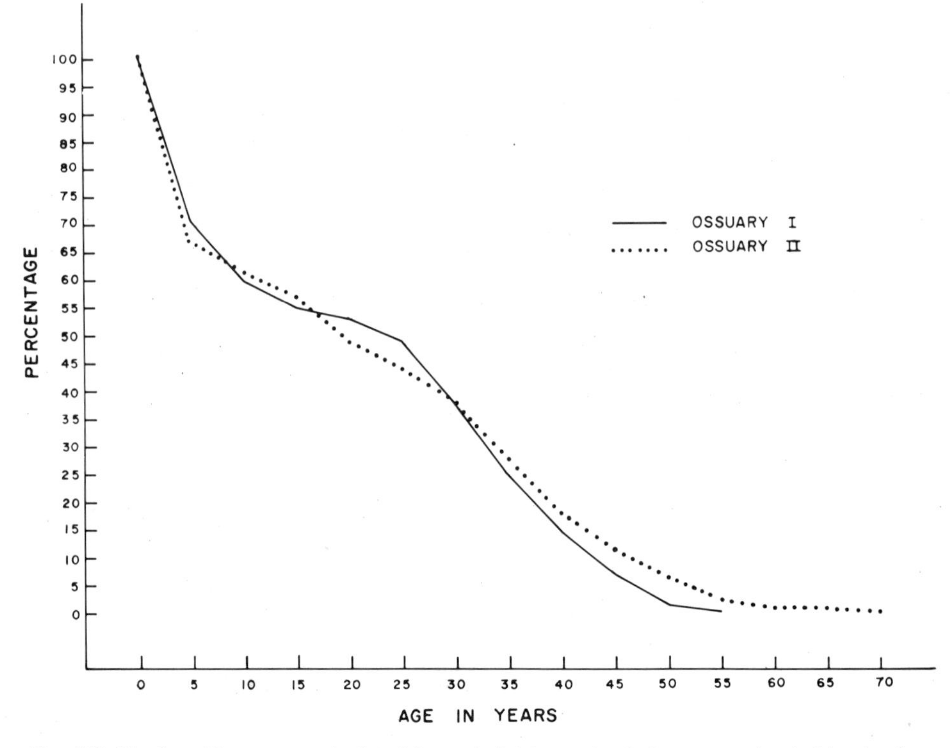

Fig. 153. Survivorship curves calculated from skeletal remains in two ossuaries in Maryland. These curves show the percentage of a theoretical original population of 100 persons still living at the end of each five-year period.

populations after 15 years. A higher mortality rate in population II between the ages of 20 and 25 years created a greater survivorship in population I, but the death rate accelerated in population I between the ages of 25 and 30 so that the percentage of survivors was again equal by the age of 30. After 30, the longevity is greater in population II. The curve of survivorship accentuates the differences between the two populations, showing a higher infant and adolescent mortality but greater adult longevity in population II.

Life Table. A third form of expressing demographic data is the life table. This is the table used by life insurance companies to summarize population statistics and estimate life expectancy, with two important differences: (1) life tables for living populations are usually divided by one-year intervals, whereas tables for prehistoric populations use five-year intervals to compensate for errors in estimating age, and (2) modern life tables are constructed using known ages obtained from a statistically valid sample of a larger population, whereas prehistoric life tables are based on estimated ages at death of all available members of a population. All life tables present averages that do not express the internal variation that always exists in time and space.

Table 31 is an example of a life table calculated from a prehistoric population, in this case Ossuary II in Maryland. The age interval (x) is five years to accommodate most of the error in estimating age at death. This five-year interval begins at age x and ends just before the following x value (that is, 0–4 years, 5–9 years, etc.). The number of deaths (Dx) is the number of skeletons in the sample with ages that fall within the limits of x (The totals are slightly higher in the intervals between 20 and 55 years than those on Table 12 because corrections were made to include individuals not represented by femora). The percentages of the deaths (dx) is the figure Dx expressed as a percentage of the total number of individuals in the ossuary (188).

The "survivors entering" column (1x) presents the data depicted by the survivorship curve (Fig. 153). The number of survivors in each interval is the percentage of the original population still alive at the beginning of the interval. It is calculated by subtracting the percentage of deaths (dx) during the preceding interval from the percentage of survivors (1x) in the same interval. The probability of death (qx) is calculated by dividing the percentage of deaths (dx) during an interval by the number of survivors entering that interval (1x).

The column Lx shows the total number of years lived by all individuals during each interval. This value is obtained by the formula

$$Lx = \frac{5\,(lx + lo)}{2}$$

where lx is the number of survivors entering interval

Table 31. Life table reconstructed from skeletons in Ossuary II in Maryland.

Age Interval (x)	No. of Deaths (Dx)	% of Deaths (dx)	Survivors Entering (lx)	Probability of Death (qx)	Total Years Lived Between X and X + 5 (Lx)	Total Years Lived After Lifetime (Tx)	Life Expectancy (e°x)
0	56	29.79	100.00	.2979	425.525	2297.900	22.98
5	12	6.38	70.21	.0909	335.100	1872.375	26.67
10	7	3.72	63.83	.0583	309.850	1537.275	24.08
15	14	7.45	60.11	.1239	281.925	1227.425	20.42
20	9	4.79	52.66	.0910	251.325	945.500	17.95
25	12	6.38	47.87	.1333	223.400	694.175	14.50
30	21	11.17	41.49	.2692	179.525	470.775	11.35
35	20	10.64	30.32	.3509	125.000	291.250	9.61
40	13	6.91	19.68	.3511	81.125	166.250	8.45
45	11	5.85	12.77	.4581	49.225	85.125	6.67
50	8	4.26	6.92	.6156	23.950	35.900	5.19
55	4	2.13	2.66	.8008	7.975	11.950	4.49
60	0	0.00	0.53	.0000	2.650	3.975	7.50
65	1	0.53	0.53	1.0000	1.325	1.325	2.50
70	0	0.00	0.00	.0000	0.000	0.000	0.00

x and lo is the number of survivors entering the following interval. The column Tx indicates the total number of years remaining in the lifetimes of all the individuals entering each age interval. It is calculated by adding the values in the Lx column for that interval and all succeeding intervals.

The last column, life expectancy ($e^{o}x$), represents the average number of years an individual entering age interval x can expect to continue to live. It is derived by the formula

$$e^{o} = \frac{Tx}{lx}$$

The life expectancy at birth of the individuals from Ossuary II was ascertained by dividing 2297.900 (Tx) by 100.00 (1x), which is 22.98 years. The table shows that a person who survived to the age of five could expect to live 27.67 more years. Life expectancy declines steadily thereafter.

The utility of a life table depends on the accuracy of the data on which it is based. Inadequate samples and erroneous ages make it meaningless. The data can also be influenced by fertility rates and rates of population growth and decline.

Crude Mortality Rate. The crude mortality rate of a population is the average number of individuals that die per thousand per year. Assuming that the rate of deaths is constant, the crude mortality rate may be calculated directly from a life table by the formula

$$M = \frac{1000}{e^{o}_{o}}$$

where M is the crude mortality rate and e^{o}_{o} is life expectancy at birth. Applying this formula to the life table for Ossuary II (Table 31) gives a crude mortality rate of 43.52. This means that about 43 persons died out of each 1000 in the population each year. Comparisons of the crude mortality rates among different groups can reveal important differences and thus suggest problems for investigation.

Population Size

The crude mortality rate offers a basis for reconstructing the size of the population to which it applies. Since it specifies the number of individuals per thousand who died each year, one needs only to know the total number of deaths and the length of time a cemetery was used to obtain an estimate of the total population size. The formula is

$$P = \frac{1000\ N}{MT}$$

where P is the size of the population, N is the number of deaths represented by the skeletal remains, M is the crude mortality rate, and T is the number of years the cemetery was in use. The values of N and M can be calculated from the life table for Ossuary II. N is the sum of the Dx column or 188 individuals. M is 43.52 years. T is estimated at 3 years from the relative amount of articulation of the bones (see p. 31). Inserting these values into the formula gives a population of 1441. When allowance is made for errors in the figures incorporated into the calculation, an estimate between 1300 and 1600 seems reasonable.

This estimate can be evaluated using archeological and ethnohistorical information. A habitation site adjacent to the ossuary was too small to have accommodated so many people, suggesting that several villages may have buried their dead communally. The average population of a village can be inferred from the life table and Captain John Smith's estimate that in 1608 a village contained an average of 50 warriors, if two assumptions are made: (1) all males between the ages of 15 and 40 were warriors, and (2) half the individuals in Ossuary II assigned these ages were male. Seventy-six skeletons in Ossuary II were between the ages of 15 and 40 (Table 31). If half were males, the warrior count would be 38. This is about 20 percent of the total number of individuals in the ossuary (188). Applying this ratio of one warrior to four non-warriors to the living population, and using Smith's estimate of 50 warriors per village provides an estimated village population of about 250.

This figure can be evaluated by information provided by a map made by John Smith in 1612, which lists 28 villages in the region of the ossuaries (Arber 1884: 384–5). Five are indicated as containing "chief's houses," implying that a chief governed five or six villages. If these groups of villages joined together to bury their dead periodically, then the estimate of between 1300 and 1600 generated from the life table would represent the combined populations of five or six villages each containing about 250 persons. This agrees with the calculation obtained from Smith's warrior count and with the evidence provided by the area of the archeological habitation site.

The preceding discussion illustrates the kinds of population estimates that can be made when the duration of a cemetery is known and the representation of deceased members of the community is essentially complete. Unfortunately, these two conditions are rarely met. Although a reliable life table can be generated from most large and well documented cemetery samples, the two most important variables in estimating population size—the total number of individuals who died and the number of years the cemetery was in use—can seldom be determined accurately. Errors in these figures can produce grossly distorted estimates.

Demographic reconstruction from skeletal remains is not difficult mathematically. The skill rests in acquiring a suitable sample and providing the necessary documentation. When the prerequisites can be met, extensive information of considerable importance results. When they cannot be met, an erroneous and misleading picture will be obtained.

RECONSTRUCTING DIET

Prehistorians have long been interested in reconstructing the diets of past human populations because of the important role of subsistence changes in biological and cultural evolution. Traditionally, most evidence has come from the analysis of faunal and floral remains, settlement patterns, and subsistence-related artifacts. The recent discovery that certain aspects of the chemical composition of human bone are determined by the types of food consumed provides a basis for obtaining more direct, specific, and detailed information. Two principal kinds of indicators are being explored: (1) isotopic ratios and (2) trace-element frequencies.

Isotopic Ratios

Carbon Isotopes. Plants can be classified into two categories according to the manner in which they metabolize carbon dioxide during photosynthesis. One process produces a compound with three carbon atoms (C3 plants) and the other a compound with four carbon atoms (C4 plants). These photosynthetic pathways equate with different proportions of the stable carbon isotopes, carbon-12 and carbon-13. Since the carbon in bone collagen comes from foods consumed, the isotopic ratios reflect the proportions of C3 and C4 plants ingested. What makes this distinction significant for reconstructing prehistoric human diets is the fact that maize, sorghum, and millet are C4 plants (van der Merwe 1982).

A dramatic application of this correlation is provided by a series of skeletons representing populations of the north-central United States during five millennia prior to European contact. The isotopic measurements showed relatively constant carbon-13 values until about A.D. 500 (Fig. 154). Then,

> During the period around A.D. 1000–1200, these values changed from about −21.4 to −12, which means that the proportion of carbon from C4 plants in the collagen went from zero to more than 70 percent. . . . This is not quite the same as saying that maize formed more than 70 percent of the diet by A.D. 1200, but close enough for the purpose of this discussion. The rapid acceptance of agriculture in exchange for a former hunter-gatherer way of life which had endured for thousands of years brought other revolutiontary changes in its wake (van der Merwe 1982:602).

In regions where several edible C4 plants are available, conclusions may be less definitive. A further complication is that high levels of carbon-13 occur in certain kinds of marine animals exploited by humans

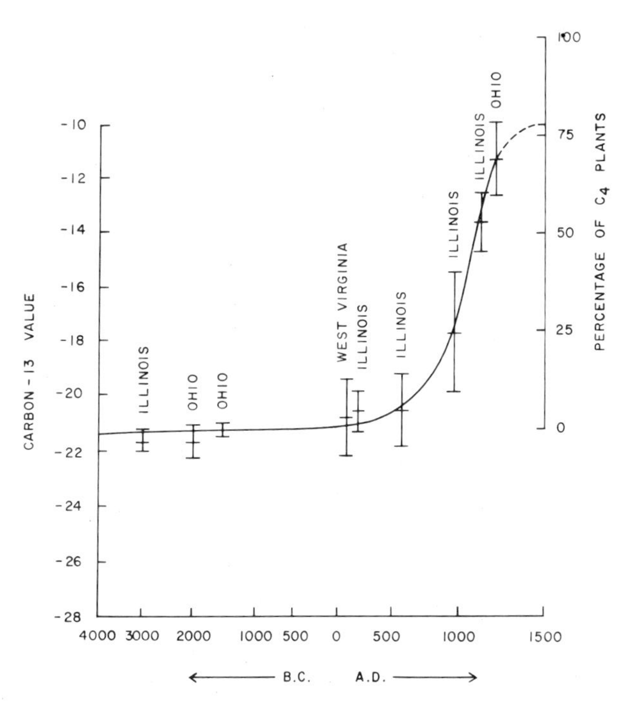

Fig. 154. Carbon-13 values in prehistoric North American populations dating between 3000 B.C. and European contact. The pronounced change beginning about A.D. 500 implies a dietary shift from local C3 plants to maize, a C4 plant (After van der Merwe 1982: Fig. 6).

for food. The chemical results must, therefore, be assessed in the context of the total potential subsistence resources available to a prehistoric population. Some of the experimental problems and results are reviewed by DeNiro and Epstein (1978), Burleigh and Brothwell (1978), van der Merwe (1982), and Vogel and van der Merwe (1977).

Nitrogen Isotopes. Like carbon, nitrogen occurs in two isotopic states, nitrogen-14 and nitrogen-15. Experiments indicate that the N-15 ratios are related to diet, but vary in different kinds of tissues within a single animal and according to location and season of collection of the foods ingested (DeNiro and Epstein 1981). The method appears most useful at present for establishing the relative amounts of marine and terrestrial foods, and of legumes and non-legumes in the diet.

This method is particularly relevant for reconstructing human diets in the New World because legumes (beans) were a staple cultigen. Increase in the ratio of N-15 should, therefore, reflect increasing dependence on agriculture. When this hypothesis was tested using estimates of dietary composition during the prehistory of the Tehuacán Valley, Mexico, the results were unexpected. Whereas the data obtained from plant remains and coprolite composition led MacNeish to infer increasing consumption of beans from 6000 B.C. to European contact, the N-15 values derived from bone collagen implied a decrease (DeNiro and Epstein 1981: Fig. 5). Both methods are subject to numerous potential sources of error and incorporate assumptions that may affect their reliability. The contradictory results call attention to the complexity of the problem of inferring diet from archeological samples and suggest that isotopic analysis may offer a more reliable basis for estimating the proportions of available foods that were actually consumed.

In another study, Schoeninger, DeNiro, and Tauber (1983) established that N-15/N-14 ratios in bone collagen are higher among laboratory animals feeding on marine foods than on terrestrial foods. Applying this finding to four human populations whose principal dietary resources were either marine (Alaskan Eskimos; Haida and Tlingit Indians) or terrestrial (New Mexican and Colombian agriculturalists) also showed significant differences, indicating that "stable nitrogen isotope ratios of bone collagen can be used in reconstructing the relative amounts of marine and terrestrial food sources in diets of historic and prehistoric human populations" (op. cit.: 1381).

Trace-mineral Analysis

Very small quantities of certain minerals in the skeleton are potential clues to dietary composition. Strontium enters the land-based food chain from the soil and ground water via the roots of plants. The amount of strontium gradually decreases moving up the food chain. Thus, humans who consume root crops should have higher levels of strontium than those who eat leafy vegetables and grains. The lowest levels should occur among groups living principally on meat.

As with carbon isotopes, strontium levels differ in marine and terrestrial foods, creating potential sources of error in interpretation. Post-mortem alterations of strontium levels may occur in certain environmental circumstances, causing further complications. The method appears promising, but more research is required to identify and quantify the variables involved. Sillen and Kavanaugh (1982) provide an excellent review of its status.

BIOCULTURAL INTERPRETATIONS

This chapter has highlighted some of the information that can be gleaned about individuals and populations from skeletal analysis. Demography, disease, stature, diet, and other aspects reconstructed from skeletal remains provide an integrated view of biology and health that can be correlated with cultural factors (Blakey 1977). Saul (1972) has labelled this approach "osteobiographic analysis." He applied it to 90 human skeletons from the Mayan ceremonial site of Altar de Sacrificios, tabulating data on stature, sex and age distribution, deformation, and disease, and comparing them with other population samples from the Maya area. This analysis led him to conclude that a temporal shift in stature was a consequence of environmental factors rather than the influx of a new population of somewhat different genetic composition.

Buikstra (1976) integrated data from the study of skeletons from the lower Illinois Valley with information provided by archeologists on settlement patterns, mortuary customs, subsistence, and other cultural behavior. She found patterns of age and sex in the mortuary contexts that suggested status was inherited. Correlating demographic profiles with archeological evidence for settlement size and pattern permitted inferences about the size and composition of the Hopewell community and the biocultural factors

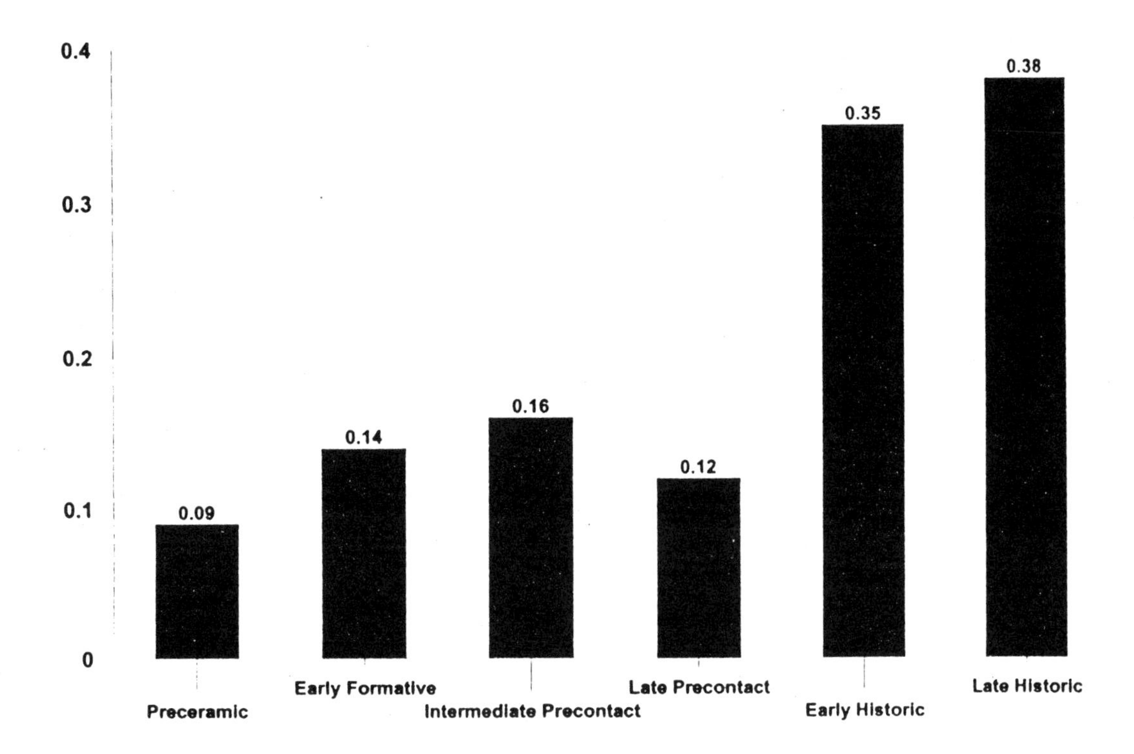

Fig. 155. The ratio of the number of bones with abnormal periosteal lesions to the number of adults in Ecuadorian samples, showing a marked increase after European contact. The type of lesions described here can be produced by infection, although other pathological conditions can not be ruled out.

influencing representation in the cemeteries excavated.

Prehistorians have long been interested in the process of plant domestication and the impact of agriculture on human populations. The identification of chemical differences and the development of techniques for extracting this information from skeletons promise to make possible more precise and comprehensive reconstructions of prehistoric diets. Well designed research on human remains representing several time periods in the same general region can show the biological impacts of shifts in subsistence and other kinds of cultural factors. For example, Lallo et al (1977) found that the frequencies of categories of disease correlated with temporal changes in subsistence among prehistoric populations in Illinois. Larsen (1980a, 1980b) observed an increase in dental caries and other indicators of biological stress with the introduction and growing dependence on maize agriculture among prehistoric Indians from Georgia. A series of samples from Ecuador has produced similar results (Ubelaker 1980a, 1980c, 1981, 1983, 1984). The samples extend from the preceramic and preagricultural period about 6000 B.C. to the late Integration Period, just prior to Spanish contact. Dramatic increases occur in nearly all indicators of biological stress about the beginning of the Regional Developmental Period, a few centuries prior to the Christian Era. These include frequencies of dental caries, tooth loss, dental hypoplasia, trauma, and infectious disease (Fig. 155). Dental caries and dental hypoplasia continue to increase thereafter, whereas the other skeletal indicators of stress show moderate declines. These data suggests the shift of primary dependence on agriculture had a negative effect on the health of the population.

These examples are multiplied in the rapidly expanding literature on biocultural interpretations of human skeletal remains. Their accuracy is determined by the sophistication of the archeological research design, the quality of the excavation, and the methods of processing the human remains. I hope that this volume contributes to the development of more reliable data and the intellectual growth of this fascinating and important subdiscipline of anthropology.

Appendix 1 Tables

Table 1. Expected maximum stature of U.S. White males estimated from maximum long-bone lengths (After Trotter 1970, Appendix Table 28).

Hum	*Rad*	*Ulna*	*Stature*		*Fem*	*Tib*	*Fib*	*Fem + Tib*
mm	*mm*	*mm*	*cm*	*in***	*mm*	*mm*	*mm*	*mm*
265	193	211	152	59^7	381	291	299	685
268	196	213	153	60^2	385	295	303	693
271	198	216	154	60^5	389	299	307	701
275	201	219	155	61	393	303	311	708
278	204	222	156	61^3	398	307	314	716
281	206	224	157	61^6	402	311	318	723
284	209	227	158	62^2	406	315	322	731
288	212	230	159	62^5	410	319	326	738
291	214	232	160	63	414	323	329	746
294	217	235	161	63^3	419	327	333	753
297	220	238	162	63^6	423	331	337	761
301	222	240	163	64^1	427	335	340	769
304	225	243	164	64^5	431	339	344	776
307	228	246	165	65	435	343	348	784
310	230	249	166	65^3	440	347	352	791
314	233	251	167	65^6	444	351	355	799
317	235	254	168	66^1	448	355	359	806
320	238	257	169	66^4	452	359	363	814
323	241	259	170	66^7	456	363	367	821
327	243	262	171	67^3	461	367	370	829
330	246	265	172	67^6	465	371	374	837
333	249	267	173	68^1	469	375	378	844
336	251	270	174	68^4	473	379	381	852
339	254	273	175	68^7	477	383	385	859
343	257	276	176	69^2	482	386	389	867
346	259	278	177	69^5	486	390	393	874
349	262	281	178	70^1	490	394	396	882
352	265	284	179	70^4	494	398	400	889
356	267	286	180	70^7	498	402	404	897
359	270	289	181	71^2	503	406	408	905
362	272	292	182	71^5	507	410	411	912
365	275	294	183	72	511	414	415	920
369	278	297	184	72^4	515	418	419	927
372	280	300	185	72^7	519	422	422	935
375	283	303	186	73^2	524	426	426	942
378	286	305	187	73^5	528	430	430	950
382	288	308	188	74	532	434	434	957
385	291	311	189	74^3	536	438	437	965
388	294	313	190	74^6	540	442	441	973
391	296	316	191	75^2	545	446	445	980
395	299	319	192	75^5	549	450	449	988
398	302	321	193	76	553	454	452	995
401	304	324	194	76^3	557	458	456	1003
404	307	327	195	76^6	561	462	460	1010
408	309	330	196	77^1	566	466	463	1018
411	312	332	197	77^4	570	470	467	1026
414	315	335	198	78	574	474	471	1033

*The expected maximum stature should be reduced by the amount of 0.06 (age in years — 30) cm to obtain expected stature of individuals over 30 years of age.

**The raised number indicates the numerator of a fraction of an inch expressed in eighths, thus 59^7 should be read 59⅞ inches.

Table 2. Expected maximum stature of U.S. Black males estimated from maximum long-bone lengths (After Trotter 1970, Appendix 2, Table 28).

Hum	*Rad*	*Ulna*	*Stature*		*Fem*	*Tib*	*Fib*	*Fem + Tib*
mm	*mm*	*mm*	*cm*	*in***	*mm*	*mm*	*mm*	*mm*
276	206	223	152	59^7	387	301	303	704
279	209	226	153	60^2	391	306	308	713
282	212	229	154	60^5	396	310	312	721
285	215	232	155	61	401	315	317	730
288	218	235	156	61^3	406	320	321	739
291	221	238	157	61^6	410	324	326	747
294	224	242	158	62^2	415	329	330	756
297	226	245	159	62^5	420	333	335	765
300	229	248	160	63	425	338	339	774
303	232	251	161	63^3	430	342	344	782
306	235	254	162	63^6	434	347	349	791
310	238	257	163	64^1	439	352	353	800
313	241	260	164	64^5	444	356	358	808
316	244	263	165	65	449	361	362	817
319	247	266	166	65^3	453	365	367	826
322	250	269	167	65^6	458	370	371	834
325	253	272	168	66^1	463	374	376	843
328	256	275	169	66^4	468	379	381	852
331	259	278	170	66^7	472	383	385	861
334	262	281	171	67^3	477	388	390	869
337	264	284	172	67^6	482	393	394	878
340	267	287	173	68^1	487	397	399	887
343	270	291	174	68^4	491	402	403	895
346	273	294	175	68^7	496	406	408	904
349	276	297	176	69^2	501	411	413	913
352	279	300	177	69^5	506	415	417	921
356	282	303	178	70	510	420	422	930
359	285	306	179	70^1	515	425	426	939
362	288	309	180	70^7	520	429	431	947
365	291	312	181	71^2	525	434	435	956
368	294	315	182	71^5	529	438	440	965
371	297	318	183	72	534	443	445	974
374	300	321	184	72^4	539	447	449	982
377	302	324	185	72^7	544	452	454	991
380	305	327	186	73^2	548	456	458	1000
383	308	330	187	73^5	553	461	463	1008
386	311	333	188	74	558	466	467	1017
389	314	336	189	74^3	563	470	472	1026
392	317	340	190	74^6	567	475	476	1034
395	320	343	191	75^2	572	479	481	1043
398	323	346	192	75^5	577	484	486	1052
401	326	349	193	76	582	488	490	1061
405	329	352	194	76^3	586	493	495	1069
408	332	355	195	76^6	591	498	499	1078
411	335	358	196	77^1	596	502	504	1087
414	337	361	197	77^4	601	507	508	1095
417	340	364	198	78	605	511	513	1104

*The expected maximum stature should be reduced by the amount of 0.06 (age in years — 30) cm to obtain expected stature of individuals over 30 years of age.

**The raised number indicates the numerator of a fraction of an inch expressed in eighths, thus 59^7 should be read 59⅞ inches.

Table 3. Expected maximum stature of U.S. White females estimated from maximum long-bone lengths (After Trotter 1970, Appendix 3, Table 28).

Hum	*Rad*	*Ulna*	*Stature*		*Fem*	*Tib*	*Fib*	*Fem + Tib*
mm	*mm*	*mm*	*cm*	*in***	*mm*	*mm*	*mm*	*mm*
244	179	193	140	55^1	348	271	274	624
247	182	195	141	55^4	352	274	278	632
250	184	197	142	55^7	356	277	281	639
253	186	200	143	56^2	360	281	285	646
256	188	202	144	56^6	364	284	288	653
259	190	204	145	57^1	368	288	291	660
262	192	207	146	57^4	372	291	295	668
265	194	209	147	57^7	376	295	298	675
268	196	211	148	58^2	380	298	302	682
271	198	214	149	58^5	384	302	305	689
274	201	216	150	59	388	305	309	696
277	203	218	151	59^4	392	309	312	704
280	205	221	152	59^7	396	312	315	711
283	207	223	153	60^2	400	315	319	718
286	209	225	154	60^5	404	319	322	725
289	211	228	155	61	409	322	326	732
292	213	230	156	61^3	413	326	329	740
295	215	232	157	61^6	417	329	332	747
298	217	235	158	62^2	421	333	336	754
301	220	237	159	62^5	425	336	340	761
304	222	239	160	63	429	340	343	768
307	224	242	161	63^3	433	343	346	776
310	226	244	162	63^6	437	346	349	783
313	228	246	163	64^1	441	350	353	790
316	230	249	164	64^5	445	353	356	797
319	232	251	165	65	449	357	360	804
322	234	253	166	65^3	453	360	363	812
324	236	256	167	65^6	457	364	366	819
327	239	258	168	66^1	461	367	370	826
330	241	261	169	66^4	465	371	373	833
333	243	263	170	66^7	469	374	377	840
336	245	265	171	67^3	473	377	380	847
339	247	268	172	67^6	477	381	384	855
342	249	270	173	68^1	481	384	387	862
345	251	272	174	68^4	485	388	390	869
348	253	275	175	68^7	489	391	394	876
351	255	277	176	69^2	494	395	397	883
354	258	279	177	69^5	498	398	401	891
357	260	282	178	70^1	502	402	404	898
360	262	284	179	70^4	506	405	407	905
363	264	286	180	70^7	510	409	411	912
366	266	289	181	71^2	514	412	414	919
369	268	291	182	71^5	518	415	418	927
372	270	293	183	72	522	419	421	934
375	272	296	184	72^4	526	422	425	941

*The expected maximum stature should be reduced by the amount of 0.06 (age in years — 30) cm to obtain expected stature of individuals over 30 years of age.

**The raised number indicates the numerator of a fraction of an inch expressed in eighths, thus 55^1 should be read $55\frac{1}{8}$ inches.

Table 4. Expected maximum stature of U.S. Black females estimated from maximum long-bone lengths (After Trotter 1970, Appendix 4, Table 28).

Hum	*Rad*	*Ulna*	*Stature*		*Fem*	*Tib*	*Fib*	*Fem + Tib*
mm	*mm*	*mm*	*cm*	*in***	*mm*	*mm*	*mm*	*mm*
245	165	195	140	55^{1}	352	275	278	637
248	169	198	141	55^{4}	356	279	282	645
251	173	201	142	55^{7}	361	283	286	653
254	176	204	143	56^{2}	365	287	290	661
258	180	207	144	56^{6}	369	291	294	669
261	184	210	145	57^{1}	374	295	298	677
264	187	213	146	57^{4}	378	299	302	685
267	191	216	147	57^{7}	383	303	306	693
271	195	219	148	58^{2}	387	308	310	701
274	198	222	149	58^{5}	391	312	314	709
277	202	225	150	59	396	316	318	717
280	205	228	151	59^{4}	400	320	322	724
284	209	231	152	59^{7}	405	324	326	732
287	213	235	153	60^{2}	409	328	330	740
290	216	238	154	60^{5}	413	332	334	748
293	220	241	155	61	418	336	338	756
297	224	244	156	61^{3}	422	340	342	764
300	227	247	157	61^{6}	426	344	346	772
303	231	250	158	62^{2}	431	348	350	780
306	235	253	159	62^{5}	435	352	354	788
310	238	256	160	63	440	357	358	796
313	242	259	161	63^{3}	444	361	362	804
316	245	262	162	63^{6}	448	365	366	812
319	249	265	163	64^{1}	453	369	370	820
322	253	268	164	64^{5}	457	373	374	828
326	256	271	165	65	462	377	378	836
329	260	274	166	65^{3}	466	381	382	843
332	264	277	167	65^{6}	470	385	386	851
335	267	280	168	66^{1}	475	389	390	859
339	271	283	169	66^{4}	479	393	394	867
342	275	286	170	66^{7}	484	397	398	875
345	278	289	171	67^{3}	488	401	402	883
348	282	292	172	67^{6}	492	406	406	891
352	285	295	173	68^{1}	497	410	410	899
355	289	298	174	68^{4}	501	414	414	907
358	293	301	175	68^{7}	505	418	418	915
361	296	304	176	69^{2}	510	422	422	923
365	300	307	177	69^{5}	514	426	426	931
368	304	310	178	70^{1}	519	430	430	939
371	307	313	179	70^{4}	523	434	434	947
374	311	316	180	70^{7}	527	438	438	955
378	315	319	181	71^{2}	532	442	442	963
381	318	322	182	71^{5}	536	446	446	970
384	322	325	183	72	541	450	450	978
387	325	328	184	72^{4}	545	454	454	986

*The expected maximum stature should be reduced by the amount of 0.06 (age in years — 30) cm to obtain expected stature of individuals over 30 years of age.

**The raised number indicates the numerator of a fraction of an inch expressed in eighths, thus 55^{1} should be read 55⅛ inches.

Table 5. Norms of root resorption of deciduous mandibular canines and molars of males (top) and females (bottom) (After Moorrees, Fanning, and Hunt 1963a, Figs. 6 and 7).

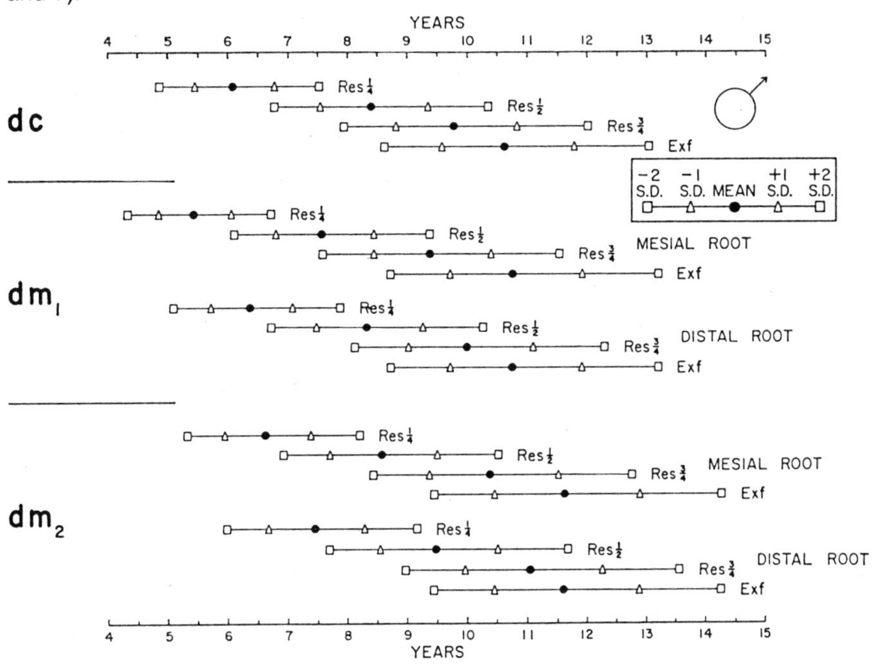

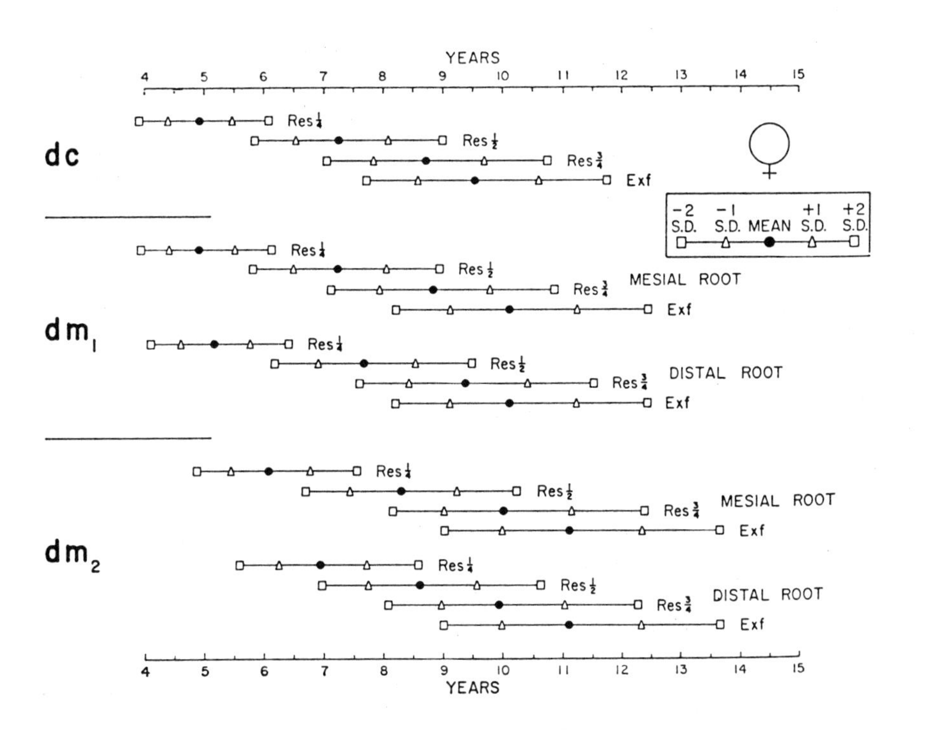

Table 6. Norms of formation of permanent mandibular canines, premolars, and molars of males (After Moorrees, Fanning, and Hunt 1963b, Fig. 5).

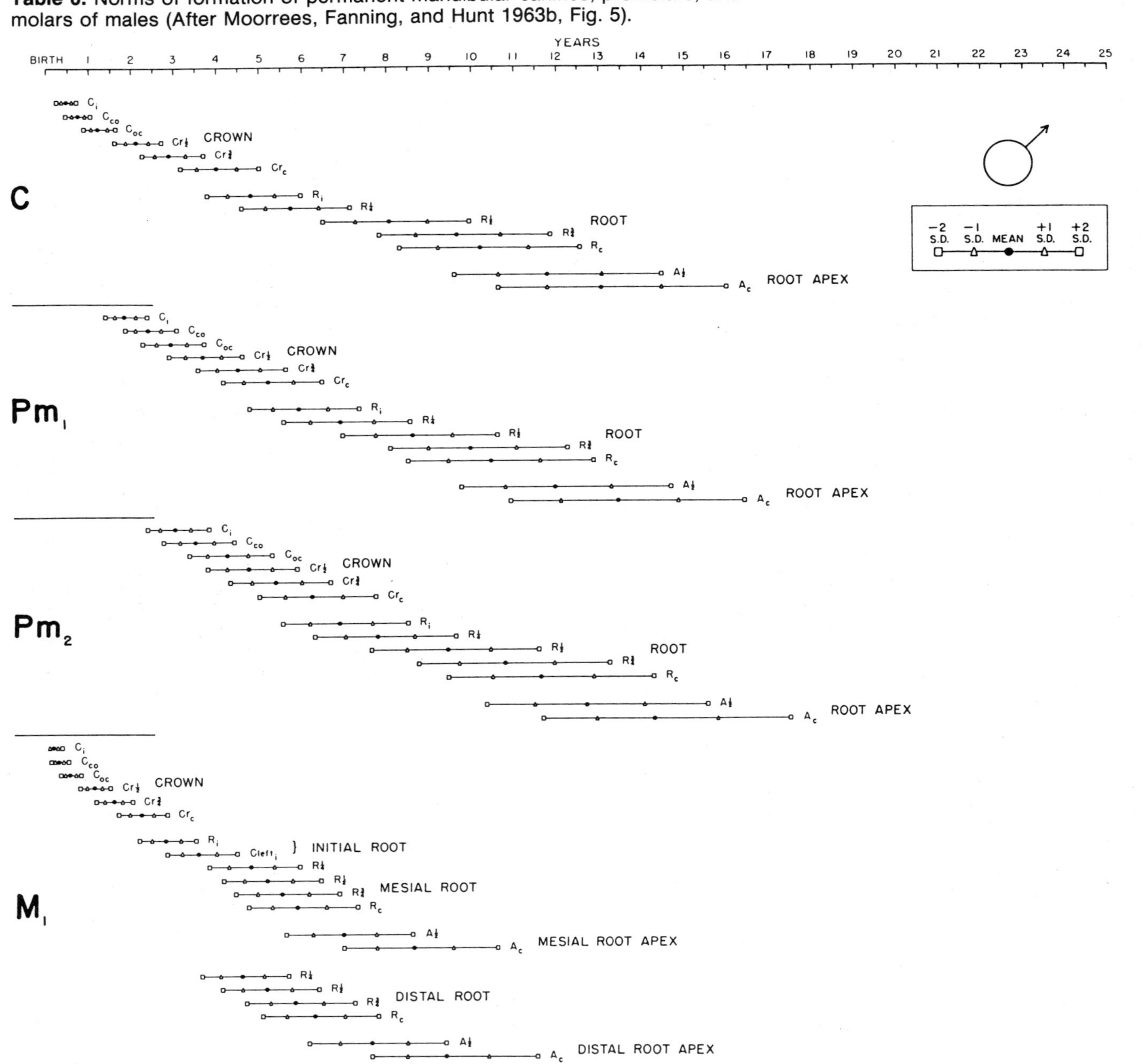

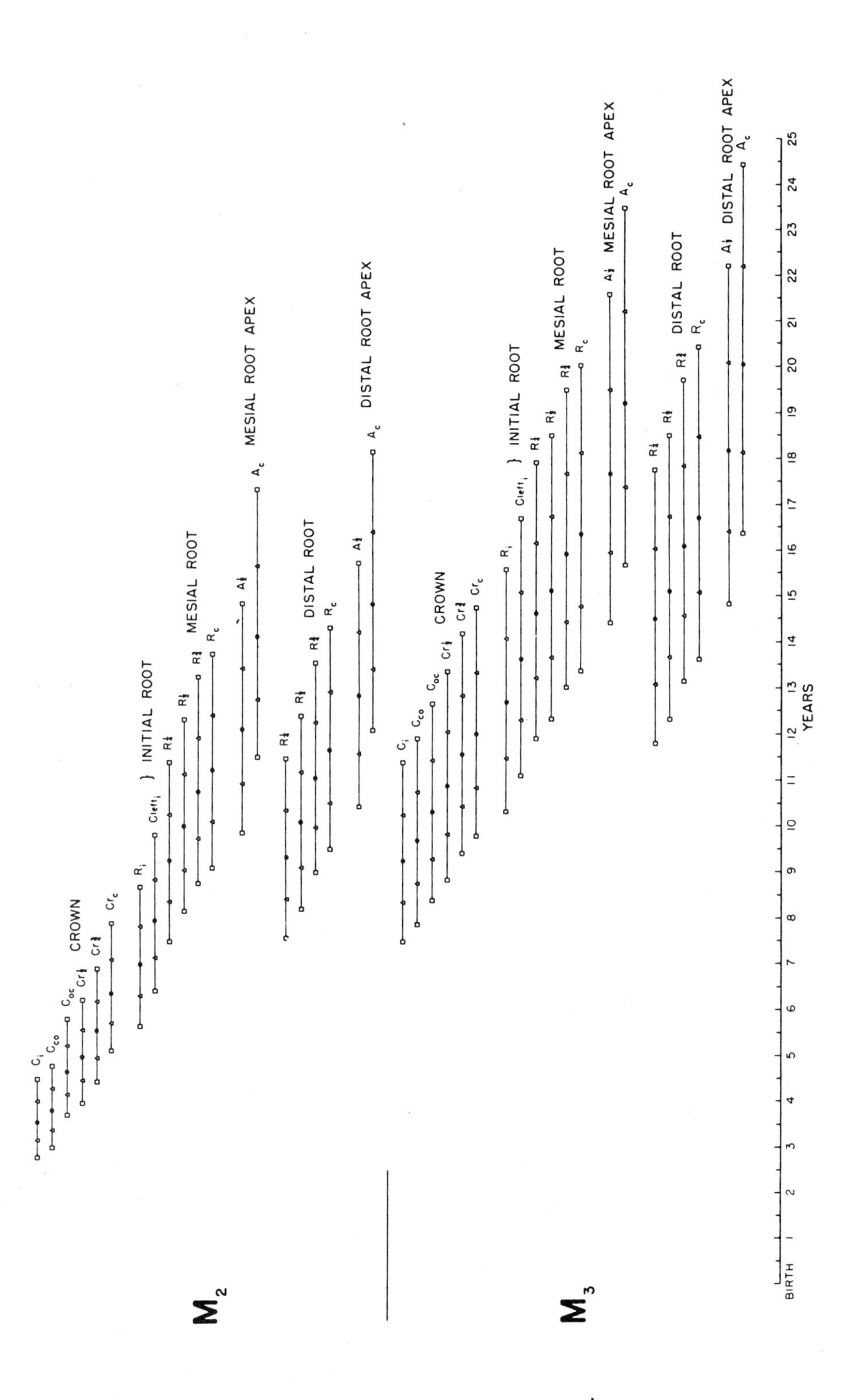
M2
CROWN
INITIAL ROOT
MESIAL ROOT
DISTAL ROOT
MESIAL ROOT APEX
DISTAL ROOT APEX
M3
CROWN
INITIAL ROOT
MESIAL ROOT
DISTAL ROOT
MESIAL ROOT APEX
DISTAL ROOT APEX
BIRTH
YEARS

Table 7. Norms of formation of permanent mandibular canines, premolars, and molars of females (After Moorrees, Fanning, and Hunt 1963b, Fig. 6).

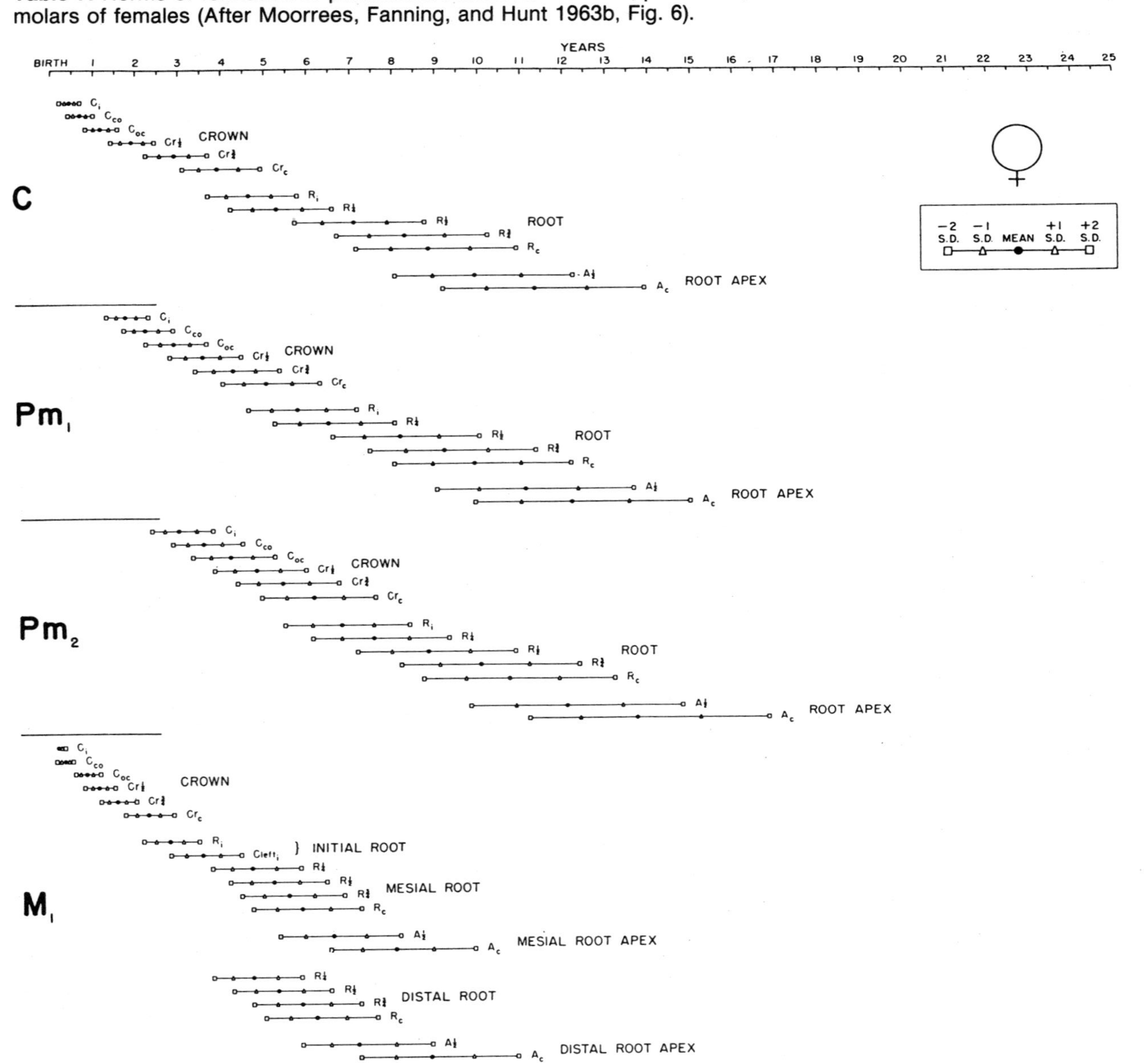

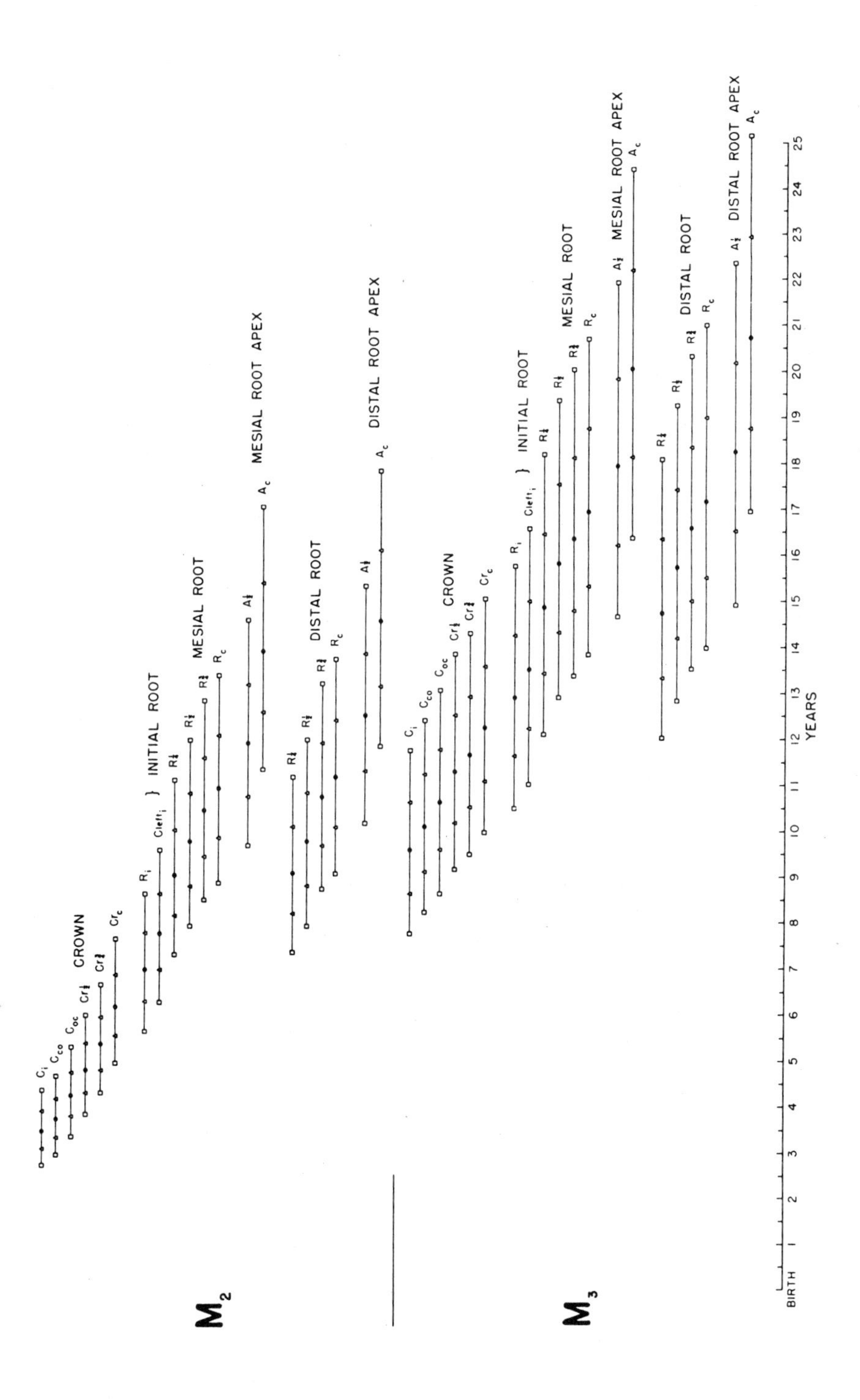
M2
CROWN
INITIAL ROOT
MESIAL ROOT
MESIAL ROOT APEX
DISTAL ROOT
DISTAL ROOT APEX
M3
CROWN
INITIAL ROOT
MESIAL ROOT
MESIAL ROOT APEX
DISTAL ROOT
DISTAL ROOT APEX
BIRTH
YEARS

Appendix 2 Procedure for the Preparation of Undecalcified, Ground Bone Sections for Microscopic Examination

An important aspect of the microscopic examination of prehistoric tissue is the preparation of adequate thin sections. Good summaries are available for preparing stained sections of dessicated soft tissue (Allison and Gerszten 1975); decalcified, stained sections of bone (Andersen and Jorgensen 1960); microtome sections (Salomon and Haas 1967) and staining (Frost 1959) of undecalcified bone, and ground sections of undecalcified bone (Moreland 1968, Ubelaker 1974). For most microscopic observations, especially those involved in age determination, I use undecalcified, ground, thin sections. We employ the following procedure at the Smithsonian to prepare these sections.

Initially, a cross section 5 to 10 millimeters thick is cut from the bone (Fig. 156) using a sabre saw with a fine blade (32 teeth per inch). A thinner (5 millimeters thick), parallel-sided section is removed from this initial section using an Ingram (Model 103) thin-section, cut-off saw (Fig. 157). Since most archeological bones need to be cleaned thoroughly, the specimen is next immersed in Decal solution (available from Scientific Products) for about 50 minutes. This is sufficient time to allow the solution to dislodge or destroy foreign particles within the cortex without seriously damaging or decalcifying the bone. Throughout this period, the container of Decal is suspended in water inside an ultrasonic cleaner, which helps the solution penetrate the tissue and remove the undesirable material. The section is then removed from the container and placed directly in water within the ultrasonic cleaner for an additional five minutes. Finally, it is allowed to dry overnight.

Frequently, the important outer periosteal aspect of the cortex is sufficiently fragile to require impregnation. This is done by leaving the specimen overnight in a syrupy solution consisting of equal parts of ar-

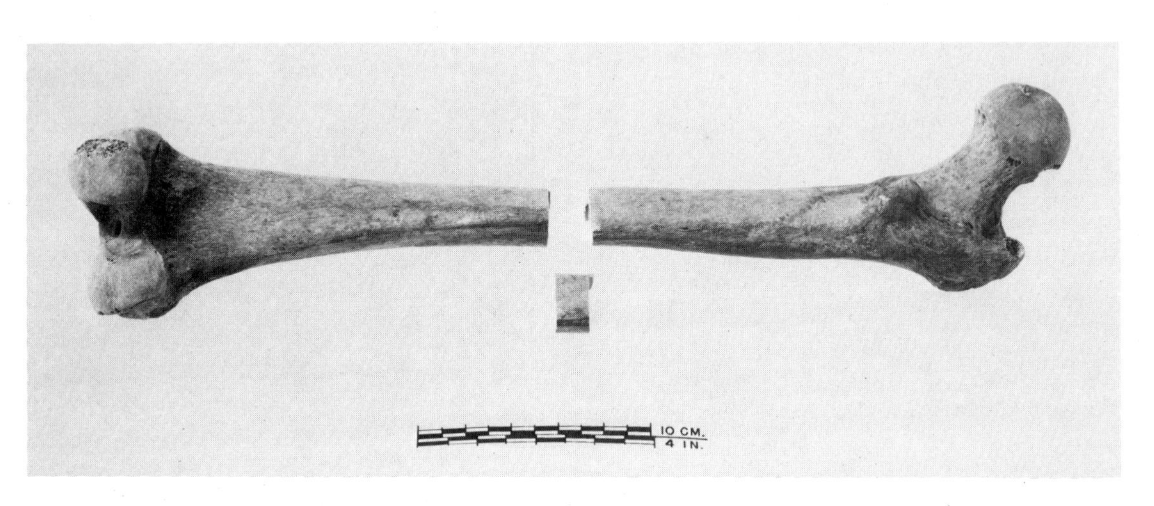

Fig. 156. A cross sectio cut from a long bone, the first step in preparing a slide for microscopic examination.

Fig. 157. Reducing the thickness of a cross section cut from a long bone. The section is mounted at the center of the picture, directly beneath the blade of the cut-off saw (protected by a plastic shield).

Fig. 158. Grinding the surface of a thin section, which has been mounted temporarily on a glass slide. The section (arrow) is moved to the left until it comes in contact with the vertical, rotating wheel.

Fig. 159. Polishing the ground surface of a thin section. The section is on the under side of the glass block held by the operator.

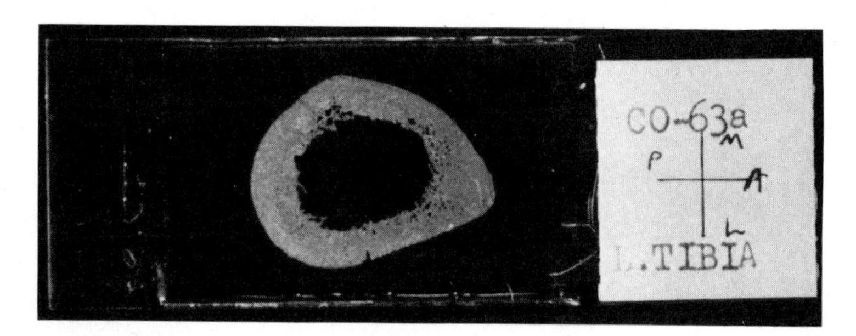

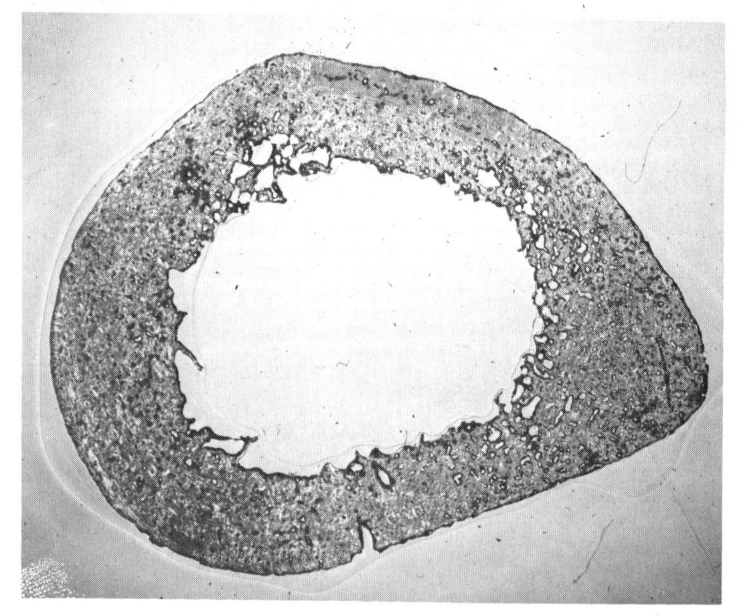

Fig. 160. The finished slide. **a,** The section mounted on a glass slide and labeled with the necessary information. **b,** An enlargement of the section shown on the slide.

aldite (AY-105) and hardener (935-F) (both available from Chemical Coating and Engineering Co., Inc.) dissolved in toluene. Impregnation is most effective when accomplished in a vacuum.

The impregnated specimen is dried overnight in an oven at a temperature of 55 degrees C and then temporarily mounted on a glass slide with an acetonebase cement. The free (upper) surface is ground (Ingram Thin Section Grinder, Model 303) until it is flat (Fig. 158). The specimen is then removed from the slide with acetone and the ground surface is polished on a polishing wheel covered by polishing compound and lubricated with A.B. polishing oil and A.B. mineral spirits (Fig. 159).

Following this procedure, the specimen is placed in acetone in the ultrasonic cleaner for one minute to remove the oil and the polishing compound. The same araldite mixture described above, but without the toluene, is then used to attach the polished surface of the specimen to a glass slide. A metal clamp holds the slide and the specimen together tightly and with even pressure. Prior to attachment of the bone to the slide, all the materials involved (araldite, hardener, bone, slide, and clamp) should be heated to about 93 degrees C on a hot plate. The specimen should be firmly attached after 20 minutes.

The specimen is then reduced to a thickness of about 75 microns by using the cut-off saw and grinder on the unpolished side. This surface must then be polished and cleaned in the manner described above, after which a cover-slip is attached with permount solution. The specimen is then ready for analysis (Fig. 160).

The procedure summarized above is described in greater detail by Ubelaker (1974:54–56), with minor variations. Note that if the unprocessed bone is well preserved and free from soil, fungus, or other foreign particles, the immersion in Decal solution and impregnation will not be required.

Literature Cited

Acsádi, Gy. and J. Nemeskéri
1970 History of Human Life Span and Mortality. Académiai Kiadó. Budapest.

Aegerter, Ernest and John A. Kirkpatrick, Jr.
1975 Orthopedic Diseases. Fourth edition. W. B. Saunders Company, Philadelphia.

Ahlqvist, J. and D. Damsten
1969 A Modification of Kerley's Method for the Microscopic Determination of Age in Human Bone. Journal of Forensic Sciences, Volume 14, Number 2, pp. 205–212.

Albrook, David
1961 The Estimation of Stature in British and East African Males. Journal of Forensic Medicine, Volume 8, pp. 15–28.

Allison, M. J. and E. Gerszten
1975 Paleopathology in Peruvian Mummies. Virginia Commonwealth University, Richmond.

Andersen, Helge and J. Balslev Jørgensen
1960 Decalcification and Staining of Archaeological Bones, with Histochemical Interpretation of Metachromasia. Stain Technology, Volume 35, Number 2, pp. 91–96.

Anderson, D. L., G. W. Thompson and F. Popovich
1976 Age of Attainment of Mineralization Stages of the Permanent Dentition. Journal of Forensic Sciences, Volume 21, Number 1, pp. 191–200.

Anderson, J. E.
1962 The Human Skeleton: A Manual for Archaeologists. National Museum of Canada, Ottawa.
1964 The People of Fairty. National Museum of Canada Bulletin 193. Ottawa.

Anderson, M. and W. T. Green
1948 Lengths of the Femur and the Tibia. American Journal of Diseases of Children, Volume 75, pp. 279–290.

Angel, J. Lawrence
1969 The Bases of Paleodemography. American Journal of Physical Anthropology, Volume 30, pp. 427–435.

Angel, J. Lawrence, Judy M. Suchey, M. Yasar Iscan, and Michael R. Zimmerman
1986 Age at Death Estimated from the Skeleton and Viscera. In Dating and Age Determination of Biological Materials, edited by M. R. Zimmerman and J. Lawrence Angel, pp. 179–220. Croom Helm, London.

Arber, E., Editor
1884 Capt. John Smith, Works, 1608–1631. The English Scholar's Library, 16, Birmingham.

Aufderheide, A.C. and C. Rodríguez-Martín
1998 The Cambridge Encyclopedia of Human Paleopathology. University Press, Cambridge.

Baby, Raymond S.
1954 Hopewell Cremation Practices. The Ohio Historical Society, Papers in Archaeology, Number 1.

Baccino, Eric., Douglas H. Ubelaker, Lee-Ann C. Hayek, and A. Zerilli
in press Evaluation of Seven Methods of Estimating Age at Death from Mature Human Skeletal Remains. Journal of Forensic Sciences.

Baccino, E. and A. Zerilli
1997 The Two Step Strategy (TSS) or the Right Way to Combine a Dental (Lamendin) and an Anthropological (Suchey-Brooks System) Method for Age Determination (abstract). Proceedings, American Academy of Forensic Sciences, Volume 3, p. 150.

Baker, R. K.
1984 The Relationship of Cranial Suture Closure and Age Analyzed in a Modern Multi-Racial Sample of Males and Females. Thesis for the Degree Master of Arts in Anthropology, California State University, Fullerton.

Banerjee, Papia and Sunetra Mukherjee
1967 Eruption of Deciduous Teeth among Bengalee Children. American Journal of Physical Anthropology, Volume 26, Number 3, pp. 357–358.

Barnett, C. H.
1954 Squatting Facets on the European Talus. Journal of Anatomy, Volume 88, Part 4, pp. 509–513.

Bass, William M.
1962 The Excavation of Human Skeletal Remains. In Field Handbook on the Human Skeleton by R. F. G. Spier, pp. 39–51. Missouri Archaeological Society, Columbia.
1971 Human Osteology: A Laboratory and Field Manual of the Human Skeleton. Special Publications, Missouri Archaeological Society, Columbia.

Bauer, W. H.
1944 Tooth Buds and Jaws in Patients with Congenital Syphilis; Correlation Between Distribution of Treponema Pallidum and Tissue Reaction. American Journal of Pathology, Volume 20, pp. 297–314.

Bergfelder, T. and B. Herrmann
1978 Zur Fertitätsschatzung an Hand geburtstraumatischer Veränderungen an Schambein. Homo, Volume 29, pp. 17–24.

Berrizbeitia, Emily L.
1987 Sex Determination with the Head of the Radius. Manuscript.

Binford, Lewis R.
1963 An Analysis of Cremations from Three Michigan Sites. Wisconsin Archeologist, Volume 44, pp. 98–110.

Black, T.K.
1978 A New Method for Assessing the Sex of Fragmentary Skeletal Remains: Femoral Shaft Circumference. American Journal of Physical Anthropology, Volume 48, pp. 227-232.

Blakely, Robert L., Editor
1977 Biocultural Adaptation in Prehistoric America. University of Georgia Press, Athens.

Boucher, Barbara J.
1955 Sex Difference in the Foetal Sciatic Notch. Journal of Forensic Medicine, Volume 2, pp. 51–54.
1957 Sex Differences in the Foetal Pelvis. American Journal of Physical Anthropology, Volume 15. pp. 581–600.

Bradbury, John
1817 Travels in the Interior of America in the Years 1809, 1810 and 1811. Smith and Galway, Liverpool.

Brooks, S. and J.M. Suchey
1990 Skeletal Age Determination Based on the Os Pubis: A Comparison of the Acsádi-Nemeskéri and Suchey-Brooks Methods. Human Evolution, Volume 5, pp. 227-238.

Brothwell, Don R.
1972 Digging Up Bones. Second Edition. British Museum of Natural History, London.

Brothwell, Don R. and A. T. Sandison, Editors
1967 Diseases in Antiquity. Charles C. Thomas, Springfield, Illinois.

Buikstra, Jane
1976 Hopewell in the Lower Illinois Valley. A Regional Study of Human Biological Variability and Prehistoric Mortuary Behavior. Northwestern University Archeological Program, Scientific Papers, Number 2. Evanston.

Burleigh, R. and D. Brothwell
1978 Studies on Amerindian Dogs; Carbon Isotopes in Relation to Maize in the Diet of Domestic Dogs from Early Peru and Ecuador. Journal of Archaeological Science, Volume 5, pp. 355–362.

Burns, K. R. and W. R. Maples
1976 Estimation of Age from Individual Adult Teeth. Journal of Forensic Sciences, Volume 21, pp. 343–356.

Bushnell, David I.
1920 Native Cemeteries and Forms of Burial East of the Mississippi. Bureau of American Ethnology, Bulletin 71. Smithsonian Institution, Washington, D.C.
1927 Burials of the Algonquian, Siouan, and Caddoan Tribes West of the Mississippi. Bureau of American Ethnology, Bulletin 83. Smithsonian Institution, Washington, D.C.

Christensen, Gordon J. and Bertram S. Kraus
1965 Initial Calcification of the Human Permanent First Molar. Journal of Dental Research, Volume 44, Number 6, pp. 1338–1342.

Coughlin, John W. and Gordon J. Christensen
1966 Growth and Calcification in the Prenatal Human Primary Molars. Journal of Dental Research, Volume 45, Number 5, pp. 1541–1547.

Cybulski, Jerome S.
1974 Tooth Wear and Material Culture: Precontact Patterns in the Tsimshian Area, British Columbia. Syesis, Volume 7, pp. 31–35.

Dahlberg, Albert and Renée M. Menegaz-Bock
1958 Emergence of the Permanent Teeth in Pima Indian Children. Journal of Dental Research, Volume 37, pp. 1123–1140.

Das, A. C.
1959 Squatting Facets of the Talus in U.P. Subjects. Journal of Anatomy, Society of India, Volume 8, pp. 90–92.

Dechaume, M., L. Dérobert and J. Payen
1960 De la Valeur de la Détermination de l'Age par l'Examen des Dents en Coupes Minces. Annales de Médecine Légale, Volume 40. pp. 165–167. Paris.

Demisch, Arthur and Peter Wartmann
1956 Calcification of the Mandibular Third Molar and its Relation to Skeletal and Chronological Age in Children. Child Development, Volume 27, Number 4, pp. 459–473.

DeNiro, Michael J. and Samuel Epstein
1978 Influence of Diet on the Distribution of Carbon Isotopes in Animals. Geochimica et Cosmochimica Acta, Volume 42, pp. 495–506.
1981 Influence of Diet on the Distribution of Nitrogen Isotopes in Animals. Geochimica et Cosmochimica Acta, Volume 45, pp. 341–351.

De Terra, Helmut
1949 Early Man in Mexico. In Tepexpan Man by Helmut De Terra, Javier Romero, and T. D. Stewart. Viking Fund Publications in Anthropology, Number 11, pp. 11–86. New York.

Dupertuis, C. W. and J. A. Hadden, Jr.
1951 On the Reconstruction of Stature from Long Bones. American Journal of Physical Anthro-

pology, Volume 9, pp. 15–54.

Dwight, Thomas
1894 Methods of Estimating the Height from Parts of the Skeleton. Medical Record, Volume 46, pp. 293–296.
1904–1905 The Size of the Articular Surfaces of the Long Bones as Characteristic of Sex; An Anthropological Study. American Journal of Anatomy, Volume 4, Number 1, pp. 19–31.

Falsetti, A.B.
1995 Sex Assessment from Metacarpals of the Human Hand. Journal of Forensic Sciences, Volume 40, pp. 774-776.

Fazekas, I. Gy. and F. Kośa
1978 Forensic Fetal Osteology. Akadémiai Kaidó, Budapest.

Frost, H. M.
1959 Staining of Fresh Undecalcified Thin Bone Sections. Stain Technology, Volume 34, pp. 135–146.

Fully, Georges
1956 Une nouvelle méthode de determination de la taille. Annales de Médecine Légale et de Criminologie, Volume 36, pp. 266–273.

Fully, G. and H. Pineau
1960 Détermination de la stature au moyen du squelette. Annales de Médecine Légale, Volume 60, pp. 145–153.

Galera, Virginia, Douglas H. Ubelaker, and Lee-Ann C. Hayek
1998 Comparison of Macroscopic Cranial Methods of Age Estimation Applied to Skeletons from the Terry Collection. Journal of Forensic Sciences, Volume 43, pp. 933-939.

Garn, S. M., A. B. Lewis, and R. M. Blizzard
1965 Endocrine Factors in Dental Development. Journal of Dental Research, Volume 43, pp. 243–258.

Garn, Stanley M., Arthur B. Lewis, and Demarest L. Polacheck
1959 Variability of Tooth Formation. Journal of Dental Research, Volume 38, pp. 135–148.

Geidel, R. A.
1981 Paleonutrition and Social Stratification: A Study of Trace Elements in Human Skeletons from the Dallas Archaeological Culture of Eastern Tennessee. Master's Thesis, Department of Anthropology, Pennsylvania State University, University Park.

Genovés, Santiago
1967 Proportionality of the Long Bones and Their Relation to Stature Among Mesoamericans. American Journal of Physical Anthropology, Volume 26, Number 1, pp. 67–77.

Ghantus, M.
1951 Growth of the Shaft of the Human Radius and Ulna During the First Two Years of Life. American Journal of Roentgenology, Volume 65, pp. 784–786.

Gilbert, B. Miles and William M. Bass
1967 Seasonal Dating of Burials from the Presence of Fly Pupae. American Antiquity, Volume 32, Number 4, pp. 534–535.

Gilbert, B. Miles and Thomas W. McKern
1973 A Method for Aging the Female Os Pubis. American Journal of Physical Anthropology, Volume 38, Number 1, pp. 31–38.

Gillman, Henry
1875 Certain Characteristics Pertaining to Ancient Man in Michigan. Annual Report of the Smithsonian Institution for 1875, pp. 234–245. Washington, D.C.

Giles, Eugene
1970 Discriminant Function Sexing of the Human Skeleton. In Personal Identification in Mass Disasters, edited by T. D. Stewart, pp. 99–109. Smithsonian Institution, Washington, D.C.

Giles, E. and O. Elliot
1962 Race Identification from Cranial Measurements. Journal of Forensic Sciences, Volume 7, pp. 147–157.

Gilster, John E., Franklin H. Smith, and Gerald K. Wallace
1964 Calcification of Mandibular Second Primary Molars in Relation to Age. Journal of Dentistry for Children, Volume 31, pp. 284–288.

Glasstone, S.
1938 A Comparative Study of the Development *in vivo* and *in vitro* of Rat and Rabbit Molars. Proceedings of the Royal Society of London, Series B, Volume 126, pp. 315–330.
1963 Regulative Changes in Tooth Germs Grown in Tissue Culture. Conference on Genetic Aspects of Oral Structures, Denver, Colorado, 1961. Journal of Dental Research, Volume 42, Number 6, pp. 1364–1368.
1964 Cultivation of Mouse Tooth Germs in a Chemically Defined Protein-Free Medium. Archives of Oral Biology, Volume 9, pp. 27–30.

Greene, David Lee
1967 Dentition of Meroitic, X-Group, and Christian Populations from Wadi Halfa, Sudan. Anthropological Papers Number 85. University of Utah Press.

Greulich, W. W. and S. I. Pyle
1950 Radiographic Atlas of Skeletal Development of the Hand and Wrist. Stanford University Press, Stanford.

Griffin, James B. and George K. Neumann
1942 A Suggested Classification and Nomenclature for Burial Location, Position, and Description. Society for American Archaeology Notebook, Volume 2, pp. 70–79. Ann Arbor.

Griffing, W. J.
1904 Committee on Explorations. Transactions of the Kansas State Historical Society, 1903–1904, Volume VIII, Topeka.

Gustafson, Gösta
1950 Age Determinations on Teeth. Journal of the American Dental Association, Volume 41, pp. 45–54.

Haglund, William D. and Marcella H. Sorg
1997 Forensic Taphonomy, The postmortem Fate of Human Remains. CRC Press, Inc. New York.

Hansen, Gerhard
1953 Die Altersbestimmung am proximalen Humerus-und Femurende in Rahmen der Identifizierung menschlicher Skelettreste. Wissenschaftliche Zeitschrift der Humboldt-Universität zu Berlin, Mathematish-Naturwissenschaftliche Reihe, Number 1, Volume 3, pp. 1–73.

Heizer, Robert F.
1958 A Guide to Archaeological Field Methods. The National Press, Palo Alto.
1974 A Question of Ethics in Archaeology—One Archaeologist's View. The Journal of California Anthropology, Volume 1, Number 2, pp. 145–151.

Hoffmann, J. M.
1979 Age Estimations from Diaphyseal Lengths: Two Months to Twelve Years. Journal of Forensic Sciences, Volume 24, pp. 461–469.
1987 Review of "Forensic Osteology: Advances in the Identification of Human Remains" by K. J. Reichs, editor, and "The Human Skeleton in Forensic Medicine" by W. M. Krogman and M. Y. Iscan. American Anthropologist, Volume 89, pp. 729–731.

Holland, Thomas Dean
1992 Estimation of Adult Stature from Fragmentary Tibias. Journal of Forensic Sciences, Volume 37, pp. 1223-1229.

Holman, D.J., and K.A. Bennett
1991 Determination of Sex from Arm Bone Measurements. American Journal of Physical Anthropology, Volume 84, pp. 421-426.

Holt, C. Adams
1978 A Re-examination of Parturition Scars on the Human Female Pelvis. American Journal of Physical Anthropology, Volume 49, pp. 91–94.

Hooton, E. A.
1922 Notes on the Skeletal Remains from the Turner Group of Earthworks, Hamilton County, Ohio. Papers of the Peabody Museum of American Archaeology and Ethnology, Harvard University, Volume 8, pp. 99–132. Cambridge, Mass.

Howard, James H.
1972 Arikara Native-Made Glass Pendants: Their Probable Function. American Antiquity, Volume 37, Number 1, pp. 93–97.

Howells, W.W.
1973 Cranial Variation in Man. Papers of the Peabody Museum, Volume 67. Cambridge.
1989 Skull Shapes and the Map. Papers of the Peabody Museum, Volume 78. Cambridge.

Hrdlička, Aleš
1905 The Painting of Human Bones Among the American Aborigines. Annual Report of the Smithsonian Institution for 1904, pp. 607-617. Washington, D.C.
1939 Practical Anthropometry. Second Edition. The Wistar Institute of Anatomy and Biology, Philadelphia.

Hunt, Edward E. and Izaac Gleiser
1955 The Estimation of Age and Sex of Preadolescent Children from Bone and Teeth. American Journal of Physical Anthropology, Volume 13, Number 3, pp. 479–487.

Hurme, V. O.
1948 Standards of Variation in the Eruption of the First Six Permanent Teeth. Child Development, Volume 19, Numbers 1–2, pp. 213–231.

Introna, Francesco, Giancarlo Di Vella, Carlo Pietro Campobasso and Michele Dragone
1997 Sex Determination by Discriminant Analysis of Calcanei Measurements. Journal of Forensic Sciences, Volume 42, pp. 725-728.

Iscan, M. Yasar and Susan R. Loth
1986 Determination of Age from the Sternal Rib in White Females: A Test of the Phase Method. Journal of Forensic Sciences, Volume 31, pp. 990–999.

Iscan, M. Yasar, S. R. Loth, and R. K. Wright
1984 Age Estimation from the Rib by Phase Analysis: White Males. Journal of Forensic Sciences, Volume 29, pp. 1094–1104.
1985 Age Estimation from the Rib by Phase Analysis: White Females. Journal of Forensic Sciences, Volume 30, pp. 853–863.
1987 Racial Variation in the Sternal Extremity of the Rib and its Effect on Age Determination. Journal of Forensic Sciences, Volume 32, pp. 452–466.

Işcan, M.Y. and P. Miller-Shaivitz
1984 Discriminant Function Sexing of the Tibia. Journal of Forensic Sciences, Volume 29, pp. 1087-1093.

Jantz, Richard L.
1973 Microevolutionary Change in Arikara Crania: A Multivariate Analysis. American Journal of Physical Anthropology, Volume 38, Number 1, pp. 15–26.
1974 The Redbird Focus: Cranial Evidence in Tribal Identification. Plains Anthropologist, Volume 19, Number 63, pp. 5–13.

Jantz, R.L., David R. Hunt and Lee Meadows
1995 The Measure and Mismeasure of the Tibia: Implications for Stature Estimation. Journal of Forensic Sciences, Volume 40, pp. 758-761.

Jantz, R. L. and P. H. Moore-Jansen
1987 A Data Base for Forensic Anthropology. Final Report to the National Institute of Justice. University of Tennessee, Knoxville.

Jit, I, and S. Singh
1956 Estimation of Stature from the Clavicles. Indian Journal of Medical Research, Volume 44, pp. 137–155.

Johnston, Francis E.
1962 Growth of the Long Bones of Infants and Young Children at Indian Knoll. American Journal of Physical Anthropology, Volume 20, Number 3, pp. 249–254.

Kalmey, Jonathan K. and Ted A. Rathbun
1996 Sex Determination by Discriminant Function Analysis of the Petrous Portion of the Temporal Bone. Journal of Forensic Sciences, Volume 41, pp. 865-867.

Kate, B. R. and S. L. Robert
1965 Some Observations on the Upper End of the Tibia in Squatters. Journal of Anatomy of London, Volume 99, Number 1, pp. 137–141.

Keen, E. N.
1953 Estimation of Stature from the Long Bones: A Discussion of its Reliability. Journal of Forensic Medicine, Volume 1, pp. 46–51.

Kerley, Ellis R.
1965 The Microscopic Determination of Age in Human Bone. American Journal of Physical Anthropology, Volume 23, Number 2, pp. 149–163.
1970 Estimation of Skeletal Age: After About Age 30. In Personal Identification in Mass Disasters, edited by T. D. Stewart, pp. 57–70. Smithsonian Institution, Washington, D.C.

Kerley, E. R. and D. H. Ubelaker
1978 Revisions in the Microscopic Method of Estimating Age at Death in Human Cortical Bone. American Journal of Physical Anthropology, Volume 49, pp. 545–546.

Kidd, Kenneth E.
1952 Sixty Years of Ontario Archeology. In Archeology of Eastern United States, edited by J. B. Griffin, pp. 71–82. University of Chicago Press, Chicago.
1953 The Excavation and Historical Identification of a Huron Ossuary. American Antiquity, Volume 18, Number 4, pp. 359–379.

Kraus, Bertram S.
1959 Calcification of the Human Deciduous Teeth. Journal of the American Dental Association, Volume 59, Number 5, pp. 1128–1136.

Krogman, Wilton Marion
1962 The Human Skeleton in Forensic Medicine. Charles C. Thomas, Springfield.

Krogman, Wilton Marion and M. Y. Iscan
1986 The Human Skeleton in Forensic Medicine. Second Edition. Charles C. Thomas, Springfield.

Lallo, J. W., G. J. Armelagos, and R. P. Mensforth
1977 The Role of Diet, Disease, and Physiology in the Origin of Porotic Hyperostosis. Human Biology, Volume 49, pp. 471–483.

Lamendin, H., E. Baccino, J.F. Humbert, J.C. Tavernier, R.M. Nossintchouk, and A. Zerilli
1992 A Simple Technique for Age Estimation in Adult Corpses: The Two Criteria Dental Method. Journal of Forensic Sciences, Volume 37, 1373-1379.

Lane, Rebecca A. and Audrey J. Sublett
1972 Osteology of Social Organization: Residence Pattern. American Antiquity, Volume 37, Number 2, pp. 186–201.

Larsen, C. S.
1980a Dental Caries, Experimental and Biocultural Evidence. Tennessee Anthropological Association, Miscellaneous Papers, Volume 5, pp. 75–80.
1980b Prehistoric Human Biological Adaptation: Case Study from the Georgia Coast. Ph.D. Dissertation, Department of Anthropology, University of Michigan, Ann Arbor.

Lasker, Gabriel W. and Marjorie M. C. Lee
1957 Racial Traits in the Human Teeth. Journal of Forensic Sciences, Volume 2, pp. 401–419.

Lazenby, R.A.
1994 Identification of Sex from Metacarpals: Effect of Side Asymmetry. Journal of Forensic Sciences, Volume 39, pp. 1188-1194.

Lewis, Arthur B. and Stanley M. Garn
1960 The Relationship Between Tooth Formation and Other Maturational Factors. The Angle Orthodontist, Volume 30, pp. 70–77.

Lovejoy, C. Owen
1985 Dental Wear in the Libben Population: Its Functional Pattern and Role in the Determination of Adult Skeletal Age at Death. American Journal of Physical Anthropology, Volume 68, pp. 47–56.

Lovejoy, C. Owen, R. S. Meindl, T. R. Pryzbeck, and Robert Mensforth
1985a Chronological Metamorphosis of the Auricular Surface of the Ilium: A New Method for the Determination of Adult Skeletal Age at Death. American Journal of Physical Anthropology, Volume 68, pp. 15–28.

Lovejoy, C. Owen, R. S. Meindl, Robert Mensforth, and Thomas J. Barton
1985 Multifactorial Determination of Skeletal Age at Death: A Method and Blind Tests of its Accuracy. American Journal of Physical Anthropology, Volume 68, pp. 1–14.

Lundy, J. K.
1983 Regression Equations for Estimating Living Stature from Long Limb Bones in South African Negros. South African Journal of Science, Volume 79, pp. 337–338.

Lunt, Roger C. and David B. Law
1974 A Review of the Chronology of Calcification of Deciduous Teeth. Journal of the American Dental Association, Volume 89, pp. 599–606.

Mall, K. P.
1914 On Stages of Development of Human Embryos from 2 to 25 mm Long. Anatomischer Anzeiger, Volume 46, pp. 78–84.

Maples, W. R.
1978 An Improved Technique Using Dental Histology for the Estimation of Adult Age. Journal of Forensic Sciences, Volume 23, pp. 764–770.

Maples, W. R. and P. M. Rice
1979 Some Difficulties in the Gustafson Dental Age Estimations. Journal of Forensic Sciences, Volume 24, pp. 168–172.

Maresh, M. M.
1943 Growth of Major Long Bones in Healthy Children. American Journal of Diseases of Children, Volume 66, pp. 227–257.
1955 Linear Growth of Long Bones of Extremities from Infancy Through Adolescence. American Journal of Diseases of Children, Volume 89, pp. 725–742.

Masset, C.
1982 Estimation of l'âge au décès par les sutures crâniennes. Thèse de Sciences Naturelles, Multigraphée. Université Paris VII.

McKern, T. W. and T. D. Stewart
1957 Skeletal Age Changes in Young American Males. Headquarters, Quartermaster Research and Development Command, Technical Report EP-45. Natick, Mass.

Meindl, Richard S. and C. Owen Lovejoy
1985 Ectocranial Suture Closure: A Revised Method for the Determination of Skeletal Age at Death Based on the Lateral-anterior Sutures. American Journal of Physical Anthropology, Volume 68, pp. 57–66.

Merchant, Virginia L. and Douglas H. Ubelaker
1977 Skeletal Growth of the Protohistoric Arikara. American Journal of Physical Anthropology, Volume 46, Number 1, pp. 61–72.

Meredith, Howard V.
1946 Order and Age of Eruption for the Deciduous Dentition. Journal of Dental Research, Volume 25, Number 1, pp. 43–66.

Miles, A. E. W.
1958 The Assessment of Age from the Dentition. Proceedings of the Royal Society of Medicine, Volume 51, pp. 1057–1060.
1962 Assessment of the Ages of a Population of Anglo-Saxons from their Dentitions. Proceedings of the Royal Society of Medicine, Volume 55, pp. 881–886.
1963a Dentition in the Assessment of Individual Age in Skeletal Material. In Dental Anthropology, edited by D. R. Brothwell, pp. 191–209. Pergamon, Oxford.
1963b Dentition in the Estimation of Age. Journal of Dental Research, Volume 42, pp. 255–263.
1978 Teeth as an Indicator of Age in Man. In Development, Function, and Evolution of Teeth, edited by P. M. Butler and K. A. Joysey, pp. 455–464. Academic Press, New York.

Moorrees, Coenraad F. A.
1965 Normal Variation in Dental Development Determined with Reference to Tooth Eruption Status. Journal of Dental Research, Volume 44, Number 1, pp. 161–173.

Moorrees, Coenraad F. A., Elizabeth A. Fanning and Edward E. Hunt, Jr.
1963a Formation and Resorption of Three Deciduous Teeth in Children. American Journal of Physical Anthropology, Volume 21, pp. 205–213.
1963b Age Variation of Formation Stages for Ten Permanent Teeth. Journal of Dental Research, Volume 42, Number 6, pp. 1490–1502.

Moreland, Grover C.
1968 Preparation of Polished Thin Sections. The American Mineralogist, Volume 53, pp. 2070–2074.

Morimoto, Iwataro
1960 The Influence of Squatting Posture on the Talus in the Japanese. Medical Journal of Shinshu University, Volume 5, Number 3, pp. 159–166.

Morse, Dan
1969 Ancient Disease in the Midwest. Reports of Investigations, Number 15, Illinois State Museum, Springfield, Illinois.
1983 Studies on the Deterioration of Associated Death Scene Materials. Appendix A in Handbook of Forensic Archaeology and Anthropology, edited by D. Morse, J. Duncan, and J. Stoutamire. Bill's Book Store, Tallahassee.

Morse, Dan, Jack Duncan, and James Stoutamire, Editors
1983 Handbook of Forensic Archaeology and Anthropology. Bill's Book Store, Tallahassee.

Murad, Turhon A. and Margie A. Boddy
1987 A Case with Bear Facts. Journal of Forensic Sciences, Volume 32, pp. 1819–1826.

Musgrave, J. H. and N. K. Harneja
1978 The Estimation of Adult Stature from Metacarpal Bone Length. American Journal of Physical Anthropology, Volume 48, pp. 113–119.

Nalbandian, J.
1959 Age Changes in Human Teeth. Journal of Dental Research, Volume 38, pp. 681–682.

Nalbandian, J., and R. F. Sognnaes
1960 Structural Age Changes in Human Teeth. In Aging, edited by Nathan W. Shock. American Association for the Advancement of Science. Volume 65, pp. 367–382. Washington, D.C.

Nat, B. S.
1931 Estimation of Stature from Long Bones in Indians of the United Provinces; Medico-legal Inquiry in Anthropometry. Indian Journal of Medical Research, Volume 18, pp. 1245–1253.

Nemeskéri, J.
1972 Die archaeologischen und antropologischen Voraussetzungen palaeodemographischer Forschungen. Praehistorische Zeitschrift, Volume 47, pp. 5–46.

Niswander, J. D., and C. Sujaku
1965 Permanent Tooth Eruption in Children with Major Physical Defect and Disease. Journal of Dentistry for Children, Volume 32, pp. 266–268.

Nolla, Carmen M.
1960 The Development of the Permanent Teeth. Journal of Dentistry for Children, Volume 27. pp. 254–266.

Olivier, G.
1963 L'Estimation de la stature par les os longs des membres. Bulletin de la Société d'Anthropologie de Paris, Volume 4 (XIe série), pp. 433–449.
Olivier, G. and H. Pineau
1958 Determination de l'age du foetus et de l'embryon. Archives d'Anatomie Pathologique, Volume 6, pp. 21–28.
1960 Nouvelle determination de la taille foetale d'après les longueurs diaphysaires des os longs. Annales de Medecine Legale, Volume 40, pp. 141–144.
Ortner, Donald J.
1968 Description and Classification of Degenerative Bone Changes in the Distal Joint Surfaces of the Humerus. American Journal of Physical Anthropology, Volume 28, Number 2, pp. 139–155.
Ortner, Donald J. and Walter G. J. Putschar
1981 Identification of Pathological Conditions in Human Skeletal Remains. Smithsonian Contributions to Anthropology, Number 28. Washington, D.C.
Ousley, S.D. and R.L. Jantz
1996 Fordisc 2.0: Personal Computer Forensic Discriminant Function. The University of Tennessee, Knoxville.
Paynter, K. J. and R. M. Grainger
1961 Influence of Nutrition and Genetics on Morphology and Caries Susceptibility. Journal of the American Medical Association, Volume 177, pp. 306–309.
1962 Relationships of Morphology and Size of Teeth to Caries. International Dental Journal, Volume 12, pp. 147–160.
Phenice, T. W.
1969a A Newly Developed Visual Method of Sexing the Os Pubis. American Journal of Physical Anthropology, Volume 30, Number 2, pp. 297–301.
1969b An Analysis of the Human Skeletal Material from Burial Mounds in North Central Kansas. University of Kansas Publications in Anthropology, Number 1. Lawrence.
Prieto Carrero, José Luis
1993 Parámetros histomorfométricos óseos normales en una población infantojuvenil española. Tesis Doctoral. Universidad Complutense de Madrid, Facultad de Medicina, Madrid.
Putschar, W.
1931 Entwicklung, Wachstum and Pathologie des Beckenverbindungen des Menschen. Gustav Fischer Verlag, Jena.
Pyle, S. I. and N. L. Hoerr
1955 Radiographic Atlas of Skeletal Development of the Knee. Charles C. Thomas, Springfield.
Rhine, J. S. and H. R. Campbell
1980 Thickness of Facial Tissues in American Blacks. Journal of Forensic Sciences, Volume 25, pp. 847–858.
Rhine, J. S., C. Elliott Moore II, and J. T. Weston
1982 Facial Reproduction. Tables of Facial Tissue Thicknesses of American Caucasoids in Forensic Anthropology. Maxwell Museum Technical Series 1. Maxwell Museum of Anthropology, Albuquerque.
Robinow, M., T. W. Richards and Margaret Anderson
1942 The Eruption of Deciduous Teeth. Growth, Volume 6, pp. 127–133.
Robling, Alexander G. and Douglas H. Ubelaker
1997 Sex Estimation from the Metatarsals. Journal of Forensic Sciences, Volume 42, pp. 1062-1069.
Roche, M. B.
1957 Incidence of Osteophytosis and Osteoarthritis in 419 Skeletonized Vertebral Columns. American Journal of Physical Anthropology, Volume 15, pp. 433–434.
Rodriguez, W. C. and W. M. Bass
1983 Insect Activity and its Relationship to Decay Rates of Human Cadavers in East Tennessee. Journal of Forensic Sciences, Volume 28, pp. 423–432.
Romans, Bernard
1961 A Concise Natural History of East and West Florida. Pelican Publishing Company, New Orleans.
Romero, Javier
1970 Dental Mutilation, Trephination, and Cranial Deformation. In Handbook of Middle American Indians edited by Robert Wauchope, Volume 9, pp. 50–67. University of Texas Press, Austin.
Salomon, Carl D. and Nicu Haas
1967 Histological and Histochemical Observations on Undecalcified Sections of Ancient Bones from Excavations in Israel. Israel Journal of Medical Science, Volume 3, pp. 747–754.
Saul, Frank P.
1972 The Human Skeletal Remains of Altar de Sacrificios. Papers of the Peabody Museum of Archaeology and Ethnology, Harvard University, Volume 63, Number 2.
Saunders, S.R., C. Fitzgerald, T. Rogers, C. Dudar, and H. McKillop
1992 A Test of Several Methods of Skeletal Age Estimation Using a Documented Archaeological Sample. Journal of the Canadian Society of Forensic Science, Volume 25, pp. 97-118.
Scammon, R. E.
1937 Two Simple Nomographs for Estimating the Age and Some of the Major External Dimensions of the Human Fetus. The Anatomical Record, Volume 68, pp. 221–225.
Scammon, R. E. and L. A. Calkins
1923a New Empirical Formulae for Determining the Age of the Human Fetus. Anatomical Record, Volume 25, pp. 148–149.
1923b Simple Empirical Formulae for Expressing the Lineal Growth of the Human Fetus. Proceedings of the Society for Experimental Biology (New York), Volume 21, pp. 353–356.
1925 Crown-heel and Crown-rump Length in the Fetal Period and at Birth. Anatomical Record, Volume 29, pp. 372–373.

1929 The Development and Growth of the External Dimensions of the Human Body in the Fetal Period. University of Minnesota Press, Minneapolis.

Scheuer, J.L., and N.M. Elkington
1993 Sex Determination from Metacarpals and the First Proximal Phalanx. Journal of Forensic Sciences, Volume 38, pp. 769-778.

Scheuer, J. L., J. H. Musgrave, and S. P. Evans
1980 The Estimation of Late Fetal and Perinatal Age from Limb Bone Length by Linear and Logarithmic Regression. Annals of Human Biology, Volume 7, pp. 257–265.

Schoeninger, Margaret J., Michael J. De Niro, and Henrik Tauber
1983 Stable Nitrogen Isotope Ratios of Bone Collagen Reflect Marine and Terrestrial Components of Prehistoric Human Diet. Science, Volume 220, pp. 1381–1383.

Schour, I. and M. Massler
1944 Chart—"Development of the Human Dentition." Second edition. American Dental Association, Chicago.

Schranz, D.
1959 Age Determinations from the Internal Structure of the Humerus. American Journal of Physical Anthropology, Volume 17, Number 4, pp. 273–278.

Schuller, A.
1921 Das Rontgenogramm der Stirnhole - Ein Hilfsmitte für de Identitatsbestimmung von Schadelen. Monatsschrift für Ohrenhbilkunde, Volume 55, pp. 1617–1620.

Schulz, Peter D.
1977 Task Activity and Anterior Tooth Grooving in Prehistoric California Indians. American Journal of Physical Anthropology, Volume 46, Number 1, pp. 87–91.

Shapiro, H. L.
1928 A Correction for Artificial Deformation of Skulls. Anthropological Papers of the American Museum of Natural History, Volume 30, Part I. New York.

Sillen, Andrew and Maureen Kavanagh
1982 Strontium and Paleodietary Research: A Review. Yearbook of Physical Anthropology, Volume 25, pp. 67–90.

Singh, Inderbir
1959 Squatting Facets on the Talus and Tibia in Indians. Journal of Anatomy, Volume 93, pp. 540–550. London.
1963 Squatting Facets on the Talus and Tibia in the Rhesus Monkey. Anatomischer Anzeiger, Volume 113, Number 5, pp. 473–476.

Singh, I. J. and D. L. Gunberg
1970 Estimation of Age at Death in Human Males from Quantitative Histology of Bone Fragments. American Journal of Physical Anthropology, Volume 33, Number 3, pp. 373–381.

Smith, S.L.
1996 Attribution of Hand Bones to Sex and Population Groups. Journal of Forensic Sciences, Volume 41, pp. 469-477.

Sprague, Roderick
1968 A Suggested Terminology and Classification for Burial Description. American Antiquity, Volume 33, Number 4, pp. 479–485.

Steele, D. Gentry
1970 Estimation of Stature from Fragments of Long Limb Bones. In Personal Identification in Mass Disasters, edited by T. D. Stewart, pp. 85–97. Smithsonian Institution, Washington, D.C.
1976 The Estimation of Sex on the Basis of the Talus and Calcaneous. American Journal of Physical Anthropology, Volume 45, pp. 581-588.

Steele, D. Gentry and Claud A. Bramblett
1988 The Anatomy and Biology of the Human Skeleton. Texas A&M Press. College Station.

Steele, D. Gentry and Thomas W. McKern
1969 A Method for Assessment of Maximum Long Bone Length and Living Stature from Fragmentary Long Bones. American Journal of Physical Anthropology, Volume 31, pp. 215–227.

Steggerda, Morris and Thomas J. Hill
1942 Eruption Time of Teeth Among Whites, Negroes, and Indians. American Journal of Orthodontics and Oral Surgery, Volume 28, Number 1, pp. 361–370.

Steinbock, R. Ted
1976 Paleopathological Diagnosis and Interpretation: Bone Disease in Ancient Human Populations. Charles C. Thomas, Springfield.

Stewart, T. D.
1931 Incidence of Separate Neural Arch in the Lumbar Vertebrae of Eskimos. American Journal of Physical Anthropology, Volume 16. Number 1, pp. 51–62.
1935 Spondylolisthesis Without Separate Neural Arch (Pseudospondylolisthesis of Junghanns). The Journal of Bone and Joint Surgery. Volume XVII, Number 3, pp. 640–648.
1941 The Circular Type of Cranial Deformity in the United States. American Journal of Physical Anthropology. Volume 28, Number 3, pp. 343–351.
1947 Racial Patterns in Vertebral Osteoarthritis. American Journal of Physical Anthropology, Volume 5, Number 2, pp. 230–231.
1957 Distortion of the Pubic Symphyseal Surface in Females and its Effect on Age Determination. American Journal of Physical Anthropology, Volume 15, pp. 9–18.
1958 The Rate of Development of Vertebral Osteoarthritis in American Whites and its Significance in Skeletal Age Identification. The Leech, Volume 28, Numbers 3, 4, 5, pp. 144–151.
1968 Identification by the Skeletal Structures. In Gradwohl's Legal Medicine edited by Francis E. Camps. Second Edition, pp. 123–154.
1970 Identification of the Scars of Parturition in the Skeletal Remains of Females. In Personal Identification in Mass Disasters, edited by T. D. Stewart, pp. 127–135. Smithsonian Institution, Washington, D.C.
1972 What the Bones Tell Today. FBI Law Enforcement Bulletin, Volume 41, pp. 16–20.
1973 The People of America. Charles Scribner's Sons, New York.

1975 Cranial Dysraphism Mistaken for Trephination. American Journal of Physical Anthropology, Volume 42, Number 3, pp. 435–438.

1979 Essentials of Forensic Anthropology, Especially as Developed in the United States. Charles C. Thomas, Springfield.

Strong, William Duncan

1935 An Introduction to Nebraska Archeology. Smithsonian Miscellaneous Collections, Volume 93, Number 10. Smithsonian Institution, Washington, D.C.

Suchey, Judy M.

1979 Problems in the Aging of Females Using the Os Pubis. American Journal of Physical Anthropology, Volume 51, pp. 467–470.

Suchey, Judy M., D. V. Wiseley, R. F. Green, and T. T. Noguchi

1979 Analysis of Dorsal Pitting in the Os Pubis in an Extensive Sample of Modern American Females. American Journal of Physical Anthropology, Volume 51, pp. 517–540.

Suchey, Judy M., D. V. Wiseley, and D. Katz

1986 Evaluation of the Todd and McKern-Stewart Methods for Aging the Male Os Pubis. In Forensic Osteology, edited by K. J. Reichs, pp. 33–67. Charles C. Thomas, Springfield.

Sundick, R. I.

1972 Human Skeletal Growth and Dental Development as Observed in the Indian Knoll Population. Ph.D. dissertation, University of Toronto.

Telkkä, A.

1950 On the Prediction of Human Stature from the Long Bones. Acta Anatomica, Volume 9, pp. 103–117.

Thomas, David Hurst

1987 The Archaeology of Mission Santa Catalina de Guale: 1. Search and Discovery. Anthropological Papers, Volume 63, Part 2. American Museum of Natural History, New York.

Thompson, D. D.

1979 The Core Technique in the Determination of Age at Death in Skeletons. Journal of Forensic Sciences, Volume 44, pp. 902–915.

Thomson, Arthur

1889 The Influence of Posture on the Form of the Articular Surfaces of the Tibia and Astragalus in the Different Races of Man and the Higher Apes. Journal of Anatomy, London, Volume 23, pp. 616–639.

1899 The Sexual Differences of the Foetal Pelvis. Journal of Anatomy and Physiology, Volume 33, Number 3, pp. 359–380.

Thwaites, Reuben Gold

1896–1901 The Jesuit Relations and Allied Documents. Travels and Explorations of the Jesuit Missionaries in New France 1610–1791. The Burrows Brothers Company, Cleveland.

Tibbetts, G. L.

1981 Estimation of Stature from the Vertebral Column in American Blacks. Journal of Forensic Sciences, Volume 26, pp. 715–723.

Todd, T. W.

1920 Age Changes in the Pubic Bone: I, The Male White Pubis. American Journal of Physical Anthropology, Volume 3, Number 3, pp. 285–334.

1921 Age Changes in the Pubic Bone. American Journal of Physical Anthropology, Volume 4, Number 1, pp. 1–70.

Todd, T. W. and J. D'Errico Jr.

1928 The Clavicular Epiphyses. The American Journal of Anatomy, Volume 41, pp. 25–50.

Todd, T. W. and D. W. Lyon, Jr.

1924 Endocranial Suture Closure: Its Progress and Age Relationship. Part I, Adult Males of White Stock. American Journal of Physical Anthropology, Volume 7, Number 3, pp. 325–384.

1925a Cranial Suture Closure: Its Progress and Age Relationship. Part II, Ectocranial Closure in Adult Males of White Stock. American Journal of Physical Anthropology, Volume 8, Number 1, pp. 23–45.

1925b Cranial Suture Closure: Its Progress and Age Relationship. Part III, Endocranial Closure in Adult Males of Negro Stock. American Journal of Physical Anthropology, Volume 8, Number 1, pp. 47–71.

1925c Cranial Suture Closure: Its Progress and Age Relationship. Part IV, Ectocranial Closure in Adult Males of Negro Stock. American Journal of Physical Anthropology, Volume 8, Number 2, pp. 149–168.

Trinkaus, Erik

1975 Squatting Among the Neantertals: A Problem in the Behavioral Interpretation of Skeletal Morphology. Journal of Archaeological Science, Volume 2, pp. 327-351.

Trotter, Mildred

1970 Estimation of Stature from Intact Limb Bones. In Personal Identification in Mass Disasters, edited by T. D. Stewart, pp. 71–83. Smithsonian Institution, Washington, D.C.

Trotter, Mildred and Goldine C. Gleser

1958 A Pre-Evaluation of Estimation of Stature Based on Measurements of Stature Taken During Life and of Long Bones after Death. American Journal of Physical Anthropology, Volume 16, Number 1, pp. 79–123.

Turner, Christy G.

1976 Dental Evidence on the Origins of the Ainu and Japanese. Science, Volume 193, Number 4256, pp. 911–913.

Turner, Christy G. and James D. Cadien

1969 Dental Chipping in Aleuts, Eskimos and Indians. American Journal of Physical Anthropology, Volume 31, Number 3, pp. 303–310.

Ubelaker, Douglas H.

1966 Arikara-Made Glass Pendants. Plains Anthropologist, Volume 71, Number 32, pp. 172–173.

1974 Reconstruction of Demographic Profiles from Ossuary Skeletal Samples; A Case Study from the Tidewater Potomac. Smithsonian Contributions to Anthropology, Number 18. Washington. D.C.

1979 Skeletal Evidence for Kneeling in Prehistoric Ecuador. American Journal of Physical Anthropology, Volume 51, Number 4, pp. 679–685.

1980a Human Remains from Site OGSE-80, A Preceramic Site on the Santa Elena Peninsula, Coastal Ecuador. Journal of the Washington Academy of Sciences, Volume 70, Number 1, pp. 3–24.

1980b Positive Identification from the Radiographic Comparison of Frontal Sinus Patterns. In Human Identification, edited by T. Rashbun and J. Buikstra, pp. 399–411. Charles C. Thomas, Springfield.
1980c Prehistoric Human Remains from the Cotocollao Site, Pichincha Province, Ecuador. Journal of the Washington Academy of Sciences, Volume 70, Number 2, pp. 59–74.
1981 The Ayalan Cemetery: A Late Integration Period Burial Site on the South Coast of Ecuador. Smithsonian Contributions to Anthropology, Number 29. Washington, D.C.
1983 Human Skeletal Remains from OGSE-MA-172, an Early Guangala Cemetery Site on the Coast of Ecuador. Journal of the Washington Academy of Sciences, Volume 73, Number 1, pp. 16–27.
1984 Prehistoric Human Biology of Ecuador, Possible Temporal Trends and Cultural Correlations. In Paleopathology at the Origins of Agriculture, edited by Mark Cohen and George J. Armelagos, Chapter 19. Academic Press, New York.
1986 Estimation of Age at Death from Histology of Human Bone. In Dating and Age Determination of Biological Materials, edited by M. R. Zimmerman and J. Lawrence Angel, pp. 240–247. Croom Helm, London.
1987a Dental Alteration in Prehistoric Ecuador. A New Example from Jama-Coaque. Journal of the Washington Academy of Sciences, Volume 77, pp. 76–80.
1987b Estimating Age at Death from Immature Human Skeletons: An Overview. Journal of Forensic Sciences, Volume 32, pp. 1254–1263.

Ubelaker, Douglas H. and William M. Bass
1970 Arikara Glassworking Techniques at Leavenworth and Sully Sites. American Antiquity, Volume 35, Number 4, pp. 467–475.

Ubelaker, Douglas H., T. W. Phenice and William M. Bass
1969 Artificial Interproximal Grooving of the Teeth in American Indians. American Journal of Physical Anthropology, Volume 30, Number 1, pp. 145–150.

Ubelaker, Douglas H. and Norman D. Sperber
1988 Alterations in Human Bones and Teeth Due to Restricted Sun Exposure and Contact with Corrosive Agents. Journal of Forensic Sciences, Volume 33, pp. 540–546.

Ubelaker, D. H. and P. Willey
1978 Complexity in Arikara Mortuary Practice. Plains Anthropologist, Vol. 23, No. 79.

Ullrich, H.
1975 Estimation of Fertility by Means of Pregnancy and Childbirth Alterations at the Pubis, the Ilium and the Sacrum. Ossa, Volume 2, pp. 23–39.

Van der Merwe, Nikolaas J.
1982 Carbon Isotopes, Photosynthesis, and Archaeology. American Scientist, Vol. 70, pp. 596–606.

Van Vark, G. N.
1970 Some Statistical Procedures for the Investigation of Prehistoric Human Skeletal Material. Rijksuniversiteit te Groningen, Groningen.

Vignati, Milciades Alejo
1930 Los Cráneos Trofeo. Archivos del Museo Etnográfico, Número 1. Buenos Aires.

Vogel, J. C. and N. J. Van der Merwe
1977 Isotopic Evidence for Early Maize Cultivation in New York State. American Antiquity, Volume 42, Number 2, pp. 238–242.

Walker, P. L.
1969 The Linear Growth of Long Bones in Late Woodland Indian Children. Proceedings of the Indiana Academy of Science, Volume 78, pp. 83–87.

Walker, Robert A. and C. Owen Lovejoy
1985 Radiographic Changes in the Clavicle and Proximal Femur and their Use in the Determination of Skeletal Age at Death. American Journal of Physical Anthropology, Volume 68, pp. 67–78.

Wallace, John
1974 Approximal Grooving of Teeth. American Journal of Physical Anthropology, Volume 40, Number 3, pp. 385–390.

Weaver, David S.
1980 Sex Differences in the Ilia of a Known Sex and Age Sample of Fetal and Infant Skeletons. American Journal of Physical Anthropology, Volume 52, pp. 191–195.

White, Leslie A., Editor
1959 Lewis Henry Morgan: the Indian Journals, 1859–62. The University of Michigan Press, Ann Arbor.

White Tim D.
1991 Human Osteology. Academic Press, New York.

Willey, P. and Alan Heilman
1987 Estimating Time Since Death Using Plant Roots and Stems. Journal of Forensic Sciences, Volume 32, pp. 1264–1270.

Willoughby, Charles C.
1922 The Turner Group of Earthworks, Hamilton County, Ohio. Papers of the Peabody Museum of American Archaeology and Ethnology, Harvard University, Volume 8, Number 3. Cambridge, Mass.

Wood, W. R.
1965 The Redbird Focus and the Problem of Ponca Prehistory. Plains Anthropologist, Volume 10, Number 28, Memoir 2.

Yakubik, Jill-Karen, H. A. Franks, R. Christopher Goodwin, and Carol J. Poplin
1986 Cultural Resources Inventory of the Bennet Carre Spillway, St. Charles Parish, Louisiana. Final Report. Department of the Army, New Orleans District Corps of Engineers. New Orleans.

Yarrow, H. C.
1880 Introduction to the Study of Mortuary Customs among the North American Indians. Government Printing Office, Washington, D.C.

Yoder, Cassady
1999 Examination of Variation in Sternal End Morphology Relevant to Age Assessment. Proceedings of the American Academy of Forensic Sciences. Volume V, p. 224.

Glossary

Alidade. An instrument consisting of a telescope mounted on a scale, used in topographic mapping.

Articulation. The point where two adjacent bones are in contact; the normal anatomical arrangement of adjacent bones.

Calcification. The process of formation of bones and teeth.

Cancellous bone. The spongy or porous internal structure, particularly characteristic at the ends of long bones.

Cartilage. A tough, elastic tissue.

Circumferential lamellar bone. Original bone deposited by the periosteum in the cortex of long bones (Fig. 104c).

Collagen. Submicroscopic protein fibers found in skin, bone, and ligaments.

Cortex. The outer portion of a bone.

Cremation. The act of burning a body or the remnants of a burned individual.

Cranial suture. The junction between two bones of the skull (Figs. 160–161).

Demography. The study of vital statistics within populations.

Diaphysis. The shaft of a long bone.

Dimorphism. The occurrence of two forms in members of the same species; most commonly applied in Homo sapiens to differences between the sexes.

Distal. Farthest from the center of the mid-line of the body; in limbs, farthest from the point of attachment to the trunk.

Dorsal. Back.

Eburnation. A polished surface produced when destruction of the intervening cartilage allows the bones in a joint to rub together (Fig. 98d).

Encephalocele. Protrusion of brain substance through an opening in the skull.

Epiphysis. A bony cap at the end of a long bone (Fig. 85).

Eruption of teeth. The emergence of teeth through the gum (Fig. 71).

Femora. Plural of femur.

Haversian canal. A small canal in the bone cortex; one of the features used in estimating age at death (Fig. 104a).

Inhumation. Synonym for burial or interment.

Long bones. Collective term for the bones of the arms and legs; specifically, the humerus, radius, ulna, femur, tibia, and fibula (Fig. 159).

Malocclusion. Condition in which the upper and lower teeth do not meet.

Medullary cavity. The canal through the center of a long bone.

Metaphysis. The area between the diaphysis and epiphysis, where bone growth occurs.

Morphology. The form and structure of an object.

Necrosis. Physiological death.

Non-Haversian canal. A canal formed within the cortex of long bones during the deposition of circumferential lamellar bone (Fig. 104d).

Ossuary. A communal grave containing the secondary remains of individuals initially stored elsewhere (Figs. 26–28).

Osteoblasts. The specialized cells that produce bone.

Osteoclasts. The specialized cells that destroy bone.

Osteology. The study of bones.

Osteophytes. Abnormal bony extensions that develop on the surface of bones (Figs. 94, 98b).

Partial articulation. A condition in which two or more (but not all) bones of a skeleton remain in articulation, indicating that decomposition was incomplete at the time of burial.

Parietal thinning. Change in the skull with age, in which the inner and outer layers move closer together producing scooped-out depressions on the exterior of the parietals (Fig. 101).

Pathology. The study of disease.

Pelvis. The portion of the skeleton composed of the sacrum and the left and right innominates (Fig. 159).

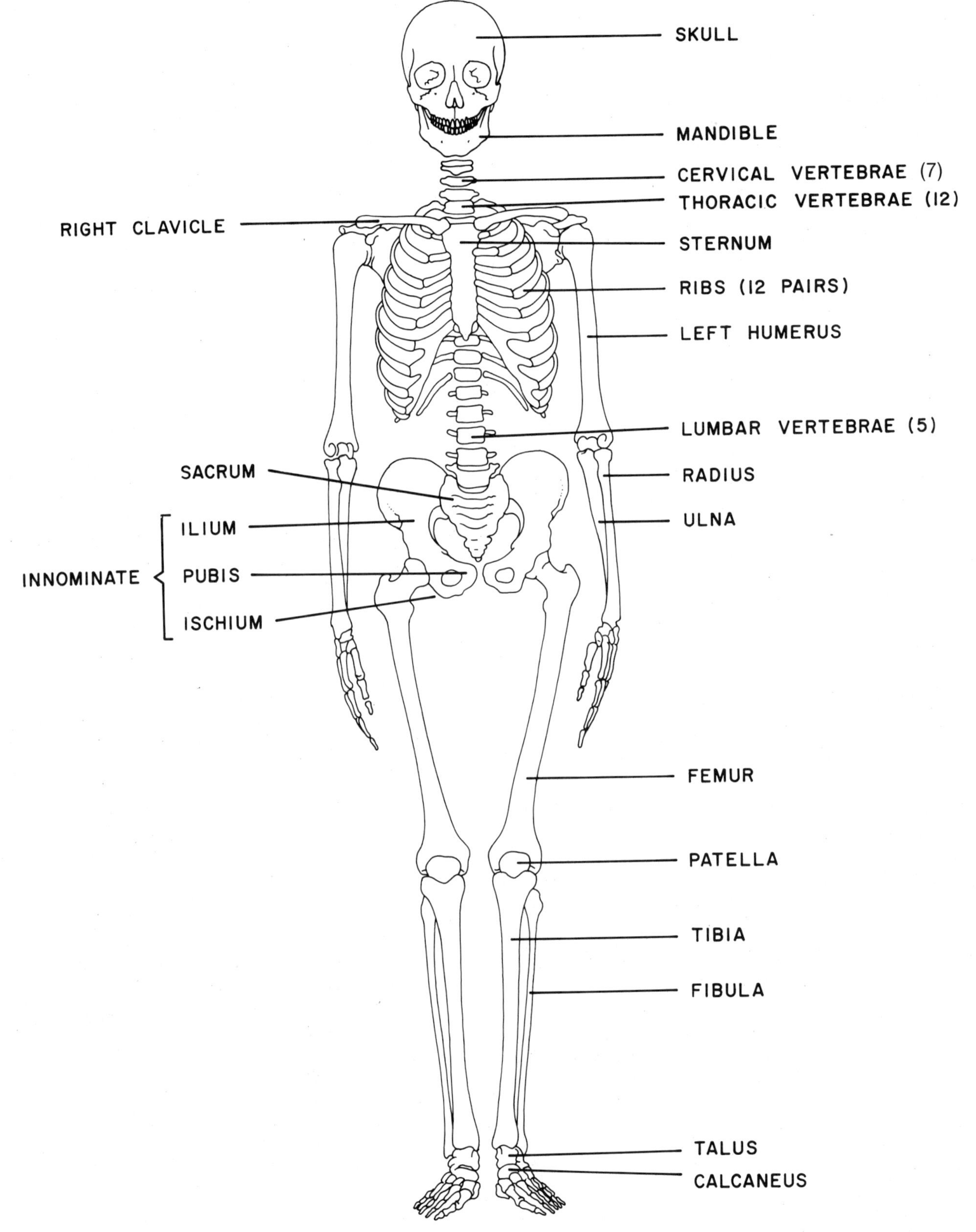

Fig. 161. The human skeleton, with the principal bones identified.

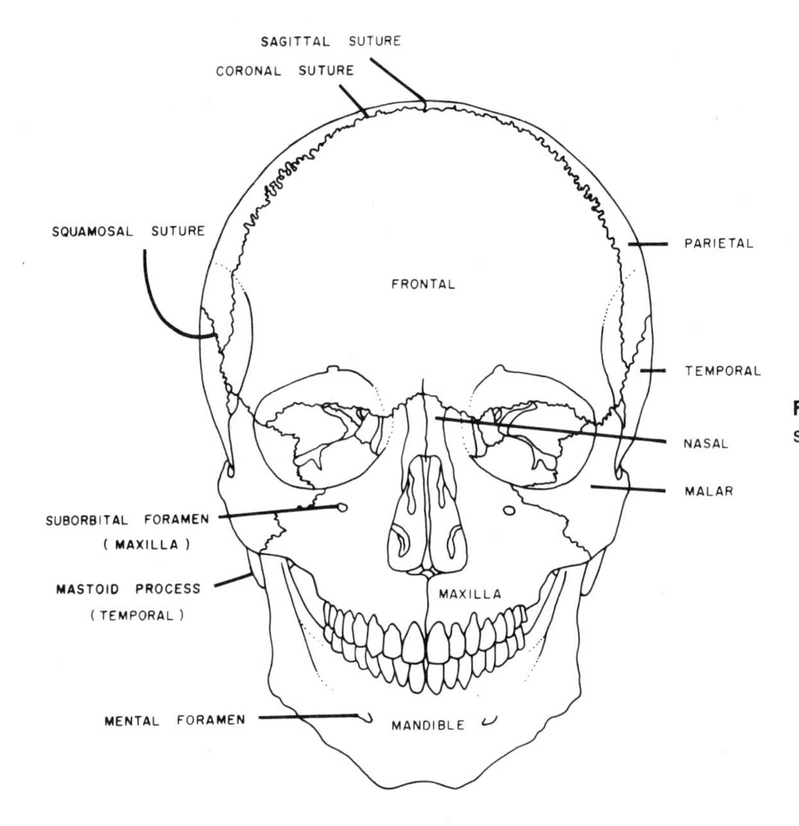

Fig. 162. Front view of the skull, with the principal bones, sutures, and other features identified.

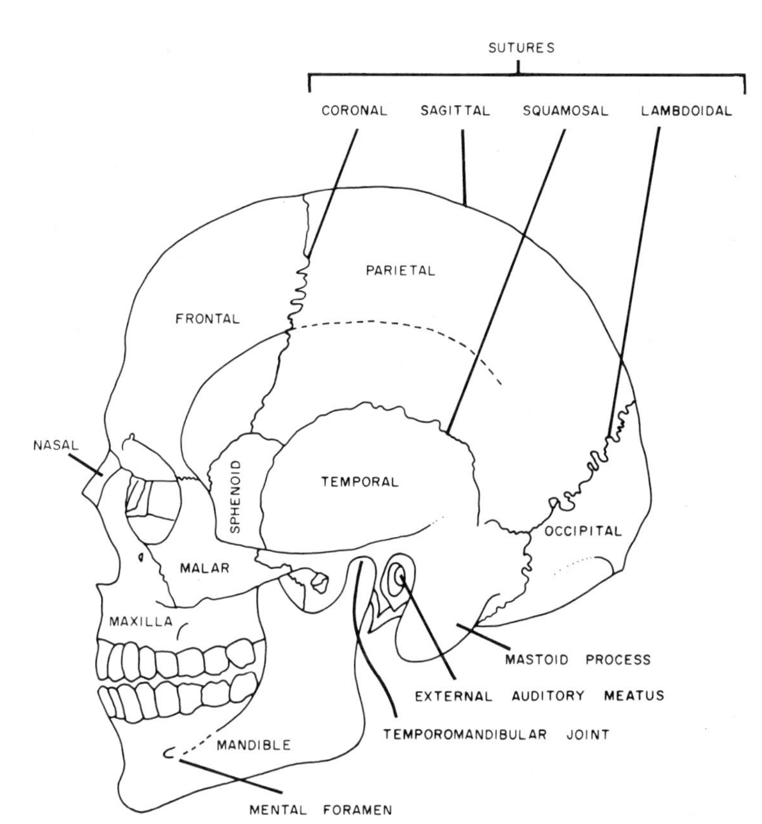

Fig. 163. Side view of the skull, with the principal bones, sutures, and other features identified.

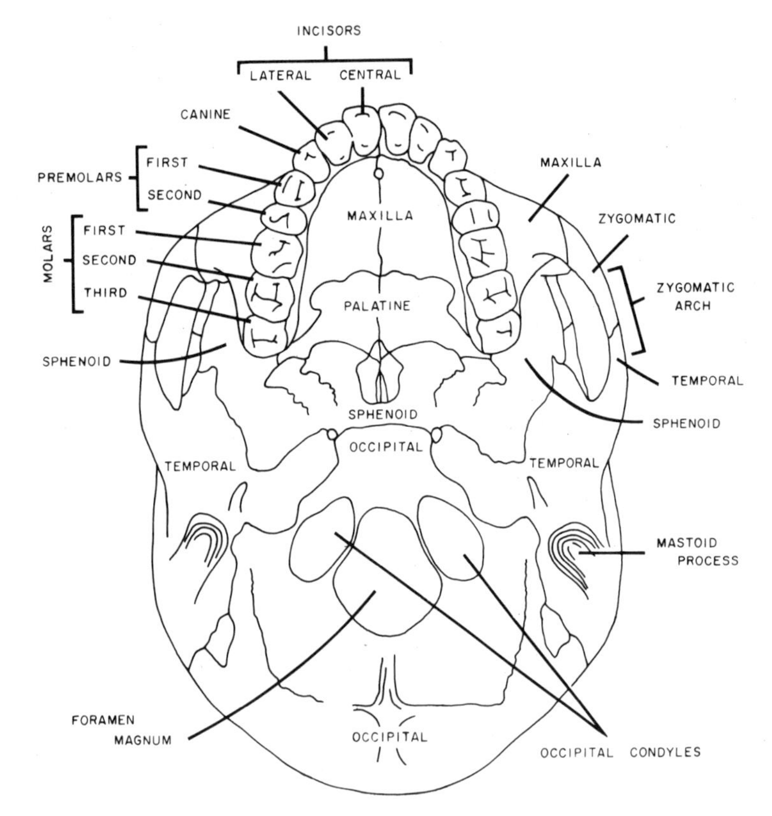

Fig. 164. Basal view of the skull, with the principal features and the teeth identified.

Peridontal disease. An abnormal condition of the tissue adjacent to the teeth.
Periosteum. The membrane of connective tissue that covers all surfaces of bones except the areas of articulation.
Post-cranial skeleton. All bones except the skull (Fig. 159).
Primary burial. An articulated skeleton, buried in the flesh (Figs. 14–22).
Proximal. Closest to the center of the mid-line of the body; in limbs, closest to the point of attachment to the trunk.
Pubic symphysis. The junction of the right and left pubic bones at the mid-line.
Resorption. The process of destruction of bone by osteoclasts.
Scaffold. A wooden platform used by some Indian groups to hold their dead (Fig. 1).
Secondary burial. An interment of disarticulated bones.
Semi-articuloses. A term proposed by Bass (1962:43) to designate the condition in which some of the bones of a skeleton are found articulated and some disarticulated (Figs. 45–46).
Stratigraphy. The superposition of layers of strata of differing geological or cultural origin.
Superiosteal. Beneath the periosteum.
Symphyseal face. The articular surface of the pubis (Figs. 86–87).
Symphyseal rim. An elevated margin that forms around the edge of the symphyseal face of the pubis (Fig. 90).
Symphysis. The area where the left and right pubes articulate.
Trabeculae. An internal network of bone fibers.
Trauma. An injury inflicted by force by some physical agent.
Trephination. A surgical procedure involving cutting a hole in the cranial vault (Figs. 114–117).
Radiogram. An image produced on photographic film by the passage of X-rays. Synomyms: Roentgenogram, radiograph.
Ventral. Front.
Ventral rampart. A ridge of bone that forms on the ventral surface of the symphyseal face of the pubis (Fig. 89).

Analytical Table of Contents